Y0-CKO-420

T 385 .P532 2009

Planchard, David C.

Certified SolidWorks 2009
[i.e. 2008] Associate CSWA
exam guide

INCLUDES CD

NEW ENGLAND INSTITUTE OF TECHNOLOGY
LIBRARY

CERTIFIED SOLIDWORKS® 2009 ASSOCIATE CSWA EXAM GUIDE

AN AUTHORIZED CSWA PREPARATION EXAM GUIDE

David C. Planchard and
Marie P. Planchard (CSWP)

DELMAR
CENGAGE Learning™

Australia • Brazil • Japan • Korea • Mexico • Singapore • Spain • United Kingdom • United States

DELMAR
CENGAGE Learning

Certified SolidWorks 2009 Associate CSWA Exam Guide:
An Authorized CSWA Preparation Exam Guide
David C. Planchard and Marie P. Planchard

Vice President, Career and Professional Editorial: Dave Garza

Director of Learning Solutions: Sandy Clark

Managing Editor: Larry Main

Senior Product Manager: John Fisher

Senior Editorial Assistant: Dawn Daugherty

Vice President, Career and Professional Marketing: Jennifer McAvey

Executive Marketing Manager: Deborah S. Yarnell

Senior Marketing Manager: Jimmy Stephens

Marketing Specialist: Mark Pierro

Production Director: Wendy Troeger

Production Manager: Stacy Masucci

Content Project Manager: Angela Iula

Senior Art Director: David Arsenault

Technology Project Manager: Christopher Catalina

Production Technology Analyst: Thomas Stover

© 2009 Delmar, Cengage Learning

ALL RIGHTS RESERVED. No part of this work covered by the copyright herein may be reproduced, transmitted, stored or used in any form or by any means graphic, electronic, or mechanical, including but not limited to photocopying, recording, scanning, digitizing, taping, Web distribution, information networks, or information storage and retrieval systems, except as permitted under Section 107 or 108 of the 1976 United States Copyright Act, without the prior written permission of the publisher.

> For product information and technology assistance, contact us at
> **Cengage Learning Customer & Sales Support, 1-800-354-9706**
> For permission to use material from this text or product,
> submit all requests online at **cengage.com/permissions**
> Further permissions questions can be emailed to
> **permissionrequest@cengage.com**

Library of Congress Control Number: 2008931249

ISBN-13: 978-1-4354-8072-8

ISBN-10: 1-4354-8072-4

Delmar
5 Maxwell Drive
Clifton Park, New York, 12065-2919
USA

Cengage Learning is a leading provider of customized learning solutions with office locations around the globe, including Singapore, the United Kingdom, Australia, Mexico, Brazil, and Japan. Locate your local office at:
international.cengage.com/region

Cengage Learning products are represented in Canada by Nelson Education, Ltd.

For your lifelong learning solutions, visit **delmar.cengage.com**

Visit our corporate website at **www.cengage.com**

Notice to the Reader
Publisher does not warrant or guarantee any of the products described herein or perform any independent analysis in connection with any of the product information contained herein. Publisher does not assume, and expressly disclaims, any obligation to obtain and include information other than that provided to it by the manufacturer. The reader is expressly warned to consider and adopt all safety precautions that might be indicated by the activities described herein and to avoid all potential hazards. By following the instructions contained herein, the reader willingly assumes all risks in connection with such instructions. The publisher makes no representations or warranties of any kind, including but not limited to, the warranties of fitness for particular purpose or merchantability, nor are any such representations implied with respect to the material set forth herein, and the publisher takes no responsibility with respect to such material. The publisher shall not be liable for any special, consequential, or exemplary damages resulting, in whole or part, from the readers' use of, or reliance upon, this material.

Printed in Canada
1 2 3 4 5 XX 10 09 08

INTRODUCTION

The Certified SolidWorks 2008 Associate CSWA Exam Guide is written to assist the SolidWorks user to pass the CSWA exam. SolidWorks Corporation offers two levels of certification representing increasing levels of expertise in 3D CAD design as it applies to engineering: Certified SolidWorks Associate CSWA, and the Certified SolidWorks Professional CSWP.

The CSWA certification indicates a foundation in and apprentice knowledge of 3D CAD design and engineering practices and principles. The main requirement for obtaining the CSWA certification is to take and pass the three hour, seven question on-line proctored exam at a Certified SolidWorks CSWA Provider, "university, college, technical, vocational, or secondary educational institution" and to sign the SolidWorks Confidentiality Agreement. Passing this exam provides students the chance to prove their knowledge and expertise and to be part of a world wide industry certification standard.

💡 Commercial customers can obtain information on taking the CSWA exam through the SolidWorks Customer Portal.

💡 At this time, the CSWA exam is given in both SolidWorks 2007 and SolidWorks 2008.

Goals of the book

The primary goal for this book is not only to help you pass the CSWA exam, but also to ensure that you understand and comprehend the concepts and implementation details of the CSWA process. The second goal of this book is to provide the most comprehensive coverage of CSWA exam related topics available, without too much coverage of topics not on the exam. The third and ultimate goal is to get you from where you are today to the point that you can confidently pass the CSWA exam. Therefore, the chapters in this book, which are outlined in the introduction, are geared toward helping you discover the topics that are on the CSWA exam, pinpoint where you have a knowledge deficiency, and determine what you need to know to master these topics.

Intended audience

The intended audience for this book and the CSWA exam is anyone with a minimum of 6 - 9 months of SolidWorks experience and basic knowledge of

Introduction

engineering fundamentals and practices. SolidWorks recommends that you review their SolidWorks Tutorials on Parts, Assemblies, Drawings, and COSMOSXpress as a prerequisite and have at least 45 hours of classroom time learning SolidWorks or using SolidWorks with basic engineering design principles and practices.

☼ The SolidWorks models in this book were created using SolidWorks 2008 SP2.1.

How the book is organized

The CSWA exam is split into five categories. Questions on the timed exam are provided in a random manor. The first chapter is a general review of the SolidWorks 2008 User Interface. The other five chapters in the book address a key exam category. The chapters provide individual short practice tutorials to review the required competencies. The complexity of the models along with the features progressively increases throughout each chapter to build knowledge and to simulate the final types of questions that would be provided on the exam.

All models for the 110 plus tutorials are provided on the enclosed book CD along with their solution. Copy the provided CD folders to your computer. Use the chapter models in each chapter folder; initial and final. Learn by doing!

The following information provides general guidelines for the content likely to be included on the exam. However, other related topics may also appear on any specific delivery of the exam. In order to better reflect the contents of the exam and for clarity purposes, the guidelines below may change at any time without notice.

Chapter 1: SolidWorks 2008 User Interface

Reviews the SolidWorks 2008 User Interface and CommandManager: *Menu bar toolbar, Menu bar menu, Drop-down menus, Context toolbars, Consolidated drop-down toolbars, System feedback icons, Confirmation Corner, Heads-up View toolbar, and more.*

Chapter 2: Basic Theory and Drawing Theory (2 Questions - Total 10 Points)

- Identify and apply basic concepts in SolidWorks
- Recognize 3D modeling techniques:
  - Understand how parts, assemblies, and drawings are related
  - Identify the feature type, parameters, and dimensions
  - Identify the correct standard reference planes: Top, Right, and Front

- Determine the design intent for a model
- Identify and understand the procedure for the following:
  - Assign and edit material to a part
  - Apply the Measure tool to a part or an assembly
  - Locate the Center of mass, and Principal moments of inertia relative to the default coordinate location, Origin.
  - Calculate the overall mass and volume of a part
- Recognize and know the function and elements of the Part and Assembly FeatureManager design tree:
  - Sketch status
  - Component status and properties
  - Display Pane status
  - Reference configurations
- Identify the default Sketch Entities from the Sketch toolbar: Line, Rectangle, Circle, etc.
- Identify the default Sketch Tools from the Sketch toolbar: Fillet, Chamfer, Offset Entities, etc.
- Identify the available SolidWorks File formats for input and export:
  - Save As type for a part, assembly, and drawing
  - Open File of different formats
- Use SolidWorks Help:
  - Contents, Index, and Search tabs
- Identify the process of creating a simple drawing from a part or an assembly:
  - Knowledge to insert and modify the 3 Standard views
  - Knowledge to add a sheet and annotations to a drawing
- Recognize all drawing name view types by their icons:
  - Model, Projected, Auxiliary, Section, Aligned Section, Detail, Standard, Broken-out Section, Break, Crop, and Alternate Position
- Identify the procedure to create a named drawing view:
  - Model, Projected, Auxiliary, Section, Aligned Section, Detail, Standard, Broken-out Section, Break, Crop, and Alternate Position

Introduction

- Specify Document Properties:
  - Select Unit System
  - Set Precision

Chapter 3: Part Modeling (1 Question - Total 30 Points)

- Read and understand an Engineering document:
  - Identify the Sketch plane, part Origin location, part dimensions, geometric relations, and design intent of the sketch and feature
- Build a part from a detailed dimensioned illustration using the following SolidWorks tools and features:
  - 2D & 3D sketch tools
  - Extruded Boss/Base
  - Extruded Cut
  - Fillet
  - Mirror
  - Revolved Base
  - Chamfer
  - Reference geometry
  - Plane
  - Axis
  - Calculate the overall mass and volume of the created part
- Locate the Center of mass for the created part relative to the Origin

Chapter 4: Advanced Part Modeling (1 Question -Total 20 Points)

- Specify Document Properties
- Interpret engineering terminology:
  - Create and manipulate a coordinate system
- Build an advanced part from a detailed dimensioned illustration using the following tools and features:
  - 2D & 3D Sketch tools
  - Extruded Boss/Base
  - Extruded Cut

- Fillet
- Mirror
- Revolved Boss/Base
- Linear & Circular Pattern
- Chamfer
- Revolved Cut

- Locate the Center of mass relative to the part Origin
- Create a coordinate system location
- Locate the Center of mass relative to a created coordinate system

Chapter 5: Assembly Modeling (1 Question - Total 30 Points)

- Specify Document Properties
- Identify and build the components to construct the assembly from a detailed illustration using the following features:
  - Extruded Boss/Base
  - Extruded Cut
  - Fillet
  - Mirror
  - Revolved Boss/Base
  - Revolved Cut
  - Linear Pattern
  - Chamfer
  - Hole Wizard
- Identify the first fixed component in an assembly
- Build a bottom-up assembly with the following Standard mates:
  - Coincident, Concentric, Parallel, Perpendicular, Tangent, Angle, and Distance
  - Aligned, Anti-Aligned options
- Apply the Mirror Component tool
- Locate the Center of mass relative to the assembly Origin
- Create a coordinate system location

Introduction

- Locate the Center of mass relative to a created coordinate system
- Calculate the overall mass and volume for the created assembly
- Mate the first component with respect to the assembly reference planes

Chapter 6: Advanced Modeling Theory and Analysis (2 Questions - Total 10 Points)

- Understand basic Engineering analysis definitions
- Knowledge of the COSMOSXpress Wizard interface
- Ability to apply COSMOSXpress to a simple part

☼ In SolidWorks 2009, COSMOSXpress is called SimulationXpress.

About the CSWA exam

The CAD world has many different certifications available. Some of these certifications are sponsored by vendors and some by consortiums of different vendors. Regardless of the sponsor of the certifications, most CAD professionals today recognize the need to become certified to prove their skills, prepare for new job searches, and to learn new skill, while at their existing jobs.

Specifying a CSWA or CSWP certification on your resume is a great way to increase your chances of landing a new job, getting a promotion, or looking more qualified when representing your company on a consulting job.

How to obtain your CSWA Certification

For Educational customers, SolidWorks Corporation requires that you take and pass the 3 hour on-line proctored exam in a secure environment at a designated CSWA Provider and to sign the SolidWorks Confidentiality Agreement. A CSWA Provider can be a university, college, technical, vocational, or secondary educational institution. Contact your local SolidWorks Value Added Reseller (VAR) or instructor for information on CSWA Providers.

☼ Commercial customers can obtain information on taking the CSWA exam through the SolidWorks Customer Portal.

There are five key categories in the CSWA exam. The minimum passing grade is 70 out of 100 points. There are two questions in both the Basic Theory and Drawing, and Advanced Modeling Theory and Analysis Categories, (multiple choice, single answer or fill in the blank) and one question in each of the Part modeling, Advanced Part Modeling and Analysis, and the Assembly Modeling categories. The single questions are on an in-depth illustrated dimension model. All questions are in a multiple choice single answer format.

How to prepare to pass the CSWA exam

Taking a SolidWorks class at a university, college, technical, vocational, or secondary educational institution or time in industry using SolidWorks does not mean that you will automatically pass the CSWA exam. In fact, the CSWA exam purposefully attempts to make the questions prove that you know the material well by making you apply the concepts in a real world situation. The CSWA exam questions tend to be a fair amount more involved than just creating a single sketch, part, or simple assembly. The exam requires that you know and apply the knowledge to different scenarios. To ensure success, use this book, along with the SolidWorks Tutorials.

How do I schedule an exam?

Contact your local SolidWorks VAR or instructor to schedule an exam. U.S. and Canada, exams usually can be scheduled up to six weeks in advance. Scheduling policies vary at international locations.

Exam day

Candidates must acknowledge the SolidWorks CSWA Certification and Confidentiality Agreement online prior to taking the exam. Candidates will not be able to proceed with the exam and a refund will not be provided.

Gather personal information prior to exam registration:

- Legal name (from government issued ID)
- CSWA Certification exam event code from the Provider
- Valid email address
- Method of payment

Students will not be able to use notes, books, calculators, PDA's, cell phones, or materials not authorized by a SolidWorks Certified Provider or SolidWorks during the exam.

The CSWA exam at this time, is provided in the following languages: English, Brazilian Portuguese, Chinese-S, Chinese-T, French, German, Italian, Japanese, Spanish, and Korean.

Introduction

Exams may contain non-scored items to collect performance data on new items. Non-scored items are not used in determining the passing score nor are reported in a subsection of the score report. All non-scored items are randomly placed in the exam with sufficient time calculated and given to complete the entire exam.

At the completion of the computer-based on-line exam, candidates receive a score report along with a score breakout by exam section and the passing score for the given exam. Note: All students are required to sign and return all supplied papers/notes that were taken during the exam to the onsite proctor before leaving the testing room.

What do I get when I pass the exam?

After a candidate passes the CSWA exam and signs the required agreements an email is sent to you to visit the following website: www.virtualtester.com/solidworks.

Certified candidates are authorized to use the appropriate CSWA SolidWorks certification logo indicating certification status. Prior to use, they must read and acknowledge the SolidWorks Certification Logo Agreement. Logos can be downloaded through the CSWA Certification Tracking System.

The CSWA Certification Tracking System provides a record of both exam and certification status. Candidates and certification holders are expected to keep contact information up to date for receiving notifications from SolidWorks.

☼ VirtualTester is a system for delivering and managing multiple choice tests over the internet. The system consists of a web service complete with company and user database, question editor and a client.

The VirtualTester website: www.virtualtester.com/solidworks provides the following menu options:

Introduction

The following dialog box with *Actions*, *Downloads*, *Search* and *More* to provided.

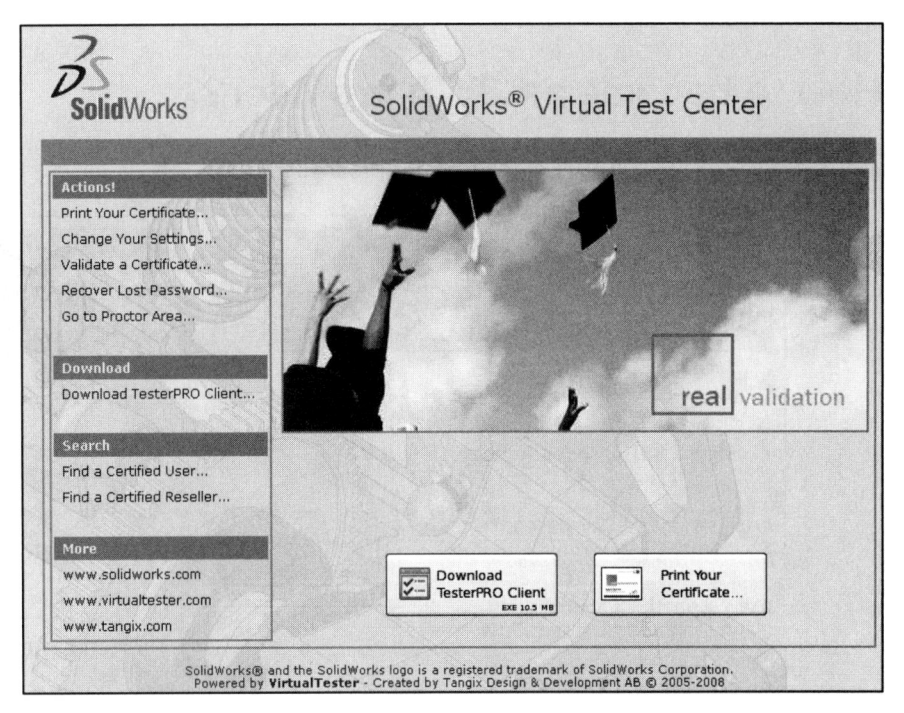

☼ Use the login credentials sent by email from virtualtester.com. If you do not receive the email, check your spam filters, junk email filters, etc. to receive the email.

A certificate suitable for framing is provided with your name and CSWA Certification ID.

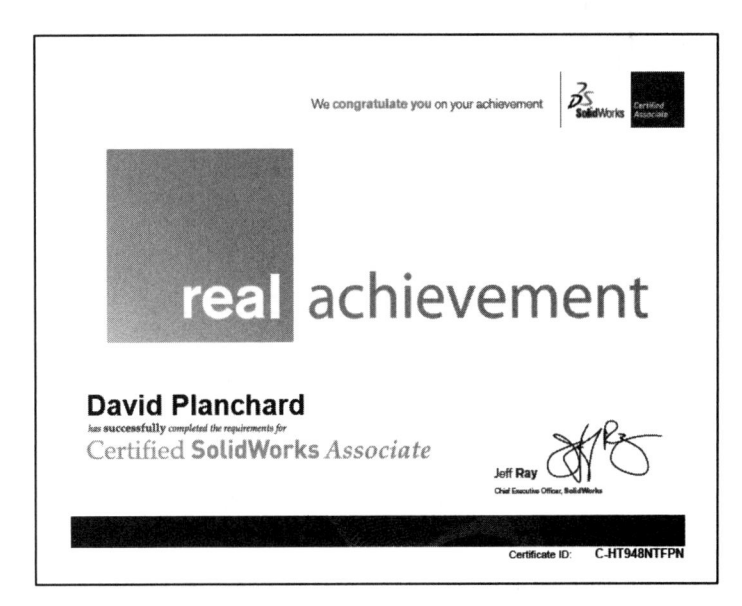

How to become a CSWA provider?

- Visit www.solidworks.com/cswa.

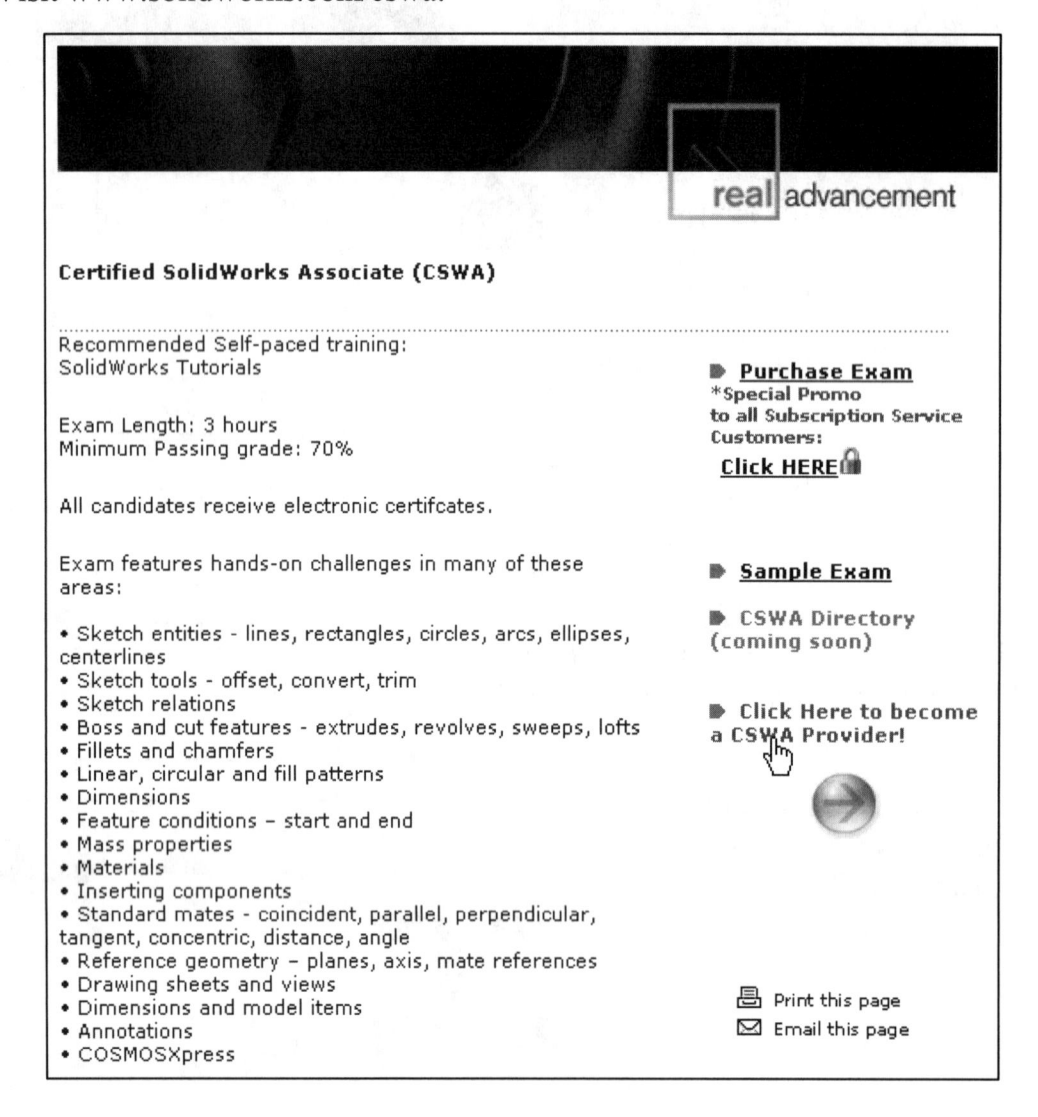

- Complete the online web application.

☼ A CSWA Provider is an educational institution, on subscription, with a CSWA Program, that administers the CSWA exam.

☼ Upon approval, the Certification department will create on online testing account for the contact person.

Syntax and organization in the book

The following conventions are used throughout this book:

1. The term document is used to refer a SolidWorks part, drawing, or assembly file.

2. The list of items across the top of the SolidWorks interface is the Main menu. Each item in the Main menu has a pull-down menu. When you need to select a series of commands from these menus, the following format is used; Click the **Plane** tool from the Reference Geometry Consolidated toolbar. The Plane PropertyManager is displayed.

3. In various chapters of the book, when a general procedure is instructed, a bold command is indicated: Create **Sketch1**, Create the **Extrude1** feature, **Assign** 1060 Alloy material to the part, etc.

4. The book is organized into six chapters. The first chapter provides an overview of the SolidWorks 2008 User Interface. The other chapters are focused on a specific category of the CSWA exam. The CD in the book provides the initial and final models for all tutorials. You can read any chapter without reading the entire book. Each chapter has stand alone tutorials to practice and reinforce the chapter subject matter and key learning objectives.

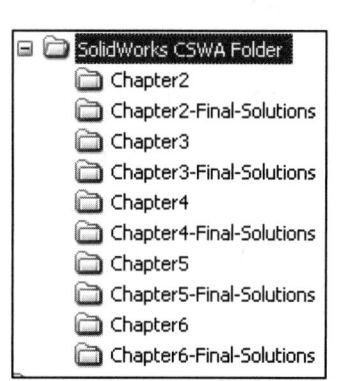

5. Copy the provided CD models and folders to your computer. Work from your hard drive. The enclosed CD files are the stand alone tutorial documents used in each chapter.

6. Compare your results with the documents in the Chapter Solutions folder. Learn by doing!

About the authors

David Planchard is the President of D&M Education, LLC. Before starting D&M Education LLC, he spent over 25 years in industry and academia holding various engineering, marketing, and teaching positions and degrees. He has five U.S. patents and one International patent. He has published and authored numerous papers on equipment design. David is also a technical editor for Cisco Press. He is a member of the New England Pro/Users Group, New England SolidWorks Users Group, and the Cisco Regional Academy Users Group. David holds a BSME and a MSM. David is a SolidWorks Solution Partner and holds the Certified SolidWorks Associate CSWA Certification.

Marie Planchard is the Director of World Education Markets at SolidWorks Corporation. Before she joined SolidWorks, Marie spent over 10 years as an engineering professor at Mass Bay College in Wellesley Hills, MA. She has 18 plus years of industry software experience and held a variety of management and engineering positions. Marie holds a BSME, MSME, and the Certified SolidWorks Professional (CSWP) Certification.

David and Marie Planchard are co-authors of the following books:

- Certified SolidWorks Associate CSWA Exam Guide 2008, 2007
- The Fundamentals of SolidWorks: Featuring the VEXplorer robot 2008, 2007
- A Commands Guide for SolidWorks 2008
- A Commands Guide Reference Tutorial for SolidWorks 2007
- Engineering Design with SolidWorks with Multimedia CD 2008, 2007, 2006, 2005, 2004, 2003, 2001Plus, 2001, and 1999
- SolidWorks Tutorial with Multimedia CD 2008, 2007, 2006, 2005, 2004, 2003, and 2001/2001Plus
- SolidWorks The Basics, with Multimedia CD 2008, 2007, 2006, 2005, 2004, and 2003
- Assembly Modeling with SolidWorks 2008, 2006, 2005-2004, 2003, and 2001Plus
- Drawing and Detailing with SolidWorks 2008, 2007, 2006, 2005, 2004, 2003, 2002, and 2001/2001Plus
- Applications in Sheet Metal Using Pro/SHEETMETAL & Pro/ENGINEER

Dedication

A special acknowledgment goes to our loving daughter Stephanie Planchard who supported us during this intense and lengthy project. Stephanie continues to support us with her patience, love, and understanding.

Contact the authors

We realize that keeping software application books up to date is important to our customers. We value the hundreds of professors, students, designers and engineers that have provided us input to enhance our books. We value your suggestions and comments. Please contact us with any comments, questions or suggestions on this book or any of our other publications. David Planchard,

D&M Education, LLC, dplanchard@verizon.net or visit our website: www.dmeducation.net.

References

- SolidWorks Users Guide, SolidWorks Corporation, 2008.
- COSMOSXpress Online help 2008.
- ASME Y14 Engineering Drawing and Related Documentation Practices.
- Beers & Johnson, Vector Mechanics for Engineers, 6th ed, McGraw Hill.
- Betoline, Wiebe, Miller, Fundamentals of Graphics Communication, Irwin, 1995.
- Hibbler, R.C, Engineering Mechanics Statics and Dynamics, 8th ed, Prentice Hall.
- Hoelscher, Springer, Dobrovolny, Graphics for Engineers, John Wiley, 1968.
- Jensen, Cecil, Interpreting Engineering Drawings, 6th ed, Delmar Thomson, 2002.
- Lockhart & Johnson, Engineering Design Communications, Addison Wesley, 1999.
- Olivo C., Payne, Olivo, T, Basic Blueprint Reading and Sketching, Delmar Thomson, 1988.
- Planchard & Planchard, Drawing and Detailing with SolidWorks, SDC Publications, 2008.
- Planchard & Planchard, Engineering Design with SolidWorks, SDC Publications, 2008.
- Madsen, David, Engineering Drawing and Design, Delmar Thomson, 2007.

TABLE OF CONTENTS

Introduction	**I-1**
Goals of the book	I-1
Intended audience	I-1
How the book is organized	I-2
About the CSWA exam	I-6
How to obtain the CSWA certification	I-6
How to prepare to pass the CSWA exam	I-6
How do I schedule the exam	I-7
Exam day	I-7
What do I get when I pass the exam	I-8
How to become a CSWA Provider	I-10
Syntax and organization of the book	I-11
About the authors	I-11
Dedication	I-12
Contact the author	I-12
References	I-13
Table of Contents	I-14
Chapter 1 – SolidWorks 2008 User Interface	**1-1**
Chapter Objective	1-1
What is SolidWorks	1-2
Basic Concepts in SolidWorks	1-2
SolidWorks User Interface (UI)	1-4
Menu Bar toolbar	1-4
Menu Bar menu	1-5
Drip-down menu	1-5
Right-click Pop-up menus	1-6
Consolidated flyout tool buttons	1-6
System Feedback	1-6
Confirmation Corner	1-7
Heads-up View toolbar	1-7
CommandManager	1-9
Task Pane	1-12
SolidWorks Resources	1-12
Design Library	1-13
File Explorer	1-13
Search	1-13
View Palette	1-14
RealView	1-14
Document Recovery	1-15
Motion Study	1-15
FeatureManager Design Tree	1-16
Fly-out FeatureManager	1-18

Summary	1-19
Key Terms	1-19

Chapter 2 – Basic Theory and Drawing Theory — **2-1**

Chapter Objective	2-1
3D Modeling techniques	2-3
How parts, assemblies, and drawings are related	2-3
Tutorial: Associativity 2-1	2-3
Identify the Feature type by icon in the FeatureManager	2-5
Tutorial: Identify model features 2-1	2-6
Tutorial: Identify model features 2-2	2-7
Tutorial: Identify model features 2-3	2-8
Tutorial: Identify model features 2-4	2-8
Tutorial: Identify model features 2-5	2-9
Identify dimensions and parameters	2-10
Tutorial: Identify dimensions and parameters 2-1	2-10
Tutorial: Identify dimensions and parameters 2-2	2-11
Tutorial: Identify dimensions and parameters 2-3	2-12
Identify the correct reference planes: Top, Right, and Front	2-13
2D sketching / reference planes	2-14
Tutorial: Reference plane 2-1	2-15
Tutorial: Reference plane 2-2	2-16
Determine the design intent for a model	2-18
Design intent in a sketch	2-18
Design intent in a feature	2-19
Design intent in a part	2-20
Design intent in an assembly	2-20
Design intent in a drawing	2-20
Tutorial: Design intent 2-1	2-21
Tutorial: Design intent 2-2	2-21
Identify material, measure, and mass properties	2-22
Assign and edit material	2-22
Tutorial: Assign and edit material 2-1	2-22
Tutorial: Assign and edit materials 2-2	2-23
Apply the Measure tool	2-23
Tutorial: Measure tool 2-1	2-24
Tutorial: Measure tool 2-2	2-24
Locate the Center of mass, and Principal moments of inertia	2-25
Tutorial: Mass properties 2-1	2-26
Tutorial: Mass properties 2-2	2-27
Identify the function and elements of a part and assembly FeatureManager	2-28
Sketch states	2-28
Component status and properties	2-30
Display Pane status	2-33
Reference configuration	2-34
Identify the default Sketch Entities from the Sketch toolbar	2-35

When do you use the appropriate Sketch Entities?	2-35
Identify the default Sketch tools from the Sketch toolbar	2-37
When do you use the appropriate Sketch tool?	2-37
SolidWorks file formats for input and export	2-38
Tutorial: File formats 2-1	2-39
Tutorial: File formats 2-2	2-40
Tutorial: File formats 2-3	2-40
Utilize SolidWorks Help topics	2-41
Tutorial: SolidWorks help 2-1	2-42
Tutorial: SolidWorks help 2-2	2-42
Identify the process of creating a simple drawing	2-42
Tutorial: Drawing process 2-1	2-43
Tutorial: Drawing process 2-2	2-45
Recognize all Drawing name view types by their icons	2-46
Tutorial: Drawing name view type 2-1	2-46
Tutorial: Drawing name view type 2-2	2-47
Identify the procedure to create a Named Drawing view	2-48
Tutorial: Drawing named procedure 2-1	2-49
Tutorial: Drawing named procedure 2-2	2-49
Tutorial: Drawing named procedure 2-3	2-49
Tutorial: Drawing named procedure 2-4	2-50
Tutorial: Drawing named procedure 2-5	2-50
Tutorial: Drawing named procedure 2-6	2-51
Tutorial: Drawing named procedure 2-7	2-51
Tutorial: Drawing named procedure 2-8	2-52
Engineering Documentation Practices	2-52
Specify Document Properties	2-53
Tutorial: Document properties 2-1	2-54
Tutorial: Document properties 2-2	2-54
Summary	2-54
Key terms	2-55
Check your understanding	2-57
Chapter 3: Part Modeling	**3-1**
Chapter Objective	3-1
Read and understand an Engineering document	3-2
Tutorial: Simple part 3-1	3-2
Build a simple part from a detailed dimensioned illustration	3-4
Tutorial: Simple part 3-2	3-4
Tutorial: Simple part 3-3	3-6
Tutorial: Simple part 3-4	3-7
Tutorial: Volume / Center of mass 3-1	3-9
Tutorial: Volume / Center of mass 3-2	3-10
Tutorial: Mass-Volume 3-3	3-13
Tutorial: Mass-Volume 3-4	3-14
Tutorial: Simple Cut 3-1	3-17

Introduction

Tutorial: Mass-Volume 3-5	3-18
Tutorial: Mass-Volume 3-6	3-20
Tutorial: Mass-Volume 3-7	3-22
2D vs. 3D Sketching	3-24
Tutorial: 3DSketch 3-1	3-24
Tutorial: Mass-Volume 3-8	3-26
Tutorial: Mass-Volume 3-9	3-28
Callout value	3-31
Tolerance type	3-31
Tutorial: Dimension text 3-1	3-32
Tutorial: Dimension text 3-2	3-32
Tutorial: Dimension text 3-3	3-33
Dimension text symbols	3-33
Tutorial: Dimension text symbols 3-1	3-34
Tutorial: Dimension text symbols 3-2	3-34
Build additional simple parts	3-35
Tutorial: Mass-Volume 3-10	3-35
Tutorial: Mass-Volume 3-11	3-37
Tutorial: Mass-Volume 3-12	3-39
Tutorial: Mass-Volume 3-13	3-40
Tutorial: Mass-Volume 3-14	3-42
Tutorial: Mass-Volume 3-15	3-43
Tutorial: Mass-Volume 3-16	3-45
Tutorial: Basic-part 3-1	3-47
Tutorial: Basic-part 3-2	3-50
Tutorial: Basic-part 3-3	3-53
Tutorial: Basic-part 3-4	3-56
Summary	3-58
Key terms	3-58
Check your understanding	3-60
Chapter 4: Advanced Part Modeling	**4-1**
Chapter Object	4-1
Build an Advanced part from a detailed dimensioned illustration	4-2
Tutorial: Advanced part 4-1	4-2
Tutorial: Advanced part 4-2	4-6
Tutorial: Advanced part 4-3	4-9
Tutorial: Advanced part 4-4	4-12
Calculate the Center of mass relative to a created coordinate system	4-17
Tutorial: Coordinate location 4-1	4-17
Tutorial: Coordinate location 4-2	4-19
Tutorial: Advanced part 4-5	4-20
Tutorial: Advanced part 4-5A	4-24
Tutorial: Advanced part 4-5B	4-25
Tutorial: Advanced part 4-6	4-27
Tutorial: Advanced part 4-6A	4-33

Introduction

Tutorial: Advanced part 4-7	4-34
Summary	4-39
Key terms	4-39
Check your understanding	4-41

Chapter 5: Assembly Modeling — **5-1**

Chapter Objective	5-1
Assembly modeling techniques	5-2
Top-down, "in-context"	5-2
Bottom-up	5-2
Mates	5-3
Standard mates	5-3
Creating mates	5-4
Tutorial: Standard mate 5-1	5-5
Tutorial: Standard mate 5-2	5-6
Tutorial: Standard mate 5-3	5-7
Tutorial: Standard mate 5-4	5-7
Build an assembly for a detailed dimensioned illustration	5-9
Tutorial: Assembly model 5-1	5-10
Tutorial: Assembly model 5-2	5-19
Tutorial: Assembly model 5-3	5-26
Mate the first component with respect to the assembly reference planes	5-36
Tutorial: Assembly model 5-4	5-36
Summary	5-39
Key terms	5-40
Check your understanding	5-42

Chapter 6: Advanced Modeling Theory and Analysis — **6-1**

Chapter Objectives	6-1
Definition review	6-1
COSMOSXpress	6-8
COSMOSXpress User Interface	6-9
Tutorial: COSMOSXpress 6-1	6-10
Tutorial: COSMOSXpress 6-2	6-17
Tutorial: COSMOSXpress 6-3	6-19
COSMOSWorks Designer	6-25
COSMOSWorks Professional	6-26
COSMOSMotion	6-27
COSMOSFloWorks	6-27
Summary	6-28
Key terms	6-28
Check your understanding	6-32

Appendix — **A-1**

Check your understanding answer key	A-1
Chapter 2	A-1

Chapter 3 A-3
Chapter 4 A-4
Chapter 5 A-5
Chapter 6 A 6

Index

Notes:

CHAPTER 1: SOLIDWORKS 2008 USER INTERFACE

Chapter Objective

SolidWorks is a design software application used to model and create 2D and 3D sketches, 3D parts and assemblies, and 2D drawings. Chapter 1 reviews the SolidWorks 2008 User Interface and CommandManager: *Menu bar toolbar, Menu bar menu, Drop-down menus, Context toolbars, Consolidated drop-down toolbars, System feedback icons, Confirmation Corner, Heads-up View toolbar,* and more.

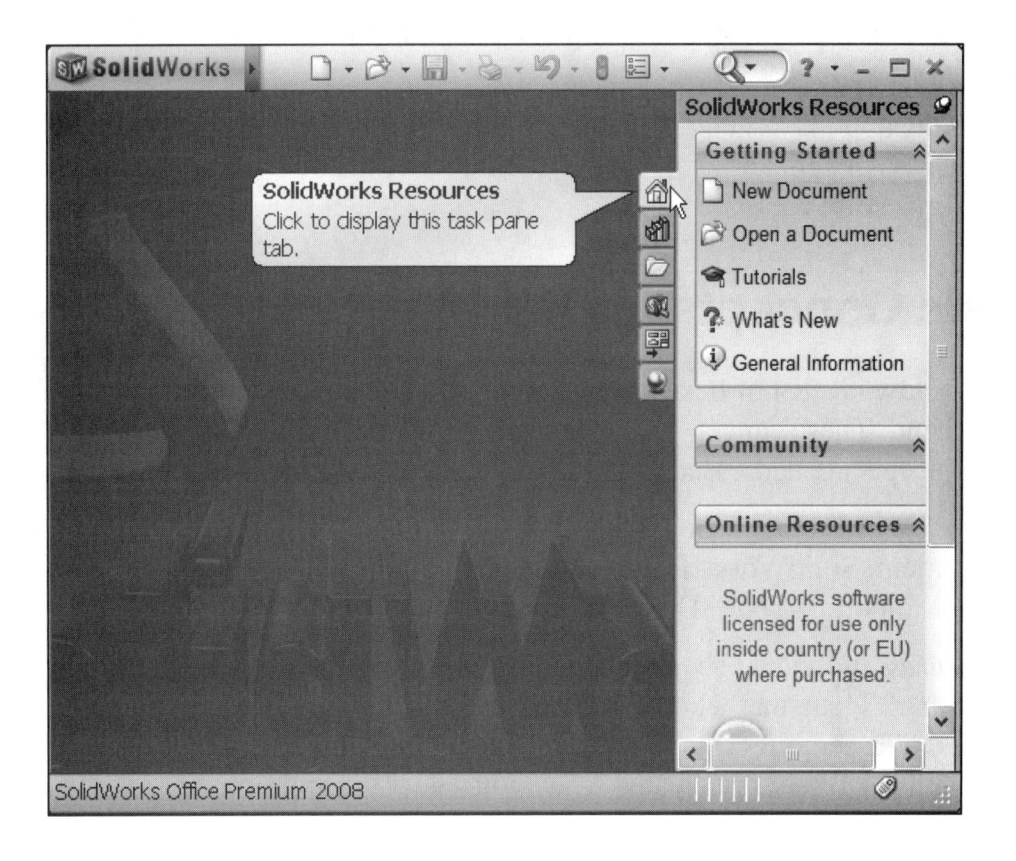

☼ The CSWA exam at this time is given in both SolidWorks 2007 and SolidWorks 2008.

What is SolidWorks?

The SolidWorks® application is a mechanical design automation software package used to create parts, assemblies, and drawings which take advantage of the familiar Microsoft Windows graphical user interface.

SolidWorks is an easy to learn design and analysis tool, (COSMOSXpress, COSMOSWorks®, COSMOSFloWorks™, and COSMOSMotion™) which makes it possible for designers to quickly sketch 2D and 3D concepts, create 3D parts and assemblies, and detail 2D drawings. In SolidWorks part, assembly, and drawing documents are all related.

☼ In SolidWorks 2009 the following name changes will occur: COSMOSXpress to SolidWorks® SimulationXpress, COSMOSWorks to SolidWorks Simulation, COSMOSFloWorks to SolidWorks Flow Simulation, and COSMOSMotion to SolidWorks Motion.

This book is targeted towards the SolidWorks user who wants to prepare for the CSWA exam and who has six or more months of engineering design and SolidWorks software experience.

Basic Concepts in SolidWorks

Below is a list of basic concepts in SolidWorks to review and to comprehend. These concepts are applicable to all versions of SolidWorks. All of these concepts are addressed in this book. They are:

- *A SolidWorks model.* Consists of 3D solid geometry in a part or assembly document. SolidWorks features start with either a 2D or 3D sketch. You can either import a 2D or 3D sketch or you can create the sketch in SolidWorks.

- *Features.* Individual shapes created by Sketch Entities tools: Lines, Circles, Rectangles, etc. that when combined, creates the part. Features can also be added to assemblies. Some features originate as sketches; other features, such as shells or fillets, are created when you select the appropriate tool or menu command and define the dimensions or characteristics that you want.

- *Base sketch.* The first sketch of a part is called the Base sketch. The Base sketch is the foundation for the 3D model. Create a 2D sketch on a default plane: Front, Top, and Right in the FeatureManager design tree, or on a created plane. You can also import a surface or solid geometry. In a 3D sketch, the Sketch Entities exist in 3D space. Sketch Entities do not need to be related to a specific Sketch plane.

SolidWorks 2008 User Interface

- *Refining the design.* The procedure of adding, editing, or reordering features in the FeatureManager design tree and in the Graphics window has been greatly enhanced. You can perform the following types of editing feature operations:

  - Rollback the part to the state it was in before a selected feature was added either with the:

    - Rollback bar in the FeatureManager.

    - Roll to Previous command from the Feature dialog box.

    - Roll to Previous command from the Pop-up shortcut toolbar.

  - Edit the definition, the sketch, or the properties of a feature by:

    - Selecting the feature or sketch in the FeatureManager. The Pop-up shortcut toolbar is displayed. Select the Edit command.

    - Selecting the feature or sketch in the FeatureManager, right-click in the Graphics window. The Feature dialog box is displayed. Select the Edit command.

To activate the Pop-up shortcut toolbar, (also know as the Context toolbar) either click the feature or sketch in the FeatureManager, right-click, or click the feature in the Graphics window.

Right-click on a feature in the FeatureManager will active the Context toolbar, Feature and Body dialog box as illustrated.

  - Control the access to selected dimensions. Click on either a feature or sketch in the FeatureManager or Graphics window. View the illustrated dimensions.

  - View the parent and child relationships of a feature.

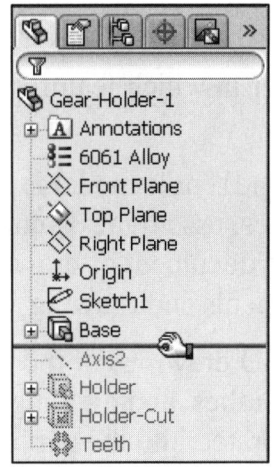

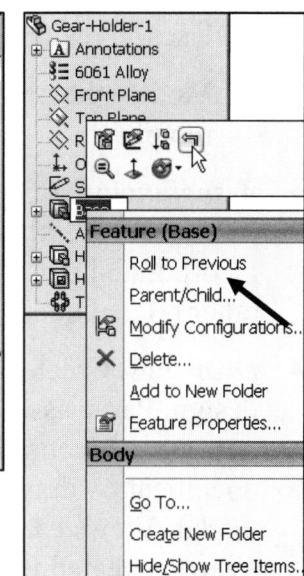

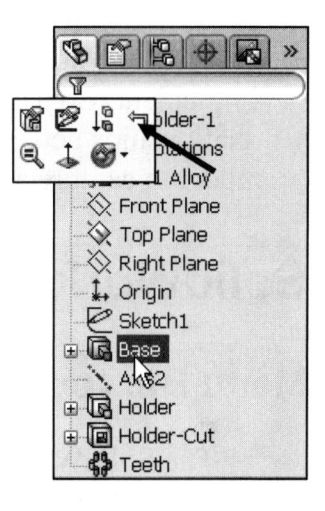

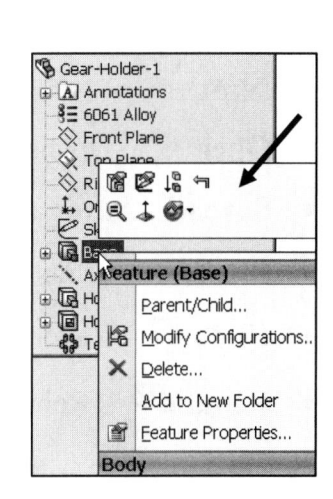

Page 1 - 3

- Use the feature handles to move and resize features.
- Modify the order in which features are reconstructed when the part is rebuilt.

- *Associativity.* A SolidWorks model is fully Associative. Associativity between parts, sub-assemblies, assemblies, and drawings assure that changes incorporated in one document or drawing view are automatically made to all other related documents and drawing views.

- *Drawings.* Create 2D drawings of the 3D solid parts and assemblies which you design. Parts, assemblies, and drawings are linked documents. This means that any change incorporated into the part or assembly changes the drawing document. A drawing generally consists of several views generated from the model. Views can also be created from existing views. Example: The Section View is created from an existing drawing view.

- *Constraints.* SolidWorks supports numerous constraints. Constraints are geometric relations such as: Perpendicular, Horizontal, Parallel, Vertical, Coincident, Concentric, etc. Apply equations to establish mathematical relationships between parameters. Insert equations and constraints to your model to capture and maintain design intent.

SolidWorks User Interface (UI)

Menu Bar toolbar

The SolidWorks 2008 (UI) is designed to make maximum use of the Graphics window area. The Menu Bar toolbar contains a set of the most frequently used tool buttons from the Standard toolbar. The available tools are:

- **New** – Creates a new document.
- **Open** – Opens an existing document.
- **Save** – Saves an active document.
- **Print** – Prints an active document.
- **Undo** – Reverses the last action.
- **Rebuild** – Rebuilds the active part, assembly or drawing.
- **Options** – Changes system options, document properties, and Add-Ins for SolidWorks.

☼ Click the down-arrow next to the tool button to expand and display the fly-out menu with additional functions.

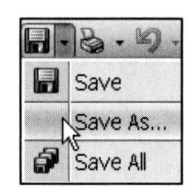

Menu Bar menu

Click SolidWorks to display the default Menu Bar menu as illustrated. SolidWorks provides a context-sensitive menu structure. The menu titles remain the same for all three types of documents, but the menu items change depending on which type of document is active. Example: The Insert menu includes features in part documents, mates in assembly documents, and drawing views in drawing documents. The display of the menu is also dependent on the work flow customization that you have selected. The default menu items for an active document are: *File, Edit, View, Insert, Tools, Window, Help,* and *Pin*.

☼ The Pin option displays the Menu Bar toolbar and the Menu Bar menu as illustrated. In future chapters, the Menu Bar menu and the Menu Bar toolbar will be referred to as just the Menu bar.

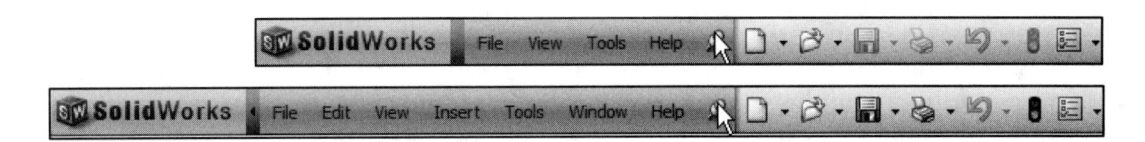

Drop-down menu

SolidWorks takes advantage of the familiar Microsoft® Windows® user interface. Communicate with SolidWorks either through the drop-down menu, Pop-up shortcut toolbar or menu, flyout consolidated toolbar or the CommandManager tab. A command is an instruction that informs SolidWorks to perform a task.

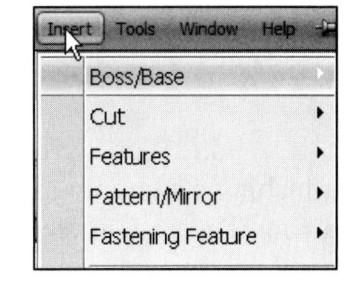

To close a SolidWorks drop-down menu, press the Esc key. You can also click any other part of the SolidWorks Graphics window, or click another drop-down menu.

Right-Click Pop-up menus

Right-click in the Graphics window on a model, or in the FeatureManager on a feature or sketch to display a context-sensitive shortcut toolbar. If you are in the middle of a command, this toolbar displays a list of options specifically related to that command.

The most commonly used tools are located in the Pop-up menu, toolbar, and CommandManager.

Consolidated fly-out tool buttons

In the Consolidated fly-out toolbar, similar commands are grouped together. Example: Variations of the rectangle tool are grouped together into a single button with a fly-out control as illustrated.

If you select the consolidated fly-out button without expanding:

- For some commands such as Sketch, the most commonly used command is performed. This command is the first listed and the command shown on the button.

- For commands such as rectangle, where you may want to repeatedly create the same variant of the rectangle, the last used command is performed. This is the highlighted command when the consolidated fly-out tool is expanded.

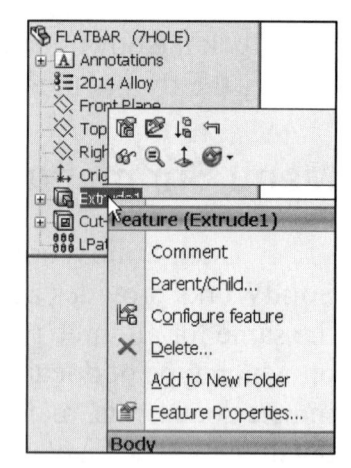

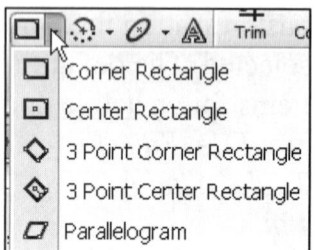

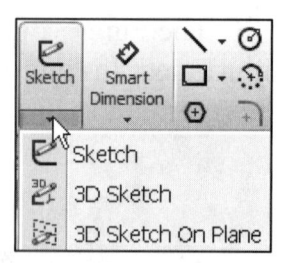

System feedback

SolidWorks provides system feedback by attaching a symbol to the mouse pointer cursor arrow. The system feedback symbol indicates what you are selecting or what the system is expecting you to select.

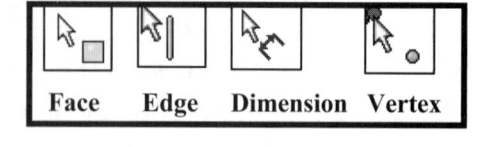

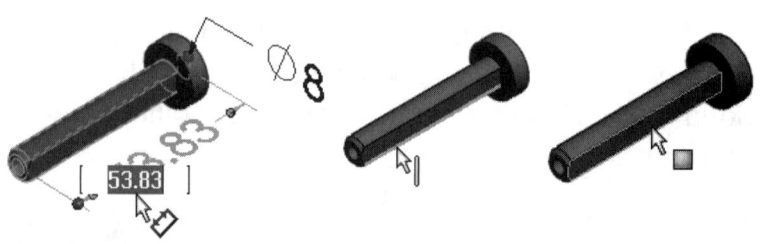

As you move the mouse pointer across your model, system feedback is provided to you in the form of symbols, riding next to the cursor arrow.

Confirmation Corner

When numerous SolidWorks commands are active, a symbol or a set of symbols are displayed in the upper right hand corner of the Graphics window. This area is called the Confirmation Corner.

When a sketch is active, the confirmation corner box displays two symbols. The first symbol is the sketch tool icon. The second symbol is a large red X. These two symbols supply a visual reminder that you are in an active sketch. Click the sketch symbol icon to exit the sketch and to saves any changes that you made.

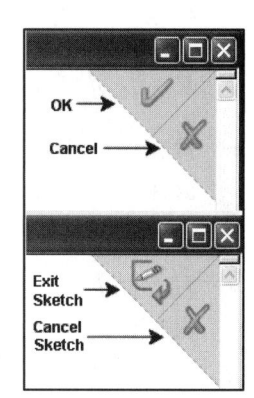

When other commands are active, the confirmation corner box provides a green check mark and a large red X. Use the green check mark to execute the current command. Use the large red X to cancel the command.

Heads-up View toolbar

SolidWorks provides the user with numerous view options from the Standard Views, View, and Heads-up View toolbar.

The Heads-up View toolbar is a transparent toolbar that is displayed in the Graphics window when a document is active. You can not hide or move the Heads-up View toolbar. You can modify it.

The following views are available: Note: The available views are document dependent.

- *Zoom to Fit* : Zooms the model to fit the Graphics window.

- *Zoom to Area* : Zooms to the areas you select with a bounding box.

- *Previous View* : Displays the previous view.

- *Section View* : Displays a cutaway of a part or assembly, using one or more cross section planes.

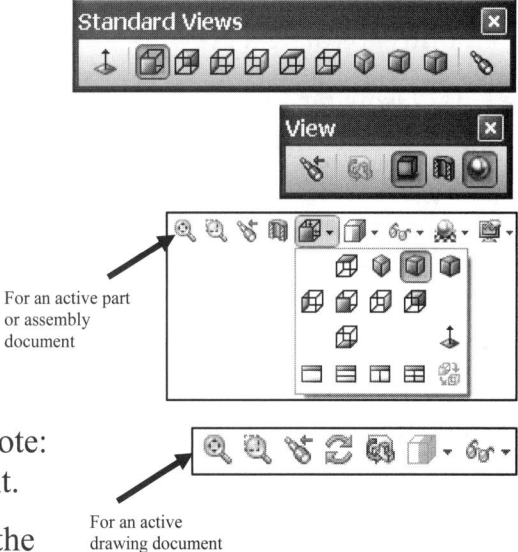

SolidWorks 2008 User Interface

- *View Orientation*: Provides the ability to select a view orientation or the number of viewports. The available options are: *Top, Isometric, Trimetric, Dimetric, Left, Front, Right, Back, Bottom, Single view, Two view - Horizontal, Two view - Vertical, Four view*.

- *Display Style*: Provides the ability to display the style for the active view: The available options are: *Wireframe, Hidden Lines Visible, Hidden Lines Removed, Shaded, Shaded With Edges*.

- *Hide/Show Items*: Provides the ability to select items to hide or show in the Graphics window. Note: The available items are document dependent.

- *Apply Scene*: Provides the ability to apply a scene to an active part or assembly document. View the available options.

- *View Setting*: Provides the ability to select the following: *RealView Graphics, Shadows in Shaded Mode*, and *Perspective*.

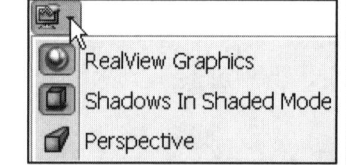

- *Rotate view*: Provides the ability to rotate a drawing view.

- *3D Drawing View*: Provides the ability to dynamically manipulate the drawing view to make a selection.

☼ The Heads-up View toolbar replaces the Reference triad in the lower left corner of the Graphics window.

☼ The default part and document setting displays the grid. To deactivate the grid, click **Options**, **Document Properties** tab. Click **Grid/Snaps**, uncheck the **Display grid** box

☼ To deactivate the planes, click **View**, uncheck **Planes** from the Menu bar.

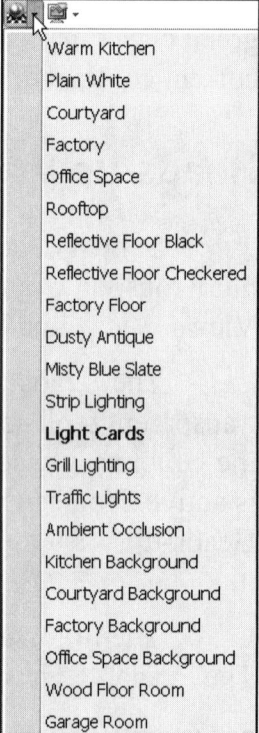

CommandManager

The CommandManager is document dependent. Drop-down tabs are located on the bottom left side of the CommandManager and display the available toolbars and features for each corresponding tab. The default Part tabs are: Features, Sketch, Evaluate, DimXpert, and Office Products.

Below is an illustrated CommandManager for a default Part document.

The Office Products toolbar display is dependent on the activated Add-Ins.. during a SolidWorks session.

If you have SolidWorks Office, SolidWorks Office Professional, or SolidWorks Office Premium, the Office Products tab is displayed in the CommandManager. The book was written with SolidWorks SP2.1.

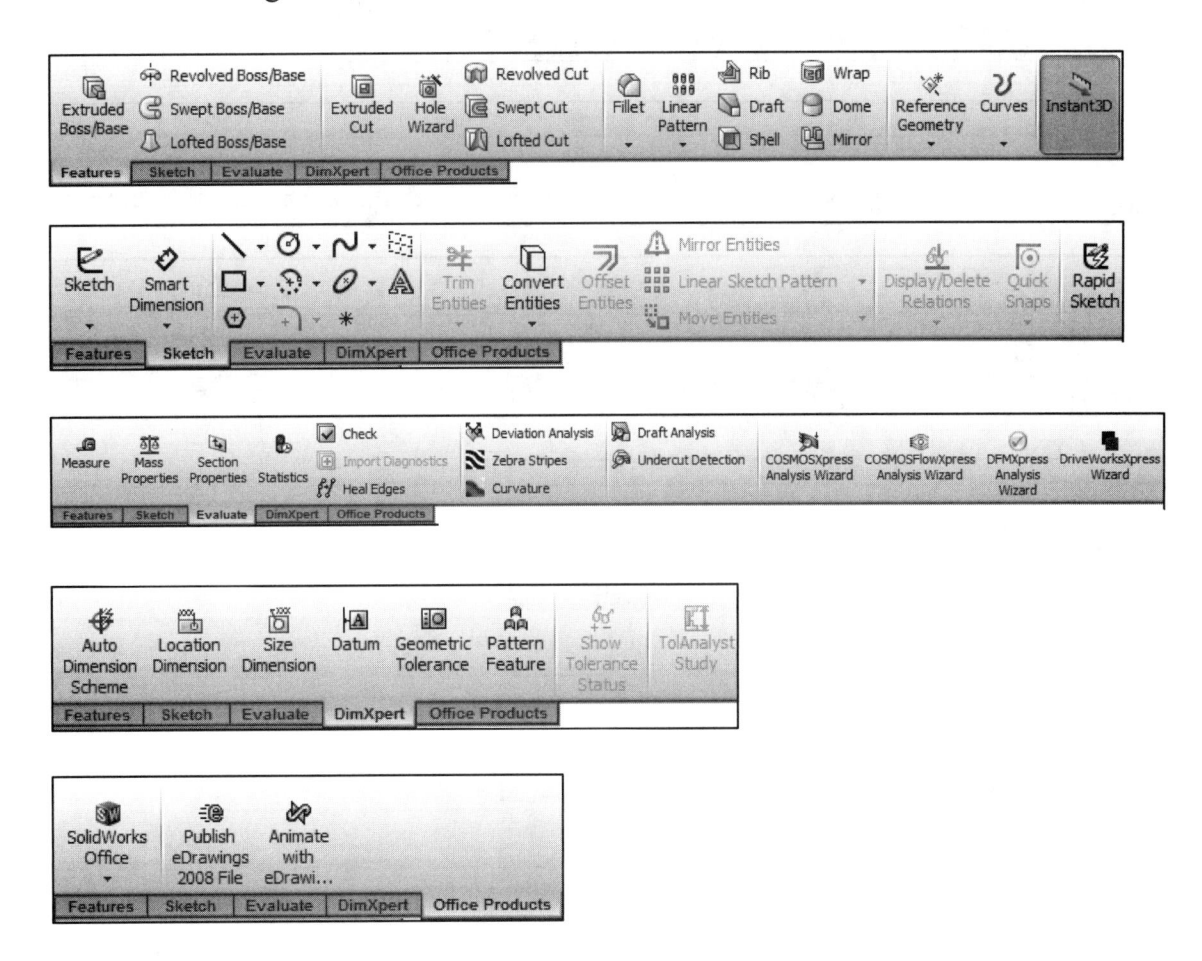

SolidWorks 2008 User Interface

Below is an illustrated CommandManager for a default Drawing document. The default Drawing tabs are: *View Layout, Annotation, Sketch, Evaluate,* and *Office Products*. Note: In older SolidWorks versions, the Annotation tab is named Annotate.

☼ The Office Products toolbar display is dependent on the activated Add-Ins.. during a SolidWorks session.

☼ If you have SolidWorks Office, SolidWorks Office Professional, or SolidWorks Office Premium, the Office Products tab is displayed in the CommandManager. The book was written with SolidWorks version SP2.1.

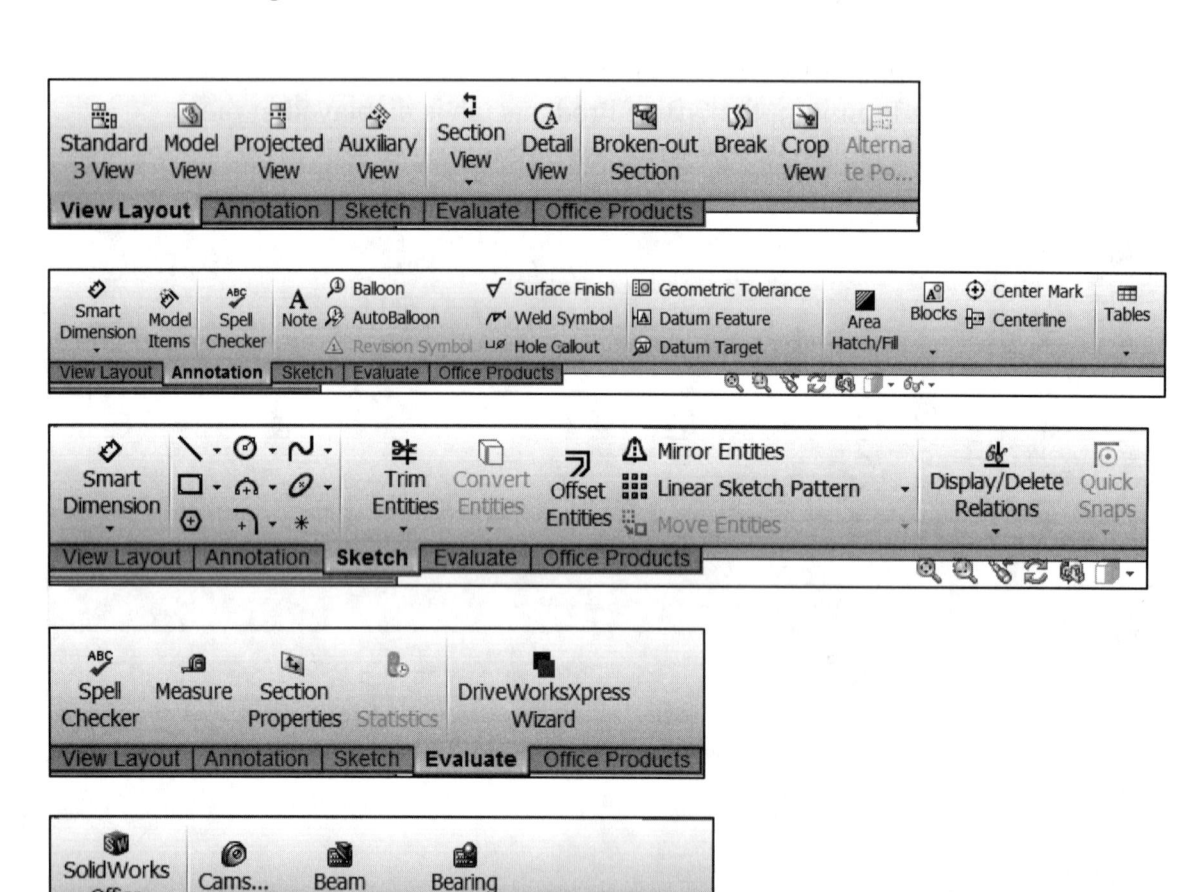

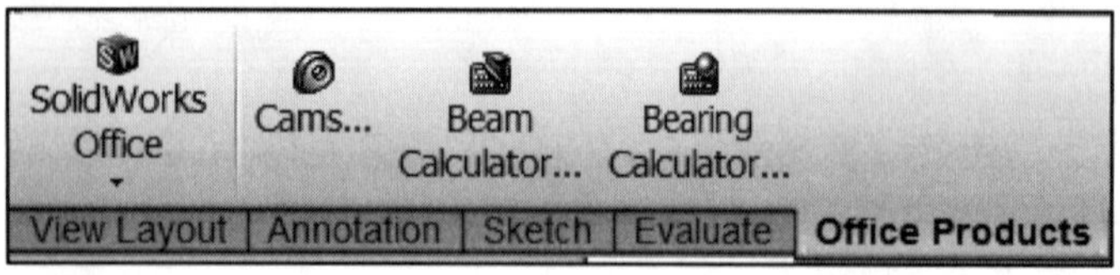

Page 1 - 10

SolidWorks 2008 User Interface

Below is an illustrated CommandManager for a default Assembly document. The default Assembly tabs are: *Assembly, Layout, Sketch, Evaluate,* and *Office Products*.

☼ The Office Products toolbar display is dependent on the activated Add-Ins.. during a SolidWorks session.

☼ If you have SolidWorks Office, SolidWorks Office Professional, or SolidWorks Office Premium, the Office Products tab is displayed in the CommandManager. The book was written with SolidWorks version SP2.1.

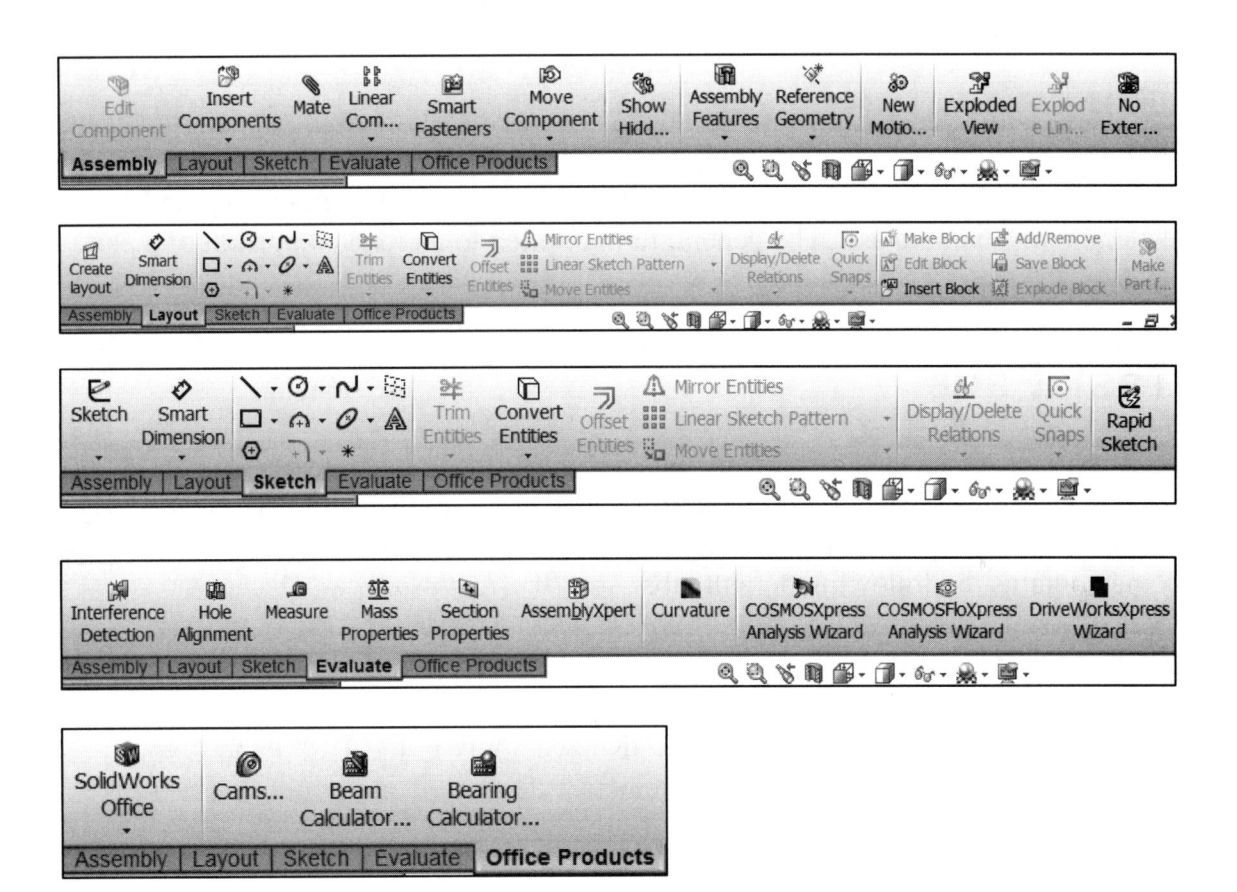

Page 1 - 11

SolidWorks 2008 User Interface

☼ The tabs replace the Control areas buttons from pervious SolidWorks versions. The tabs that are displayed by default depend on the type of open document and the work flow customization that you have selected.

☼ To customize the CommandManager tabs, **right-click** on a tab, and select the required **custom** option or select Customize CommandManager to access the Customize dialog box.

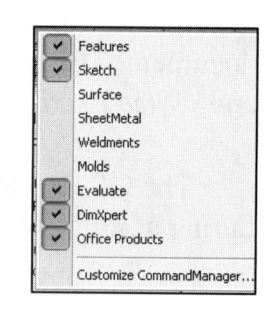

☼ DimXpert for parts provides the ability to graphically check if the model is fully dimensioned and toleranced.

☼ Both DimXpert for parts and drawings automatically recognize manufacturing features. Manufacturing features are *not SolidWorks features*. Manufacturing features are defined in 1.1.12 of the ASME Y14.5M-1994 Dimensioning and Tolerancing standard as: "The general term applied to a physical portion of a part, such as a surface, hole or slot.

Task Pane

The Task Pane is displayed when a SolidWorks session starts. The Task Pane can be displayed in the following states: visible or hidden, expanded or collapsed, pinned or unpinned, docked or floating. The Task Pane contains the following default tabs: *SolidWorks Resources*, *Design Library*, *File Explorer*, *SolidWorks Search*, *View Palette*, *RealView*, and *Document Recovery*.

☼ The Document Recovery tab is displayed only in the Task Pane if your system terminates unexpectedly with an active document and if auto-recovery is enabled in System Options.

SolidWorks Resources

The basic SolidWorks Resources menu displays the following default selections: *Getting Started*, *Community*, *Online Resources*, and *Tip of the Day*.

Other user interfaces are available: *Machine Design*, *Mold Design*, or *Consumer Products Design* during the initial software installation selection.

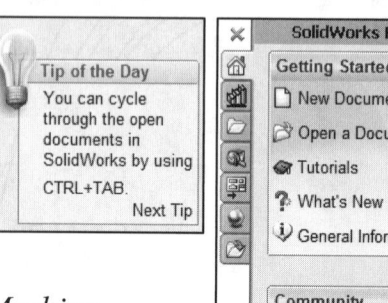

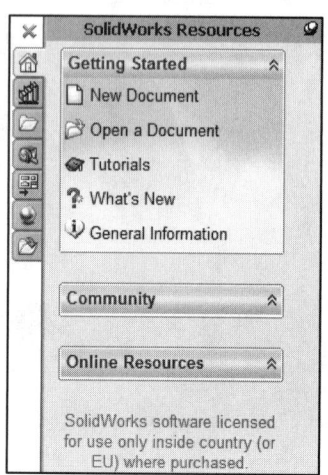

Page 1 - 12

Design Library

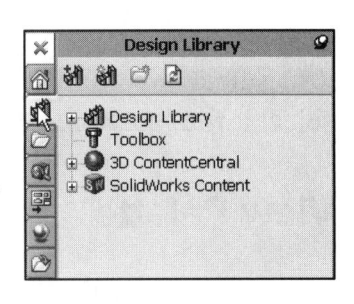

The Design Library ⌘ contains reusable parts, assemblies, and other elements, including library features.

The Design Library tab contains four default selections. Each default selection contains additional sub categories. The default selections are: *Design Library*, *Toolbox*, *3D ContentCentral*, and *SolidWorks Content*.

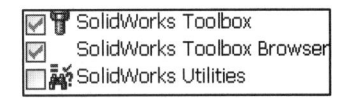

☼ Click **Tools**, **Add-Ins..**, **SolidWorks Toolbox** and **SolidWorks Toolbox Browser** to activate the SolidWorks Toolbox.

☼ At this time, there are no questions on the exam that requires knowledge of the Design Library, SolidWorks Toolbox, 3D ContentCentral or SolidWorks Content.

File Explorer

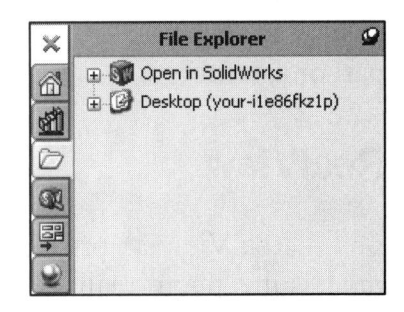

File Explorer ⌘ duplicates Windows Explorer from your local computer and displays the following directories: *Open in SolidWorks* and *Desktop*.

Search

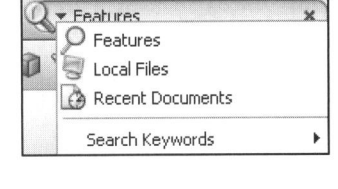

SolidWorks Search ⌘ is installed with Microsoft Windows Search and indexes the resources once before searching begins, either after installation, or when you initiate the first search.

The SolidWorks Search box is displayed in the upper right corner of the SolidWorks Graphics window. Enter the text or key words to search. Click the drop-down arrow to view the last 10 recent searches.

The Search tool ⌕ in the Task Pane searches the following default locations: *All Locations*, *Local Files*, *Design Library*, *SolidWorks Toolbox*, and *3D ContentCentral*.

☼ Select any or all of the above locations. If you do not select a file location, all locations are searched.

View Palette

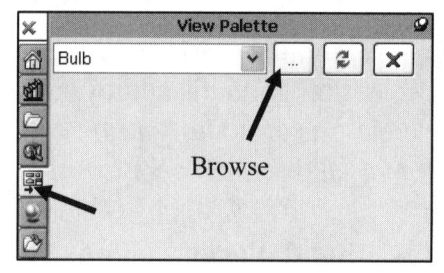

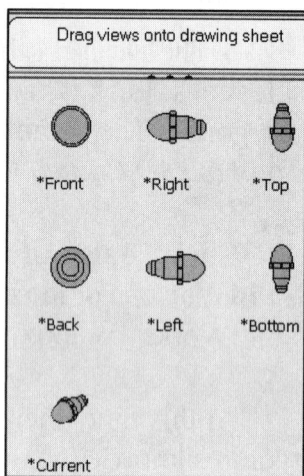

The View Palette tool located in the Task Pane provides the ability to insert drawing views of an active document, or click the Browse button to locate the desired document.

Click and drag the view from the View Palette into an active drawing sheet to create a drawing view.

☼ At this time, there are no questions on the CSWA exam that requires the user to create an actual drawing of a part or assembly.

RealView

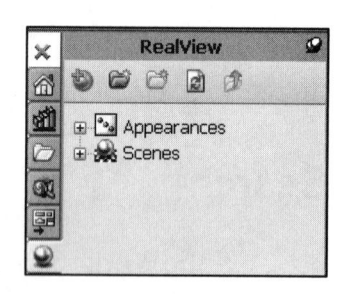

RealView provides a simplified way to display models in a photo-realistic setting using a library of appearances and scenes.

On RealView compatible systems, you can select Appearances and Scenes to display your model in the Graphics window. Drag and drop a selected appearance onto the model or FeatureManager. View the results in the Graphics window.

☼ PhotoWorks needs to be active to apply the Scenes tool.

☼ RealView graphics is only available with supported graphics cards. For the latest information on graphics cards that support RealView Graphics display, visit: www.solidworks.com/pages/services/videocardtesting.html.

Document Recovery

If auto recovery is initiated in the System Options section and the system terminates unexpectedly with an active document, the saved information files are available on the Task Pane Document Recovery tab the next time you start a SolidWorks session.

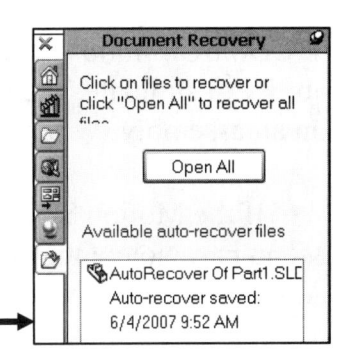

Motion Study tab

The Motion Study tab is located in the bottom left corner of the Graphics window. Motion Study uses a key frame-based interface, and provides a graphical simulation of motion for a model. Click the Motion Study tab to view the MotionManager. Click the Model tab to return to the FeatureManager design tree.

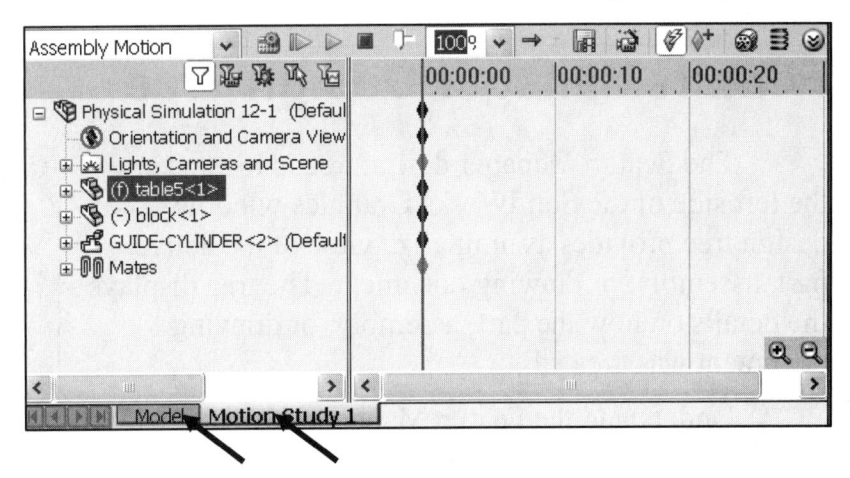

The MotionManager displays a timeline-based interface, and provides the following selections:

1. *All levels*. Provides the ability to change viewpoints, display properties, and create animations displaying the assembly in motion.

2. *Assembly Motion*. (Available in core SolidWorks.) Provides the ability to animate the assembly and to control the display at various time intervals. The Assembly Motion option computes the sequences required to go from one position to the next.

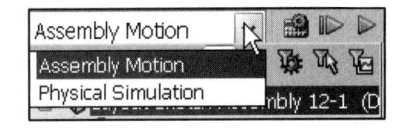

3. *Physical Simulation*. (Available in core SolidWorks.) Provides the ability to simulating the effects of motors, springs, dampers, and gravity on assemblies. This options combines simulation elements with SolidWorks tools such as mates and Physical Dynamics to move components around the assembly.

4. *COSMOSMotion*. (Available in SolidWorks Office Premium.) Provides the ability to simulate, and analyze the effects of forces, contacts, friction, and motion on an assembly.

☼ If the Motion Study tab is not visible, click **View**, **MotionManager** from the Menu bar. Note: On a model that was created before SolidWorks 2008, the Annotation tab may be displayed in the Motion Study location.

☼ To create a new Motion Study, click **Insert, New Motion Study** from the Menu bar menu.

☼ The PhotoWorks Items tab is displayed if PhotoWorks is installed. No PhotoWorks questions are on the CSWA exam at this time.

☼ No Motion Study questions are on the CSWA exam at this time.

FeatureManager Design Tree

The FeatureManager design tree is located on the left side of the SolidWorks Graphics window. The design tree provides a summarize view of the active part, assembly, or drawing document. The tree displays the details on how the part, assembly, or drawing document was created.

Understand the FeatureManager design tree to troubleshoot your model. The FeatureManager is use extensively throughout this book.

The FeatureManager consist of four default tabs:

- *FeatureManager design tree*
- *PropertyManager*
- *ConfigurationManager*
- *DimXpertManager*

☼ Select the Hide FeatureManager Tree Area arrows to enlarge the Graphics window for modeling.

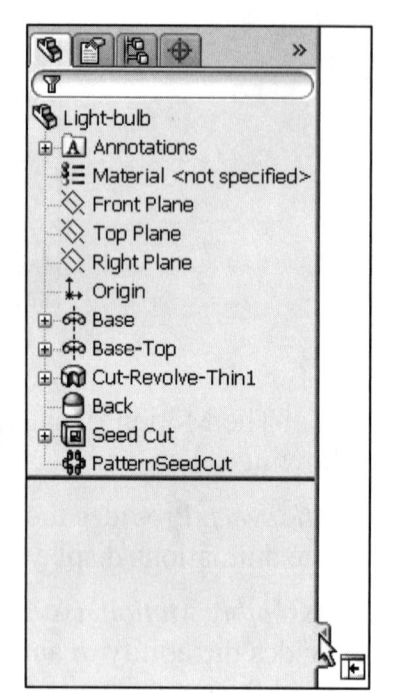

Various commands provide the ability to control what is displayed in the FeatureManager design tree. They are:

1. Show or Hide FeatureManager items.

💡 Click **Options** from the Menu bar. Click **FeatureManager** from the System Options tab. **Customize** your FeatureManager from the Hide/Show Tree Items dialog box.

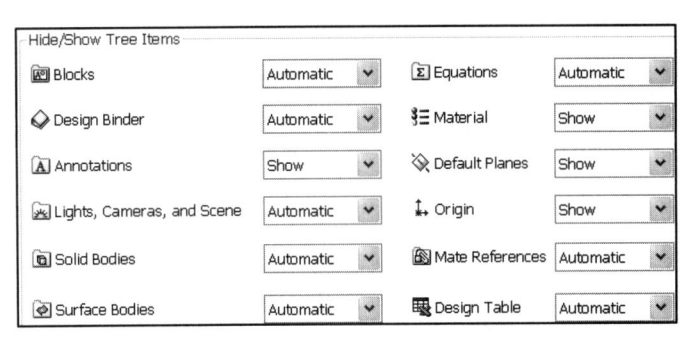

2. Filter the FeatureManager design tree. Enter information in the filter field. You can filter by: *Type of features, Feature names, Sketches, Folders, Mates, User-defined tags*, and *Custom properties*.

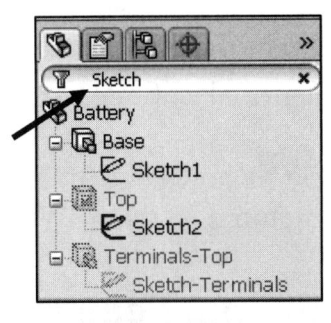

💡 Tags are keywords you can add to a SolidWorks document to make them easier to filter and to search. The Tags icon is located in the bottom right corner of the Graphics window.

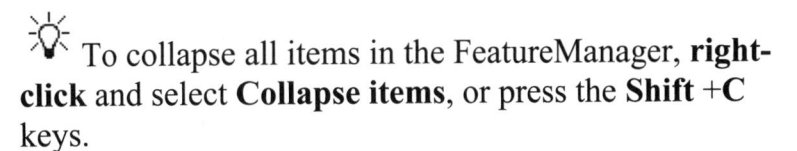

💡 To collapse all items in the FeatureManager, **right-click** and select **Collapse items**, or press the **Shift +C** keys.

The FeatureManager design tree and the Graphics window are dynamically linked. Select sketches, features, drawing views, and construction geometry in either pane.

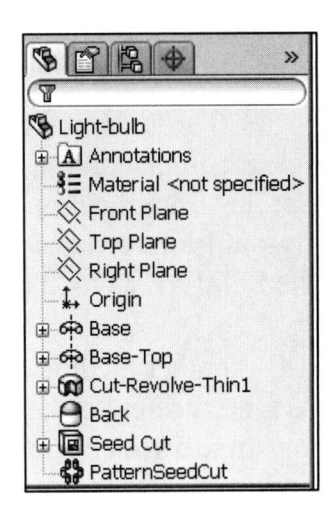

Split the FeatureManager design tree and either display two FeatureManager instances, or combine the FeatureManager design tree with the ConfigurationManager or PropertyManager.

Move between the FeatureManager design tree, PropertyManager, ConfigurationManager, and DimXpertManager by selecting the tabs at the top of the menu.

The ConfigurationManager is located to the right of the FeatureManager. Use the ConfigurationManager to create, select, and view multiple configurations of parts and assemblies.

Split the ConfigurationManager and either display two ConfigurationManager instances, or combine the ConfigurationManager with the FeatureManager design tree, PropertyManager, or a third party application that uses the panel. The icons in the ConfigurationManager denote whether the configuration was created manually or with a design table.

The DimXpertManager tab provides the ability to insert dimensions and tolerances manually or automatically. The DimXpertManager provides the following selections: Auto Dimension Scheme, Show Tolerance Status, Copy Scheme, and TolAnalyst Study.

Knoweldge of DimXpert and TolAnalyst is not required on the CSWA exam at this time.

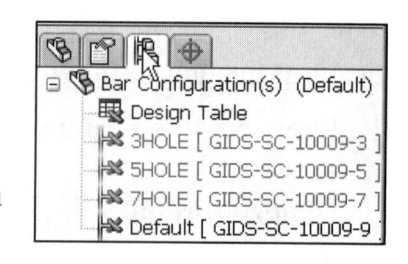

Fly-out FeatureManager

The fly-out FeatureManager design tree provides the ability to view and select items in the PropertyManager and the FeatureManager design tree at the same time.

Throughout the book, you will select commands and command options from the drop-down menus, fly-out FeatureManagers, shortcut toolbar, or from the SolidWorks toolbars.

Another method for accessing a command is to use the accelerator key. Accelerator keys are special keystrokes which activates the drop-down menu options. Some commands in the menu bar and items in the drop-down menus have an underlined character. Press the Alt key followed by the corresponding key to the underlined character activates that command or option.

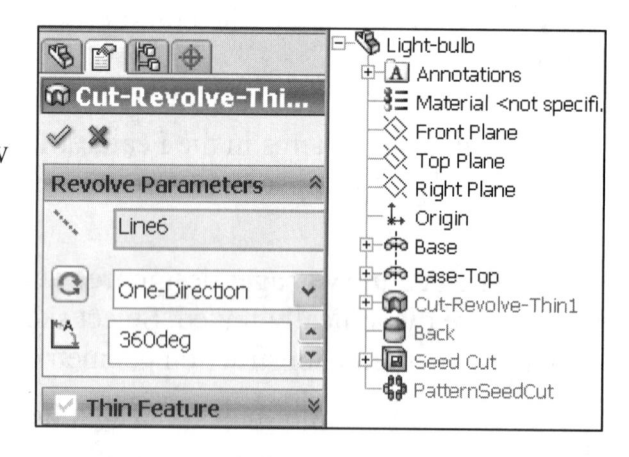

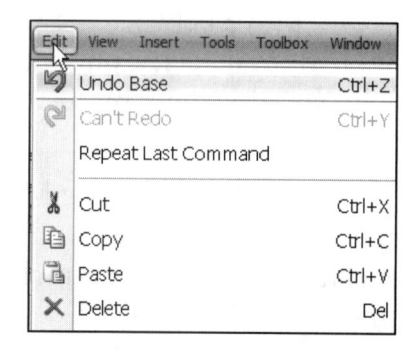

Summary

The SolidWorks 2008 User Interface and CommandManager consist of the following options: Menu bar toolbar, Menu bar menu, Drop-down menus, Shortcut toolbars, Context toolbar, Consolidated flyout menus, System feedback icons, Confirmation Corner, Heads-up View toolbar and more.

There are two modes in the New SolidWorks Document dialog box: Novice and Advanced. The Novice option is the default option with three templates. The Advanced option contains access to more templates.

The Part FeatureManager design tree consist of four default tabs: FeatureManager design tree, PropertyManager, ConfigurationManager, and DimXertManager.

The CommandManager is document dependent. The CommandManager tabs are located on the bottom left side of the CommandManager and display the available toolbars and features for each corresponding tab.

The default Part tabs are: Features, Sketch, Evaluate, DimXpert, and Office Products.

The default Drawing tabs are: View Layout, Annotation, Sketch, Evaluate, and Office Products.

The default Assembly tabs are: Assembly, Layout, Sketch, Evaluate, and Office Products. The Office Products toolbar display is dependent on the activated Add-Ins.. during a SolidWorks session.

The Task Pane contains the following default tabs: SolidWorks Resources, Design Library, File Explorer, SolidWorks Search, View Palette, RealView, and Document Recovery.

At this time, there are no questions on the CSWA exam on Motion Study, DimXpert, TolAnalyst, and PhotoWorks.

Key Terms

Assembly: An assembly is a document in which parts, features, and other assemblies (sub-assemblies) are put together. A part in an assembly is called a component. Adding a component to an assembly creates a link between the assembly and the component. When SolidWorks opens the assembly, it finds the component file to show it in the assembly. Changes in the component are automatically reflected in the assembly. The filename extension for a SolidWorks assembly file name is .SLDASM.

SolidWorks 2008 User Interface

CommandManager: The CommandManager is a Context-sensitive toolbar that dynamically updates based on the toolbar you want to access. By default, it has toolbars embedded in it based on the document type. When you click a tab below the Command Manager, it updates to display that toolbar. For example, if you click the **Sketches** tab, the Sketch toolbar is displayed.

ConfigurationManager: The ConfigurationManager is located on the left side of the SolidWorks window and provides the means to create, select, and view multiple configurations of parts and assemblies in an active document. You can split the ConfigurationManager and either display two ConfigurationManager instances, or combine the ConfigurationManager with the FeatureManager design tree, PropertyManager, or third party applications that use the panel.

Coordinate System: SolidWorks uses a coordinate system with origins. A part document contains an original Origin. Whenever you select a plane or face and open a sketch, an Origin is created in alignment with the plane or face. An Origin can be used as an anchor for the sketch entities, and it helps orient perspective of the axes. A three-dimensional reference triad orients you to the X, Y, and Z directions in part and assembly documents.

Cursor Feedback: The system feedback symbol indicates what you are selecting or what the system is expecting you to select. As you move the mouse pointer across your model, system feedback is provided.

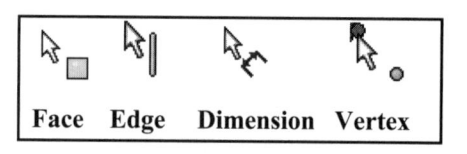

Dimension: A value indicating the size of the 2D sketch entity or 3D feature. Dimensions in a SolidWorks drawing are associated with the model, and changes in the model are reflected in the drawing, if you DO NOT USE DimXpert.

DimXpertManager: The DimXpertManager lists the tolerance features defined by DimXpert for a part. It also displays DimXpert tools that you use to insert dimensions and tolerances into a part. You can import these dimensions and tolerances into drawings. DimXpert is not associative.

Document: In SolidWorks, each part, assembly, and drawing is referred to as a document, and each document is displayed in a separate window.

Drawing: A 2D representation of a 3D part or assembly. The extension for a SolidWorks drawing file name is .SLDDRW. Drawing refers to the SolidWorks module used to insert, add, and modify views in an engineering drawing.

Feature: Features are geometry building blocks. Features add or remove material. Features are created from 2D or 3D sketched profiles or from edges and faces of existing geometry.

FeatureManager: The FeatureManager design tree located on the left side of the SolidWorks window provides an outline view of the active part, assembly, or drawing. This makes it easy to see how the model or assembly was constructed or to examine the various sheets and views in a drawing. The FeatureManager and the Graphics window are dynamically linked. You can select features, sketches, drawing views, and construction geometry in either pane.

Graphics window: The area in the SolidWorks window where the part, assembly, or drawing is displayed.

Heads-up View toolbar: A transparent toolbar located at the top of the Graphic window.

Model: 3D solid geometry in a part or assembly document. If a part or assembly document contains multiple configurations, each configuration is a separate model.

Motion Studies: Graphical simulations of motion and visual properties with assembly models. Analogous to a configuration, they do not actually change the original assembly model or its properties. They display the model as it changes based on simulation elements you add.

Origin: The model origin is displayed in blue and represents the (0,0,0) coordinate of the model. When a sketch is active, a sketch origin is displayed in red and represents the (0,0,0) coordinate of the sketch. Dimensions and relations can be added to the model origin, but not to a sketch origin.

Part: A 3D object that consist of one or more features. A part inserted into an assembly is called a component. Insert part views, feature dimensions and annotations into 2D drawing. The extension for a SolidWorks part filename is .SLDPRT.

Plane: Planes are flat and infinite. Planes are represented on the screen with visible edges.

PropertyManager: Most sketch, feature, and drawing tools in SolidWorks open a PropertyManager located on the left side of the SolidWorks window. The PropertyManager displays the properties of the entity or feature so you specify the properties without a dialog box covering the Graphics window.

RealView: Provides a simplified way to display models in a photo-realistic setting using a library of appearances and scenes. RealView requires graphics card support and is memory intensive.

Rebuild: A tool that updates (or regenerates) the document with any changes made since the last time the model was rebuilt. Rebuild is typically used after changing a model dimension.

Relation: A relation is a geometric constraint between sketch entities or between a sketch entity and a plane, axis, edge or vertex.

Rollback: Suppresses all items below the rollback bar.

Sketch: The name to describe a 2D profile is called a sketch. 2D sketches are created on flat faces and planes within the model. Typical geometry types are lines, arcs, corner rectangles, circles, polygons, and ellipses.

Task Pane: The Task Pane is displayed when you open the SolidWorks software. It contains the following tabs: SolidWorks Resources, Design Library, File Explorer, Search, View Palette, Document Recovery, and RealView/PhotoWorks.

Toolbars: The toolbars provide shortcuts enabling you to access the most frequently used commands. When you enable add-in applications in SolidWorks, you can also display their associated toolbars.

Units: Used in the measurement of physical quantities. Decimal inch dimensioning and Millimeter dimensioning are the two types of common units specified for engineering parts and drawings.

Notes:

Chapter 2: Basic Theory and Drawing Theory

Chapter Objective

Basic Theory and Drawing Theory is one of the five categories on the CSWA exam. This chapter covers the general concepts, symbols, and terminology used throughout the book and then addresses the core elements which are aligned to the exam.

There are two questions on the CSWA exam in this category. Each question is worth five points. The two questions are in a multiple choice single answer or fill in the blank format and requires general knowledge and understanding of the SolidWorks User Interface, FeatureManager, ConfigurationManager, Sketch toolbar, Feature toolbar, Drawing view methods, and basic 3D modeling techniques and engineering principles.

The knowledge of the topics covered in this chapter is directly and indirectly required for the other four categories on the exam.

On the completion of the chapter, you will be able to:

- Recognize 3D modeling techniques:
    - Understand how parts, assemblies, and drawings are related
    - Identify the feature type, parameters, and dimensions
    - Identify the correct standard reference planes: Top, Right, and Front
    - Determine the design intent for a model
- Identify and understand the procedure for the following:
    - Assign and edit material to a part
    - Apply the Measure tool to a part or an assembly
    - Locate the Center of mass, and Principal moments of inertia relative to the default coordinate location, Origin
    - Calculate the overall mass and volume of a part
- Recognize and know the function and elements of the Part and Assembly FeatureManager design tree:

Basic Theory and Drawing Theory

- - Sketch status
  - Component status and properties
  - Display Pane status
  - Reference configurations
- Identify the default Sketch Entities from the Sketch toolbar: Line, Rectangle, Circle, etc.
- Identify the default Sketch Tools from the Sketch toolbar: Fillet, Chamfer, Offset Entities, etc.
- Identify the available SolidWorks File formats for input and export:
  - Save As type for a part, assembly, and drawing
  - Open File of different formats
- Use SolidWorks Help:
  - Contents, Index, and Search tabs
- Identify the process of creating a simple drawing from a part or an assembly:
  - Knowledge to insert and modify the 3 Standard views
  - Knowledge to add a sheet and annotations to a drawing
- Recognize all drawing name view types by their icons:
  - Model, Projected, Auxiliary, Section, Aligned Section, Detail, Standard, Broken-out Section, Break, Crop, and Alternate Position
- Identify the procedure to create a named drawing view:
  - Model, Projected, Auxiliary, Section, Aligned Section, Detail, Standard, Broken-out Section, Break, Crop, and Alternate Position
- Specify Document Properties:
  - Select Unit System
  - Set Precision

☼ The models in this book were created in SolidWorks 2008 SP2.1.

☼ At this time, the CSWA exam is given in both SolidWorks 2007 and SolidWorks 2008.

3D Modeling techniques

Understand and address the following items for the CSWA exam:

- How parts, assemblies, and drawings are related
- Identify the feature type, parameters, and dimensions
- Identify the correct standard reference planes: Top, Right, and Front
- Determine the design intent for a model

In a SolidWorks application, each part, assembly, and drawing is referred to as a document. Each document is displayed in a separate Graphics window.

How parts, assemblies, and drawings are related

Parts, assemblies, and drawings are associative. This means that changes made in one area are automatically reflected in all the associated areas affected by that change. For example, changes that you make to a part are reflected in the associative assembly and drawing.

Typically you design each part, combine the parts, "components" into an assembly, and then generate drawings of the assembly for inspection and manufacturing. The associative part, assembly, and drawing share a common data base.

Questions on the CSWA exam are in a multiple choice single answer or fill in the blank format.

Tutorial: Associativity 2-1

Verify the association between a part, assembly, and drawing.

Copy the SolidWorks CSWA Folder from the CD in the book to your hard drive before you perform the tutorials in the book.

1. **Open** the Wedge-Feature part from the SolidWorks CSWA Folder\Chapter2 location as illustrated. The part is displayed in the Graphics window.

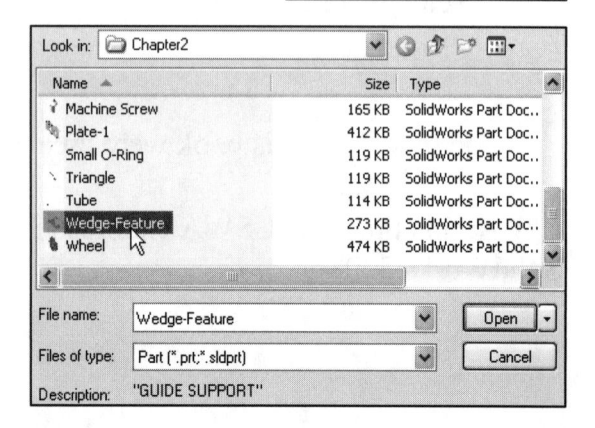

Basic Theory and Drawing Theory

2. **Open** the Wedge-Feature drawing from the SolidWorks CSWA Folder\Chapter2 location. View the FeatureManager. View the 120mm length dimension in the Front view.

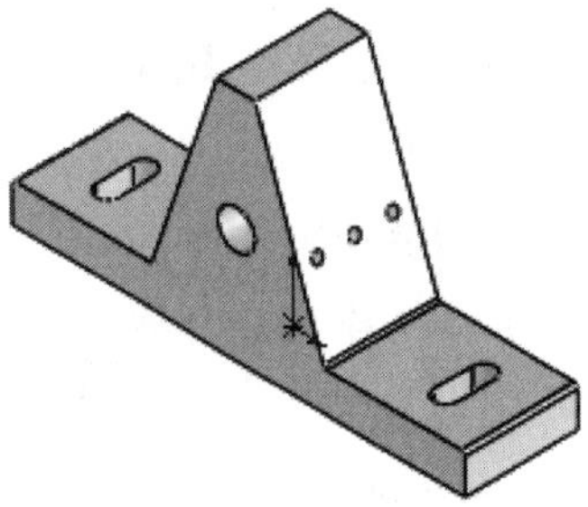

3. **Open** the Wedge-Tube assembly from the SolidWorks CSWA Folder\Chapter2 location. View the assembly in the Graphics window. The Wedge-Feature part, Wedge-Feature drawing and the Wedge-Tube assembly are active documents.

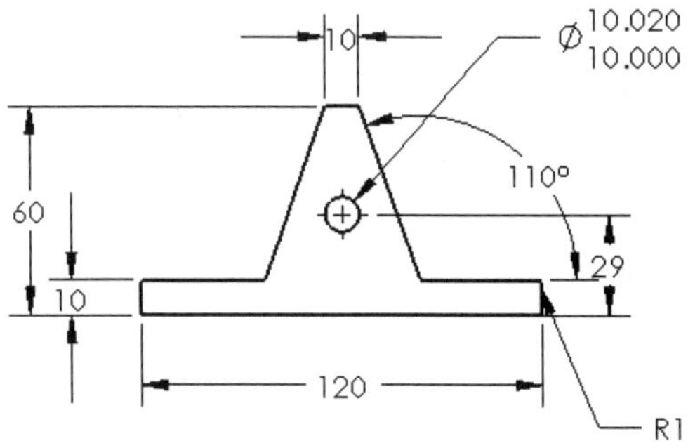

Modify the Extruded Base feature depth dimension of the part.

Return to the Wedge-Feature part.

4. Click **Window**, **Wedge-Feature** from the Menu bar menu.

5. Click **Base-Extrude** from the Wedge-Feature Part FeatureManager.

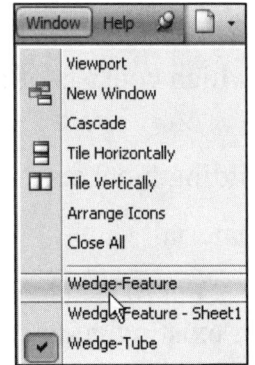

6. **View** the dimensions in the Graphics window.

Modify the Extruded Base feature length dimension.

7. Click the **120**mm length dimension in the Graphics window.

8. Enter **140**.

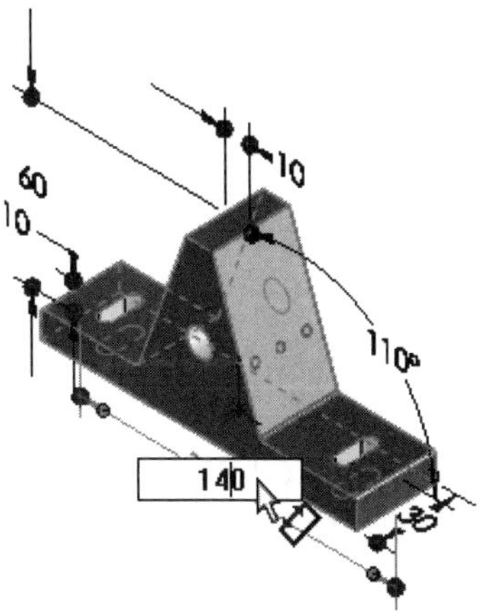

Return to the Wedge-Feature drawing.

9. Click **Window, Wedge-Feature – Sheet1** from the Menu bar menu. View the 140mm dimension in the drawing

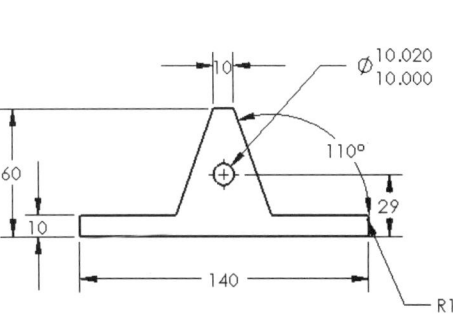

Page 2 - 4

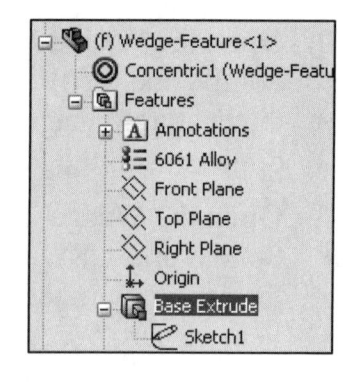

Basic Theory and Drawing Theory

View the Wedge-Tube assembly.

10. Click **Window, Wedge-Tube** from the Menu bar menu.

11. **View** the length dimension of Base Extrude in the Wedge-Feature component. The length dimension is modified from 120mm to 140mm.

12. **Close** all models. Think about the association between the part, assembly, and drawing.

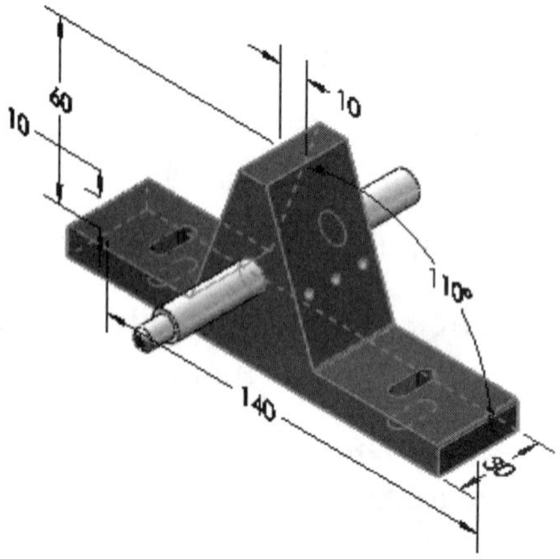

☼ Understand the procedure to open and close a part from a drawing or from an assembly. Understand the process to edit features and sketches from the FeatureManager design tree.

Identify the Feature type by icon in the FeatureManager

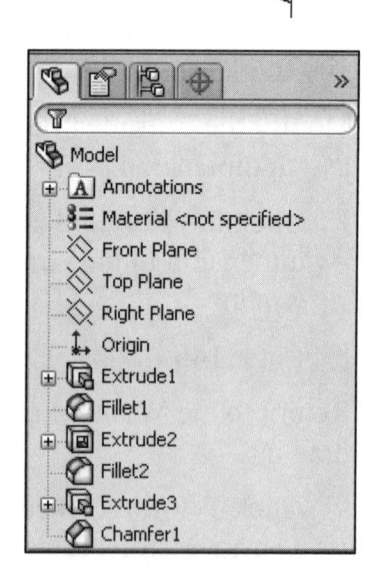

A part is a 3D model which consists of features. What are features?

- Features are geometry building blocks
- Features add or remove material
- Features are created from 2D or 3D sketched profiles or from edges and faces of existing geometry

☼ You can use the same sketch to create different features.

☼ The CSWA test is a three hour exam. Apply symmetry to decrease design time.

☼ Optimize the number of features required to build a model. You will need the knowledge to build the illustrated model in the next three chapters.

Basic Theory and Drawing Theory

All questions on the exam are in a multiple choice single answer or fill in the blank format. In the Part Modeling category, an exam question could read:

Question 1: Build the illustrated model from the provided information. Locate the Center of mass relative to the default coordinate system, Origin.

- A: X = -1.63 inches, Y = 1.48 inches, Z = -1.09 inches
- B: X = 1.63 inches, Y = 1.01 inches, Z = -0.04 inches
- C: X = 43.49 inches, Y = -0.86 inches, Z = -0.02 inches
- D: X = 1.63 inches, Y = 1.01 inches, Z = -0.04 inches

The correct answer is B.

In the Part Modeling category; Chapter 3, you are required to read and understand an engineering document, set document properties, identify the correct Sketch planes, apply the correct Sketch and Feature tools, and apply material to build a simple part.

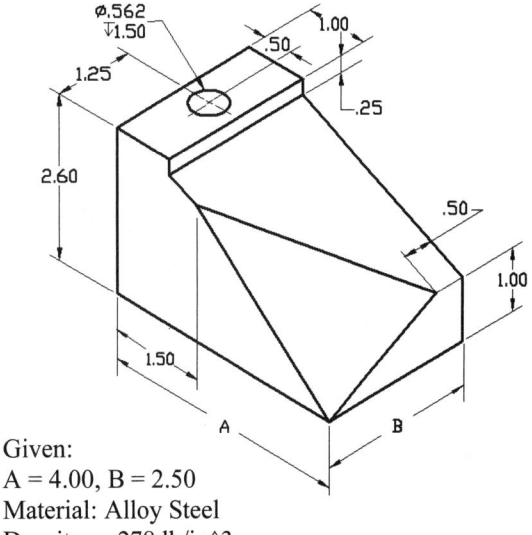

Given:
A = 4.00, B = 2.50
Material: Alloy Steel
Density = .278 lb/in^3
Units: IPS
Decimal places = 2

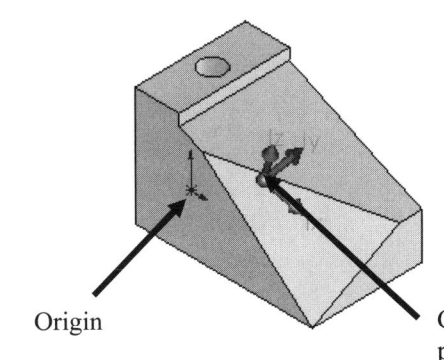

Origin

Center of mass relative to the part Origin

Mass = 4.97 pounds

Volume = 17.86 cubic inches

Surface area = 46.77 inches^2

Center of mass: (inches)
X = 1.63
Y = 1.01
Z = -0.04

Tutorial: Identify model features 2-1

Identify the features in the FeatureManager based on the feature icon and understand the procedure to build the Plate-1 part.

1. **Open** the Plate-1 part from the SolidWorks CSWA Folder\Chapter2 location. View the illustrated model and FeatureManager design tree. The features have been renamed.

    a. Base-Extrude: Extruded Base feature

    b. Mounting Holes: Extruded Cut feature for the two outside holes

    c. Small Edge Fillet: Fillet feature for the top and bottom edges

    d. Front-Back Edge Fillet: Fillet feature for the front and back side edges

    e. CBORE: Hole Wizard feature for the center hole

    f. Angle-Cut1: Chamfer feature for the left front hole

    g. Angle-Cut2: Chamfer feature for the right front hole

2. **Close** the model.

Tutorial: Identify model features 2-2

Identify the features in the FeatureManager based on the feature icon and understand the procedure to build the Wedge-Feature part.

1. **Open** the Wedge-Feature part from the SolidWorks CSWA Folder\Chapter2 location. The features have been renamed.

    a. Base Extrude: Extruded Base feature

    b. Slot Cut: Extruded Cut feature

    c. Slot Cut 2: Mirror feature

    d. Guide Hole: Extruded Cut feature

    e. M3X0.5 Tapped Hole1: Hole Wizard feature

    f. Hole Pattern: Linear Pattern feature

    g. Edges: Fillet feature

2. **Close** the model.

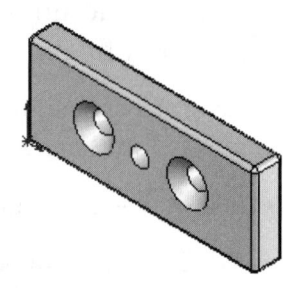

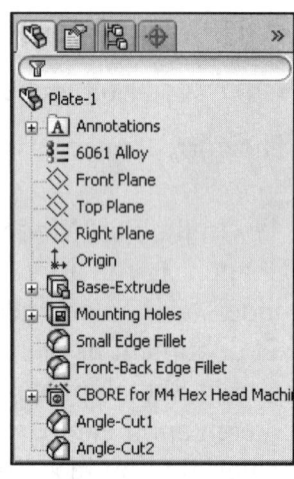

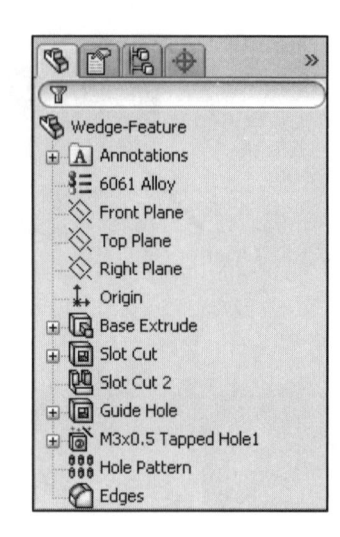

Tutorial: Identify model features 2-3

Identify the features in the FeatureManager based on the feature icon and understand the procedure to build the Light-bulb part.

1. **Open** the Light-bulb part from the SolidWorks CSWA Folder\Chapter2 location. View the illustrated model and FeatureManager. The features have been renamed.

   a. Base: Revolved Base feature

   b. Base-Top: Revolved Boss feature

   c. Cut-Revolve-Thin1: Revolved Cut feature

   d. Back: Dome feature

   e. Seed Cut: Extruded Cut feature

   f. PatternSeedCut: Circular Pattern feature

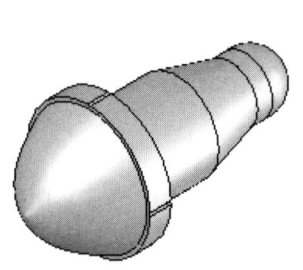

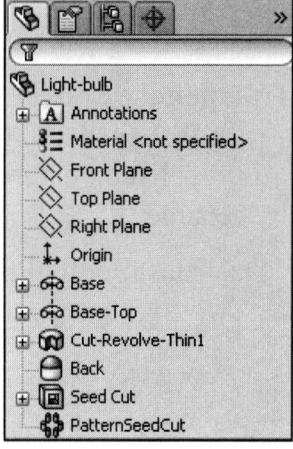

2. **Close** the model.

Edit the sketches and features in the provided FeatureManagers to explore how each model was built.

There are numerous ways to create the models in this chapter. A goal for this book is to display different design intents and modeling techniques.

Tutorial: Identify model features 2-4

Identify the features in the FeatureManager based on the feature icon and understand the procedure to build the Machine Screw part.

1. **Open** the Machine Screw part from the SolidWorks CSWA Folder\Chapter2 location. View the illustrated model and FeatureManager. The features have been renamed.

   a. Base: Revolved Base feature

   b. Slot: Extruded Cut feature

   c. Pattern: Circular Pattern feature

   d. Edge: Fillet feature

   e. End: Chamfer feature

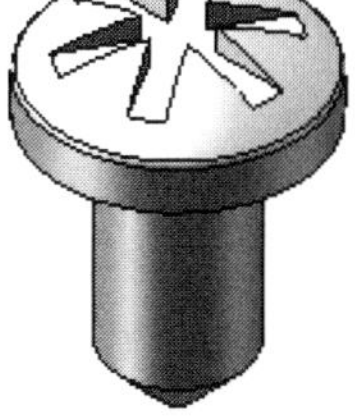

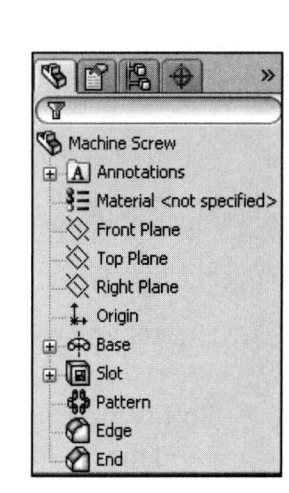

2. **Close** the model.

Tutorial: Identify model features 2-5

Identify the features in the FeatureManager based on the feature icon and understand the procedure to build the Small O-Ring part.

1. **Open** the Small O-Ring part from the SolidWorks CSWA Folder\Chapter2 location. View the illustrated model and FeatureManager. The feature has been renamed.

    a. Base: Lofted Base feature

2. **Close** the model.

    All questions on the exam are in a multiple choice single answer or fill in the blank format. In the Basic Theory and Drawing Theory category, an exam question could read:

Question 1: Identify the following Feature icon.

- A: Extruded Cut feature
- B: Extruded Boss/Base feature
- C: Fillet feature
- D: Chamfer feature

The correct answer is D.

Question 2: Identify the following Feature icon.

- A: Extruded Boss/Base feature
- B: Revolved Boss/Base feature
- C: Fillet feature
- D: Chamfer feature

The correct answer is B.

☼ SolidWorks states that the first feature you create in a part is the Base. This feature is the basis on which you create the other features for the model. The Base feature can be an extrusion, a revolve, a sweep, a loft, thickening of a surface, or a sheet metal flange.

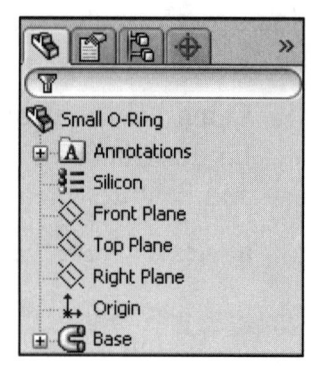

Basic Theory and Drawing Theory

☼ Use the Features toolbar from the CommandManager in SolidWorks to help identify the feature tool icon during the exam.

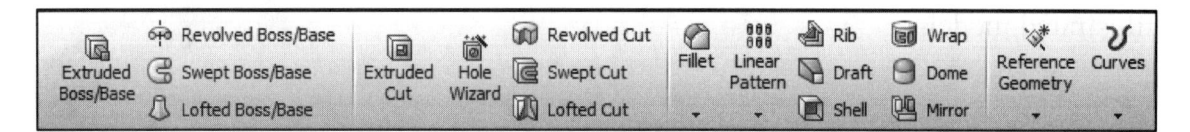

Identify dimensions and parameters

Create model dimensions as you create each part feature. Then insert the dimensions into the various drawing views. Changing a dimension in the model updates the drawing, and changing an inserted dimension in a drawing view changes the model.

☼ By default, single model dimensions are displayed in black. This includes dimensions that are blue in the part or assembly document (such as the extrusion depth). Reference dimensions are gray by default, design table dimensions are magenta by default.

Tutorial: Identify dimensions and parameters 2-1

Understand the illustrated dimensions, annotations, and parameters of the model.

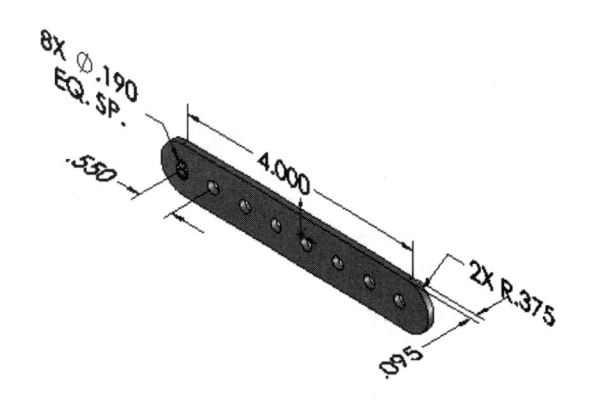

1. **Open** the Bar part from the SolidWorks CSWA Folder\Chapter2 location. View the illustrated dimensions, annotations, and parameters of the model.

2. Click **8X Ø .190 EQ. SP** in the Graphics window. Click **OK**. Click **OK**. Understand the procedure to insert dimension text from the Dimension PropertyManager.

3. **Edit** LPattern1 in the FeatureManager. View the parameters, "number of instances". The number of instances = 8.

4. **Close** the model.

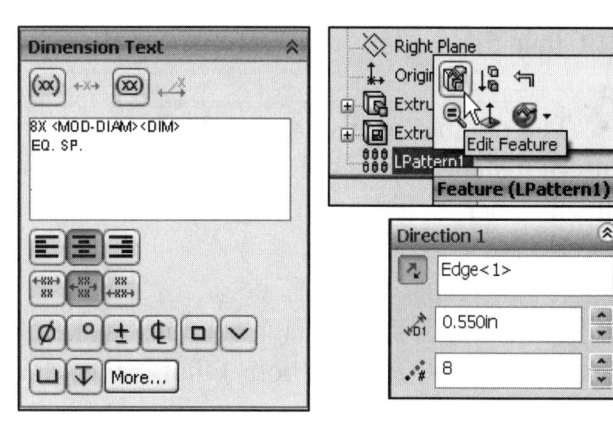

Page 2 - 10

In the Basic Theory and Drawing Theory category, an exam question could read:

Question 1: Identify the number of instances in the illustrated model.

- A: 7
- B: 5
- C: 8
- D: None

The correct answer is B.

Tutorial: Identify dimensions and parameters 2-2

Understand the illustrated dimensions and parameters of the model and the procedure to build the Triangle part.

1. **Open** the Triangle part from the SolidWorks CSWA Folder\Chapter2 location.
2. **Click** Extrude2 from the FeatureManager.
3. **View** the Extruded Cut feature dimensions.
4. **Edit** LPattern1 from the FeatureManager.
5. **View** the displayed parameters. Extrude2 is the seed feature for the Linear Pattern. D4@Sketch2 is the Pattern Direction. Spacing = 20mm.
6. **Close** the model.

The number of instances sets the number of pattern instances for Direction 1 and Direction 2. This number includes the original features or selections.

SolidWorks states that a seed is, "a sketch or an entity (a feature, face, or body) that is the basis for a pattern.

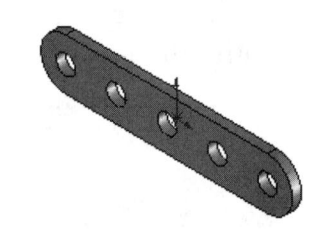

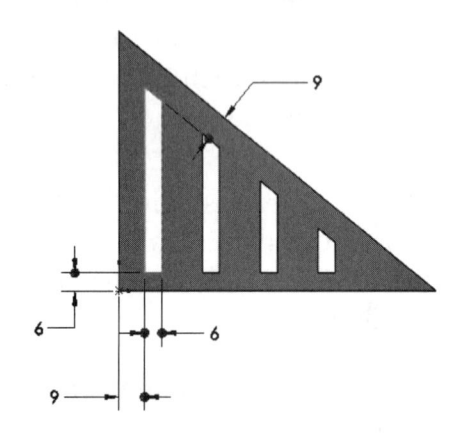

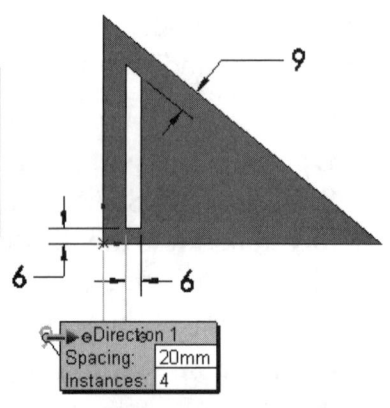

Tutorial: Identify dimensions and parameters 2-3

Understand the illustrated dimensions, and parameters of the model and the procedure to build the Wheel part.

1. **Open** the Wheel part from the SolidWorks CSWA Folder\Chapter2 location.

2. **Double-click** Extrude1 from the FeatureManager. View the Extruded Base feature dimensions.

3. **Edit** Sketch1 from the FeatureManager.

4. **Modify** the diameter dimension from 3.000in to 3.5000in.

5. **Edit** CirPattern1 from the FeatureManager. Extrude1 and Extrude2 are the seed features for the pattern. Number of instances = 8. Axis<1> is the Pattern Axis. (5) is the pattern instance to skip.

6. **Modify** the number of instances from 8 to 6.

7. **Close** the model.

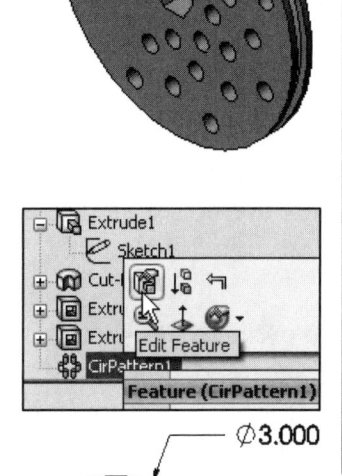

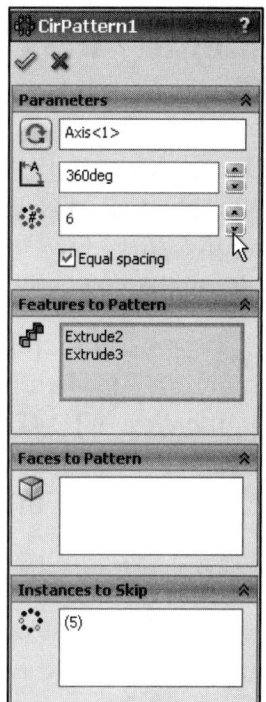

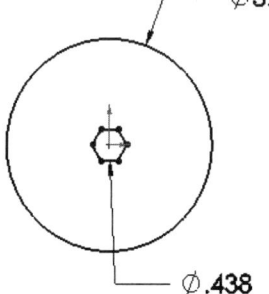

Instances to Skip, eliminates the pattern instances that you select in the Graphics window when you are creating the pattern.

In the Basic Theory and Drawing Theory category, an exam question could read:

Question 1: Identify the number of instances in the illustrated model.

- A: 7
- B: 4
- C: 8
- D: None

The correct answer is B.

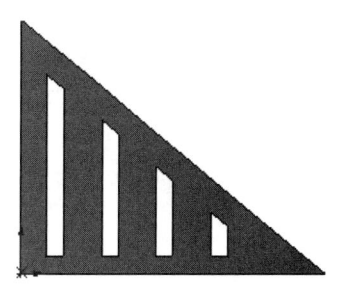

Identify the correct reference planes: Top, Right, and Front

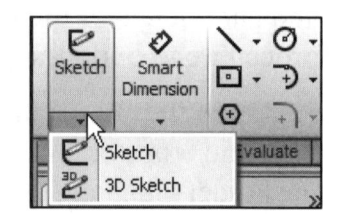

Most SolidWorks features start with a 2D sketch. Sketches are the foundation for creating features. SolidWorks provides the ability to create either 2D or 3D sketches.

A 2D sketch is limited to a flat 2D Sketch plane. A 3D sketch can include 3D elements. As you create a 3D sketch, the entities in the sketch exist in 3D space. They are not related to a specific Sketch plane as they are in a 2D sketch.

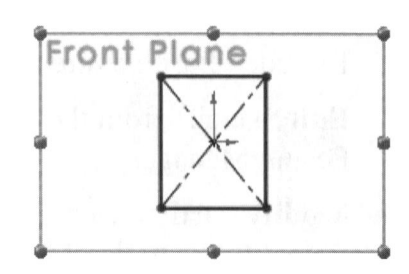

☼ You may need to create a 3D sketch on the exam. The illustrated model displays a 3D sketch, using the Sketch Line tool for an Extruded Cut feature. You will create a 3D sketch in Chapter 3.

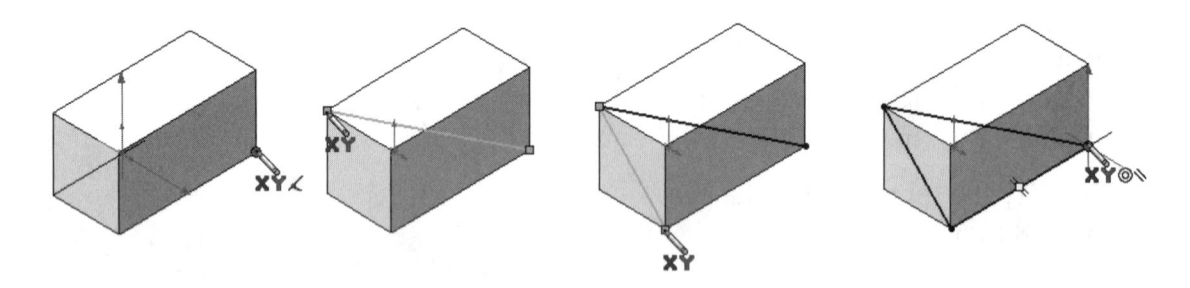

Does it matter where you start sketching a 2D sketch? Yes! When you create a new part or assembly, the three default planes are aligned with specific views.

The plane you select for your first sketch determines the orientation of the part. Selecting the correct plane is very important.

☼ When you create a new part or assembly, the three default planes are aligned with specific views. The plane you select for the Base sketch determines the orientation of the part.

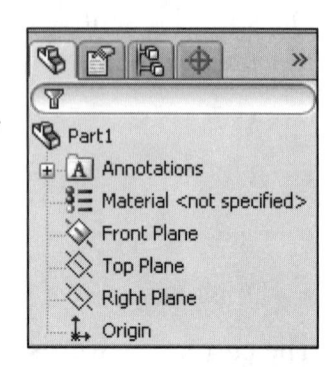

2D sketching / reference planes

The three default ⊥ reference planes, displayed in the FeatureManager design tree represent infinite 2D planes in 3D space. They are:

- Front
- Top
- Right

Planes have no thickness or mass. Orthographic projection is the process of projecting views onto parallel planes with ⊥ projectors. The default ⊥ datum planes are:

- Primary
- Secondary
- Tertiary

Use the following planes in a manufacturing environment:

- Primary datum plane: Contacts the part at a minimum of three points
- Secondary datum plane: Contacts the part at a minimum of two points
- Tertiary datum plane: Contacts the part at a minimum of one point

The part view orientation depends on the sketch plane. Compare the Front, Top, and Right Sketch planes for an L-shaped profile in the following illustration.

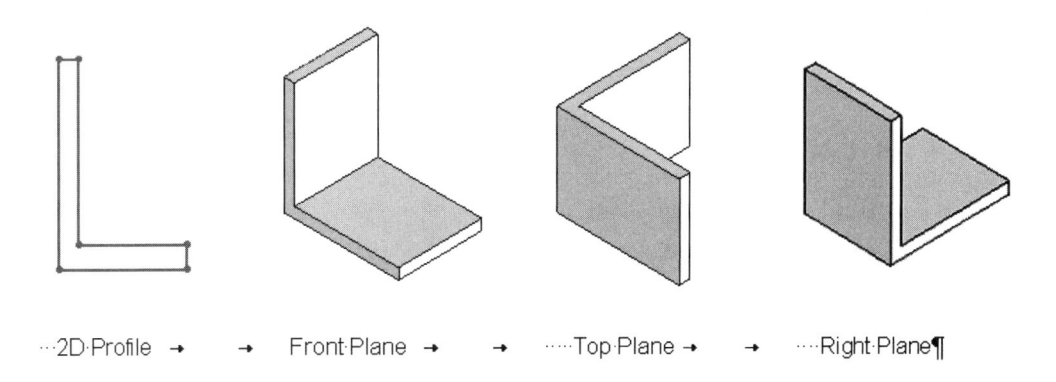

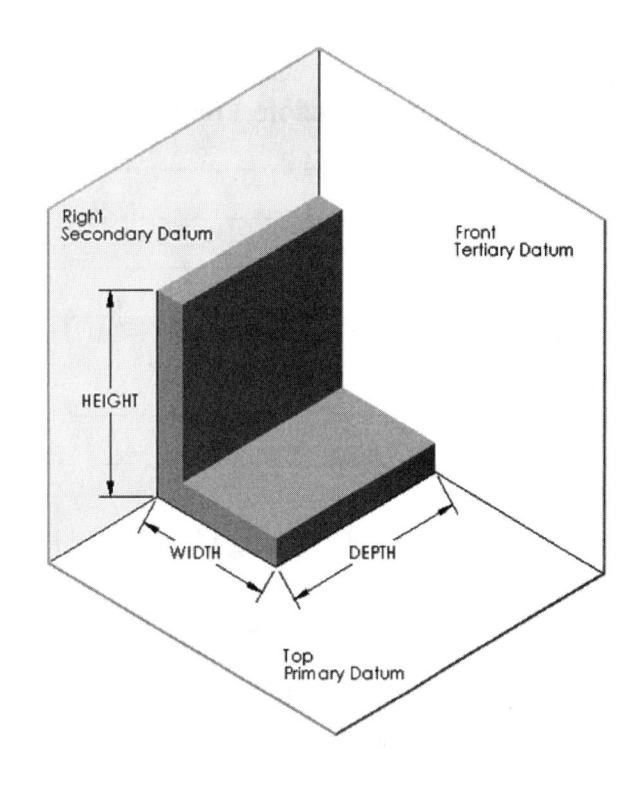

2D Profile → → Front Plane → → Top Plane → → Right Plane

The six principle views of orthographic projection listed in the ASME Y14.3M standard are: Top, Front, Right side, Bottom, Rear, and Left side. SolidWorks Standard view names correspond to these orthographic projection view names.

ASME Y14.3M Principle View Name:	SolidWorks Standard View:
Front	Front
Top	Top
Right side	Right
Bottom	Bottom
Rear	Back
Left side	Left

The standard drawing views in third angle orthographic projection are: Front, Top, Right, & Isometric.

ANSI is the default dimensioning standard used in this book.

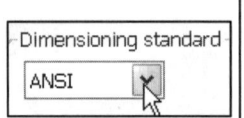

Tutorial: Reference plane 2-1

1. **Create** a New part in SolidWorks as illustrated. Think about the steps that you would take to build the model. Identify the location of the part Origin and the reference plane for Sketch1. Sketch1 is the Base sketch.

Set the document properties for the model.

2. Select **ANSI**. Select the **IPS** unit system. Select **.1** for Length decimal place. Click **OK**.

3. Create **Sketch1**. Select Front Plane as the Sketch plane. Apply the Line Sketch tool. Locate the sketch Origin in the bottom left corner. Insert the illustrated geometric relations and dimensions.

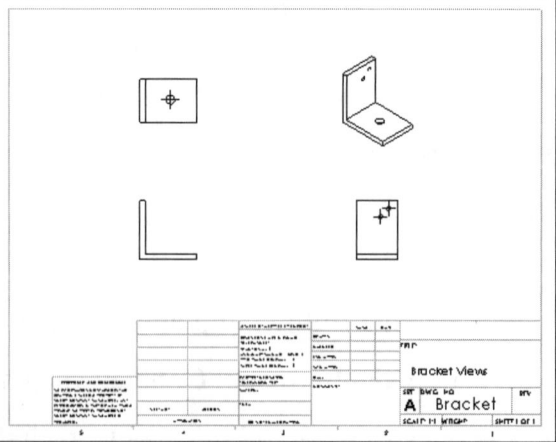

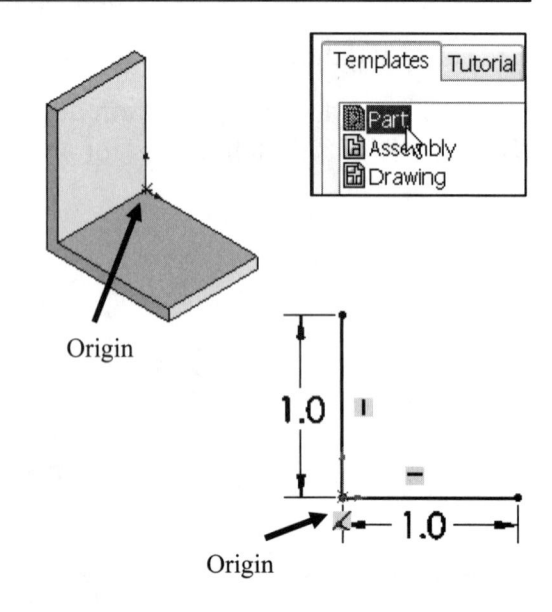

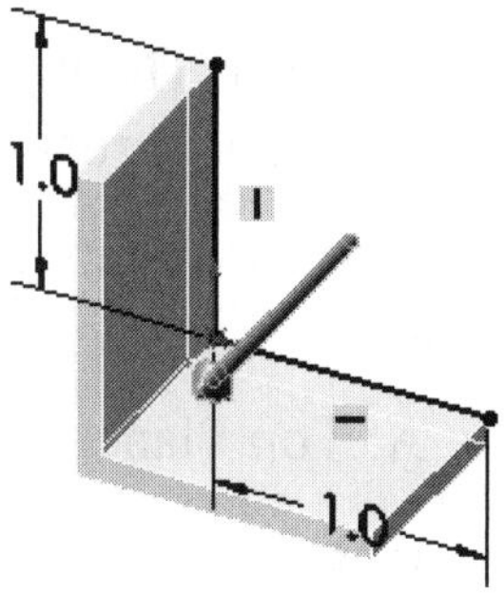

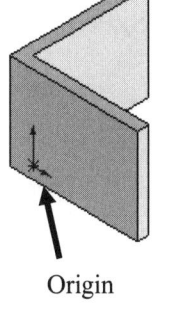

Basic Theory and Drawing Theory

4. Create the **Extrude-Thin1** feature. Blind is the default End Condition in Direction 1. Depth = .8in. One-Direction is the default type in the Thin Feature box. Thickness = .1.

5. **Save** the part and name it Bracket-FrontPlane.

6. **Close** the model.

☼ Apply the Thin Features options to control the extrude thickness, not the depth.

Tutorial: Reference plane 2-2

1. **Create** a New part in SolidWorks. Identify the location of the part Origin and the reference plane for Sketch1. Sketch1 is the Base sketch.

2. **Set** the document properties for the model. Select **ANSI**. Select the **IPS** unit system. Select **.1** for Length decimal places.

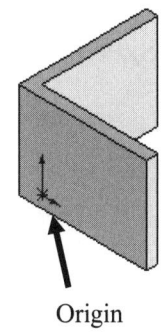

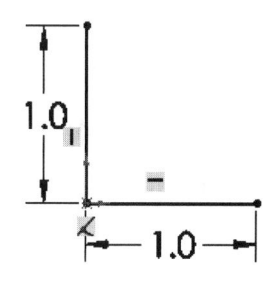

3. Create **Sketch1**. Select the Top Plane as the Sketch plane. Apply the Line Sketch tool. Locate the sketch Origin in the bottom left corner. Insert the illustrated geometric relations and dimensions.

4. Create the **Extrude-Thin1** feature. Blind is the default End Condition in Direction 1. Depth = .8in. One-Direction is the default type in the Thin Feature box. Thickness = .1in.

5. **Save** the part and name it Bracket-TopPlane.

6. **Close** the model.

☼ To modify the Sketch plane in a sketch: Right-click the **Sketch<#>** in the FeatureManager, select **Edit Sketch Plane**. Select a new plane from the fly-out FeatureManager. The name of the plane is displayed in the Sketch Plane/Face box. Click **OK** from the Sketch Plane PropertyManager.

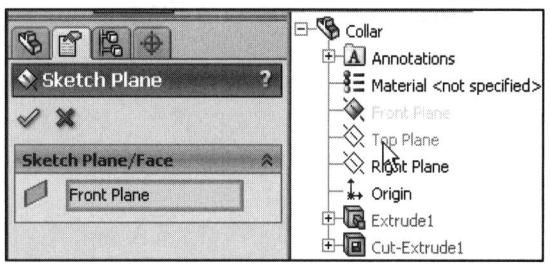

In the Basic Theory and Drawing Theory category, an exam question could read:

Question 1: Identify the Sketch plane for the Extruded Base feature.

- A: Top Plane
- B: Front Plane
- C: Right Plane
- D: Left Plane

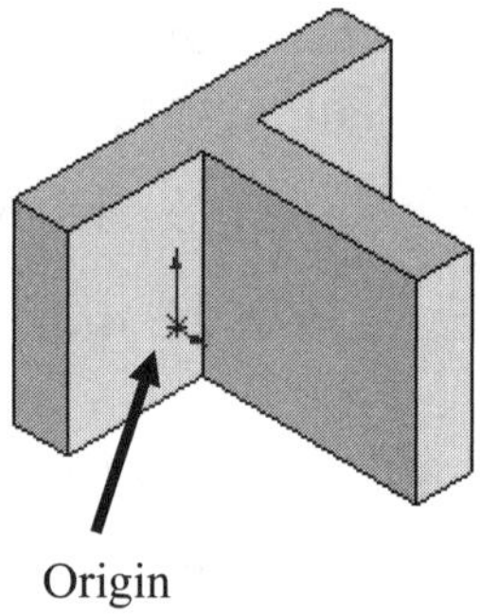

The correct answer is A.

Question 2: Identify the Sketch plane for the Extruded Base feature.

- A: Top Plane
- B: Front Plane
- C: Right Plane
- D: Left Plane

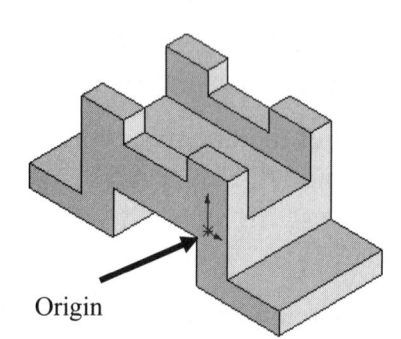

The correct answer is B.

Question 3: Identify the Sketch plane for the Extruded Base feature.

- A: Top Plane
- B: Front Plane
- C: Right Plane
- D: Left Plane

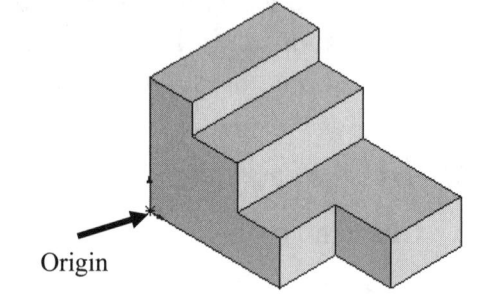

The correct answer is B.

Question 4: The _____ is the Sketch plane for the Extruded Base feature.

The correct answer is Top Plane.

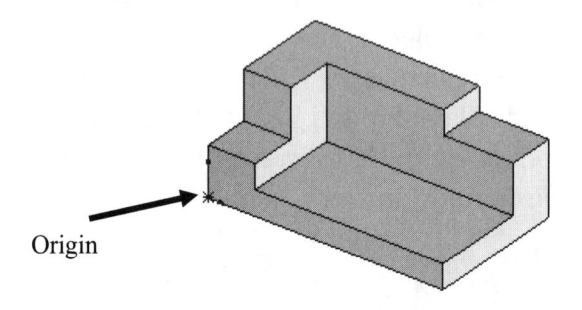

Determine the design intent for a model

What is design intent? All designs are created for a purpose. Design intent is the intellectual arrangements of features and dimensions of a design. Design intent governs the relationship between sketches in a feature, features in a part, and parts in an assembly.

The SolidWorks definition of design intent is the process in which the model is developed to accept future modifications. Models behave differently when design changes occur.

Design for change! Utilize geometry for symmetry, reuse common features, and reuse common parts. Build change into the following areas that you create: sketch, feature, part, assembly, and drawing.

Design intent in a sketch

Build design intent in a sketch as the profile is created. A profile is determined from the Sketch Entities. Example: Rectangle, Circle, Arc, Point, etc. Apply symmetry into a profile through a sketch centerline, mirror entity, and position about the reference planes and Origin.

Build design intent as you sketch with automatic geometric relations. Document the decisions made during the up front design process. This is very valuable when you modify the design later.

A rectangle contains Horizontal, Vertical, and Perpendicular automatic geometric relations. Apply design intent using added geometric relations. Example: Horizontal, Vertical, Collinear, Perpendicular, Parallel, etc.

The Properties PropertyManager and the pop-up Context toolbar is displayed when you select multiple sketch entities in the Graphics window.

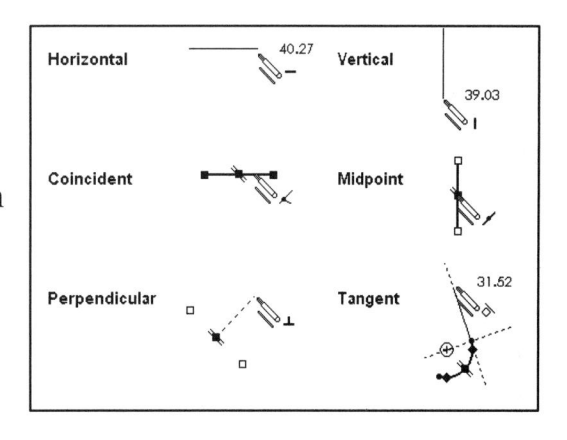

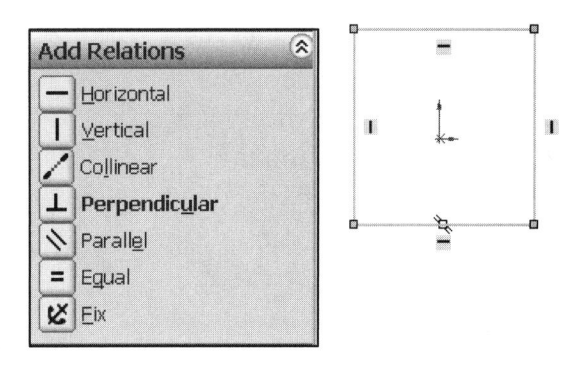

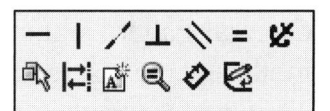

Basic Theory and Drawing Theory

Example A: Apply design intent to create a square profile. Sketch a rectangle (Corner Rectangle tool) with the Origin approximately in the center. Insert a construction reference centerline. Add a Midpoint relation. Add an Equal relation between the two vertical and horizontal lines. Insert a dimension to define the square.

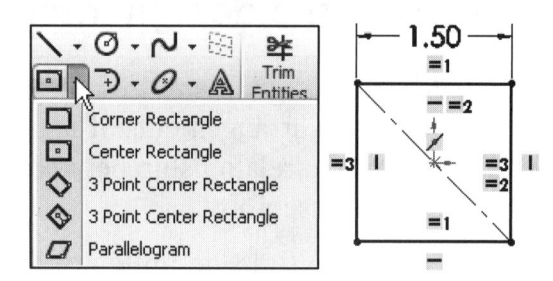

💡 The new Center Rectangle tool eliminates the need to create a centerline, Midpoint and Equal relation.

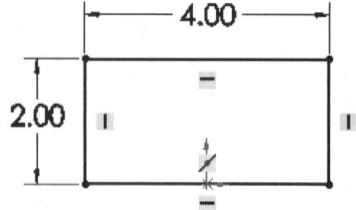

Example B: Develop a rectangular profile. The bottom horizontal midpoint of the rectangular profile is located at the Origin. Sketch a rectangle. Add a Midpoint relation between the horizontal edge of the rectangle and the Origin. Insert two dimensions to define the width and height of the rectangle.

Design intent in a feature

Build design intent into a feature by addressing symmetry, feature selection, and the order of feature creation.

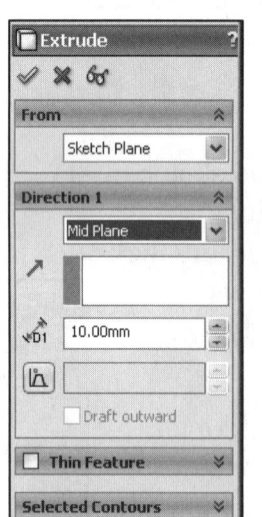

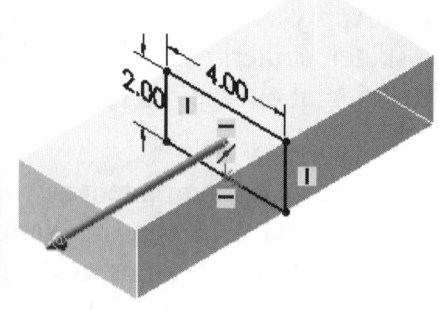

Example A: The Base Extrude feature remains symmetric about the Front Plane. Utilize the Mid Plane End Condition option in Direction 1. Modify the depth, and the feature remains symmetric about the Front Plane.

Example B: Create 34 teeth for a Circular Pattern feature. Do you create each tooth separate using the Extruded Cut feature? No. Create a single tooth and then apply the Circular Pattern feature. Modify the number of teeth from 32 to 24.

Page 2 - 19

Design intent in a part

Utilize symmetry, feature order and reusing common features to build design intent into a part.

Example A: Feature order. Is the entire part symmetric? Feature order affects the part. Apply the Shell feature before the Fillet feature and the inside corners remain perpendicular.

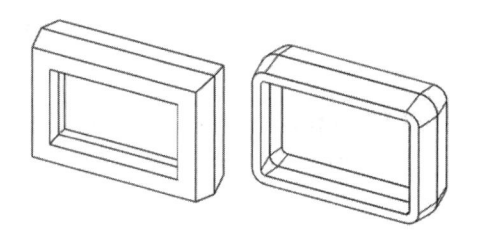

Design intent in an assembly

Utilizing symmetry, reusing common parts and using the Mate relation between parts builds the design intent into an assembly.

Example A: Reuse geometry in an assembly. The assembly contains a linear pattern of holes. Insert one screw into the first hole. Utilize the Component Pattern feature to copy the machine screw to the other holes.

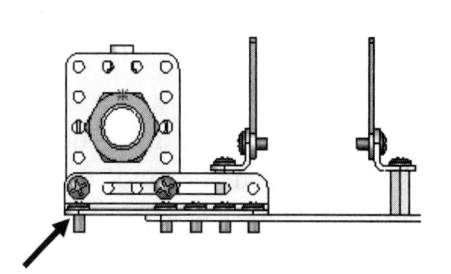

Design intent in a drawing

Utilize dimensions, tolerance and notes in parts and assemblies to build the design intent into a drawing.

Example A: Tolerance and material in the drawing. Insert an outside diameter tolerance +.000/-.002 into the Pipe part. The tolerance propagates to the drawing.

Define the Custom Property Material in the Part. The Material Custom Property propagates to your drawing.

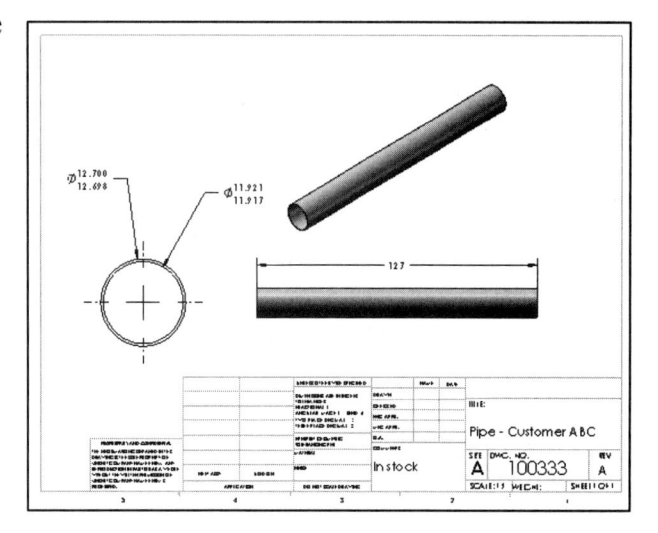

☼ Create a 2D sketch on any of the default planes: Front, Top, and Right or a created plane.

Basic Theory and Drawing Theory

Tutorial: Design intent 2-1

💡 Design intent is an important consideration when creating a SolidWorks model. Plan your sketches and features.

1. **Open** the Battery part from the SolidWorks CSWA Folder\Chapter2 location. Drag the rollback bar up under the Base feature in the FeatureManager.

2. **Edit** Sketch1 from the FeatureManager. Top Plane is the Sketch plane. View the geometric relations and dimensions. Drag the rollback bar down under Terminals-Top in the FeatureManager.

   The first feature of the Battery is Extrude1. Extrude1 is the Base feature and requires a Base sketch. The Base sketch plane determines the orientation of the Base feature. Note: Extrude1 was renamed to Base.

   The Sketch plane locates the sketch profile on any plane or face. The sketch for the feature was created using symmetry. The sketch is a rectangle profile on the Top Plane, centered at the Origin using the Sketch Centerline tool with a Midpoint and Equal relation. The Equal relation was applied to the two perpendicular line segments.

3. **Close** the model.

💡 Click **View**, **Sketch Relations** from the Menu bar menu to view the relations in a sketch.

Tutorial: Design Intent 2-2

1. **Open** the Box part from the SolidWorks CSWA Folder\Chapter2 location.

2. **Edit** Extrude1 from the FeatureManager. Extrude1 is the Base feature. Mid Plane was selected as the End Condition in Direction 1. Both sides of the Box are equal. The design intent was symmetry, and the Mid Plane option was selected.

3. **Close** the model.

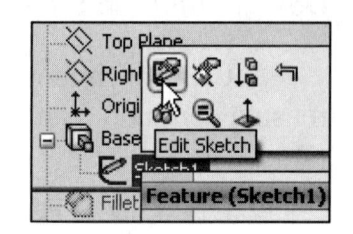

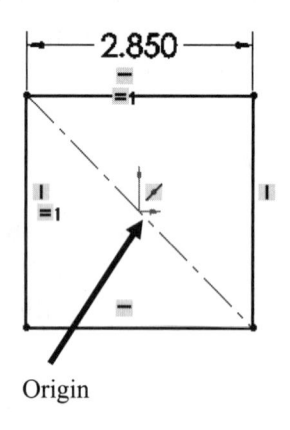

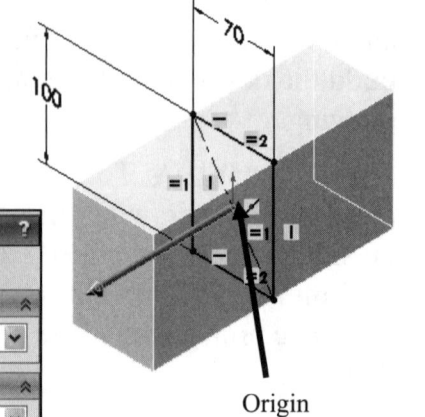

Page 2 - 21

Identify material, measure, and mass properties

Understand the process and procedure of the following:

- Assign and edit material to a part
- Apply the Measure tool to a part or assembly
- Locate the Center of mass, and Principal moments of inertia relative to the default coordinate location, Origin.
- Calculate the overall mass and volume of a part

Assign and edit material

Use the Materials Editor PropertyManager to assign or edit material in a part. Material is required to calculate the correct Mass Properties of a part or an assembly.

💡 The Material icon is displayed in the Part FeatureManager regardless of whether a material is applied. Right-click the icon to view a list of the ten most recently used materials.

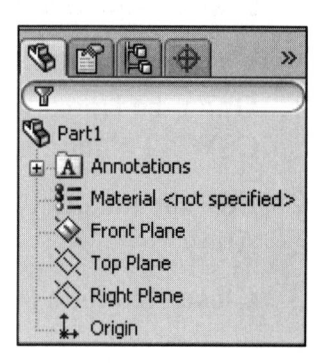

Tutorial: Assign and edit material 2-1

Assign material to the Machine Screw part.

1. **Open** the Machine Screw part from the SolidWorks CSWA Folder\Chapter2 location.
2. **Assign** 6061 Alloy to the part.
3. **View** the available materials.
4. **View** the physical properties of the 6061 Alloy in the Physical Properties box.
5. **Close** the model.

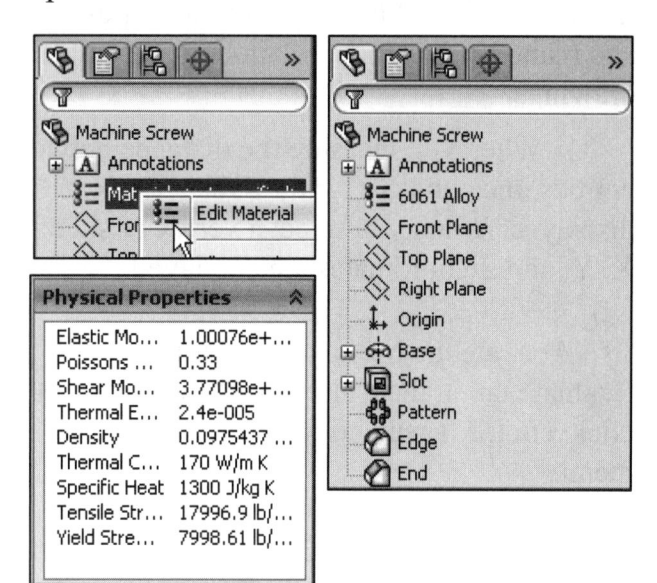

☼ Know the available default materials and their location.

An exam question could read:

Question 1: Identify the material category for 6061 Alloy.

- A: Steel
- B: Iron
- C: Aluminum Alloys
- D: Other Alloys
- E: None of the provided

The correct answer is C.

Tutorial: Assign and edit material 2-2

Edit the material for the Hook-1 part.

1. **Open** the Hook-1 part from the SolidWorks CSWA Folder\Chapter2 location.
2. **Edit** the material from 2014 Alloy to 6061 Alloy.
3. **Close** the model.

Apply the Measure tool

The Measure tool measures distance, angle, radius, and size of and between lines, points, surfaces, and planes in sketches, 3D models, assemblies, or drawings.

When you measure the distance between two entities, the delta X, Y, and Z distances can be displayed. When you select a vertex or sketch point, the X, Y, and Z coordinates are displayed.

☼ Activate the Measure tool from the Evaluate tab in the CommandManager or click **Tool**, **Measure** from the Menu bar menu.

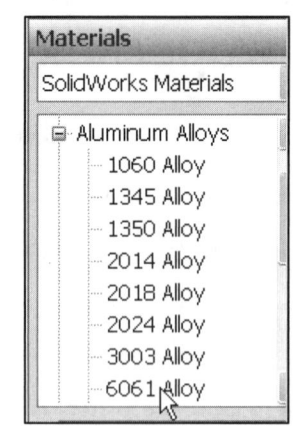

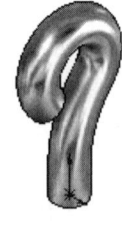

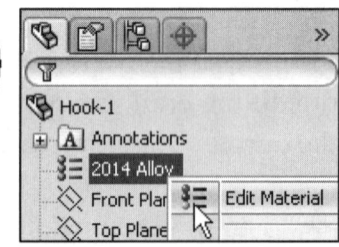

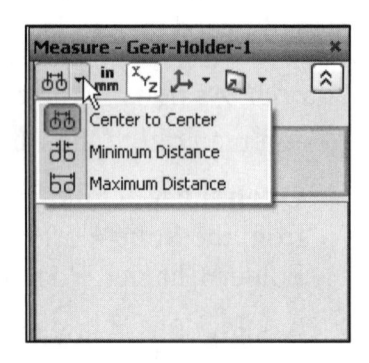

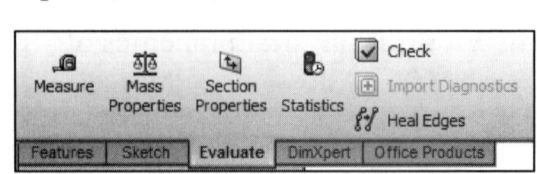

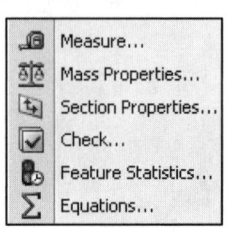

Basic Theory and Drawing Theory

☼ Select the Units/Precision option to specify custom measurement units and precision for your model.

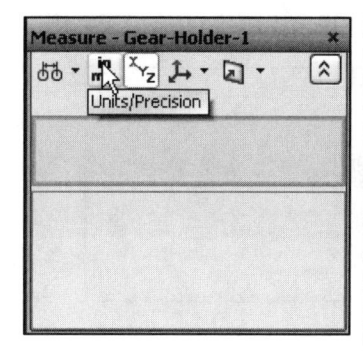

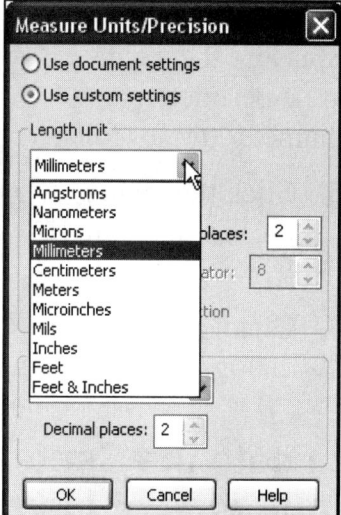

Tutorial: Measure tool 2-1

1. **Open** the Gear-Holder-1 part from the SolidWorks CSWA Folder\Chapter2 location.

   Activate the Measure dialog box.

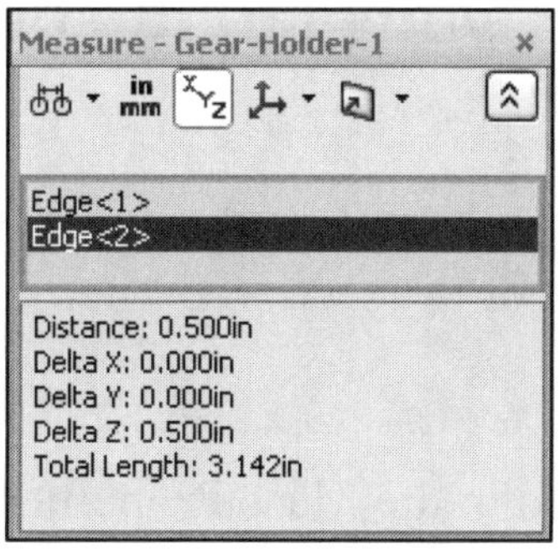

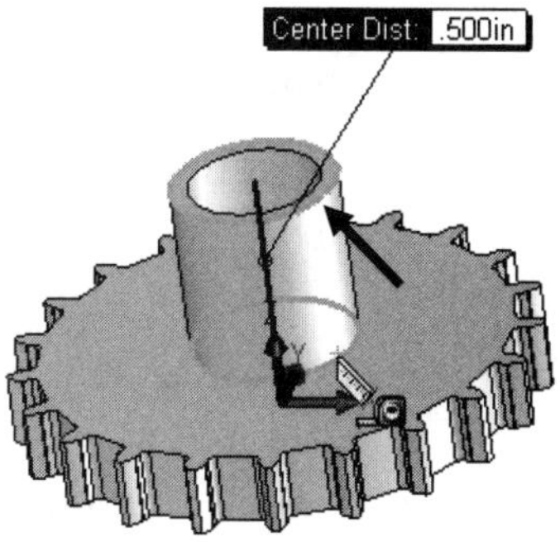

2. Click the **Measure** tool from the Evaluate tab.
   The Measure dialog box is displayed.

3. **Measure** the overall length of the Holder. Select the two edges as illustrated.

4. **Close** the Measure dialog box. **Close** the model.

Tutorial: Measure tool 2-2

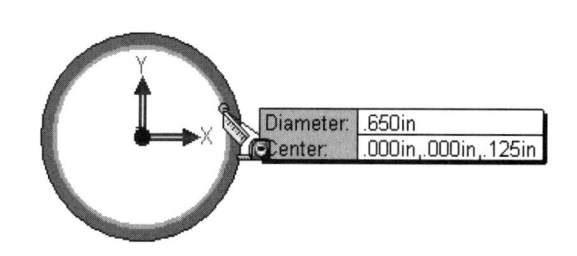

1. **Open** the Collar part from the SolidWorks CSWA Folder\Chapter2 location.

   Activate the Measure dialog box.

2. Click the **Measure** tool from the Evaluate tab. The Measure dialog box is displayed.

3. **Measure** the inside diameter of the Collar. Select the inside circular edge as illustrated. The inside diameter = .650in. Length = 2.042in.

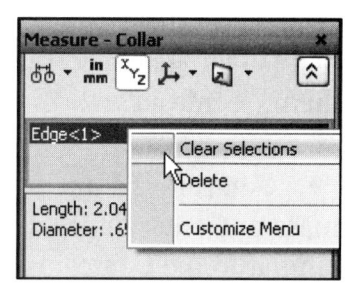

4. **Clear** all selections in the Measure - Collar dialog box.

Measure the inside area, perimeter, and diameter of the Collar.

5. Click the **inside face** as illustrated. View the results.

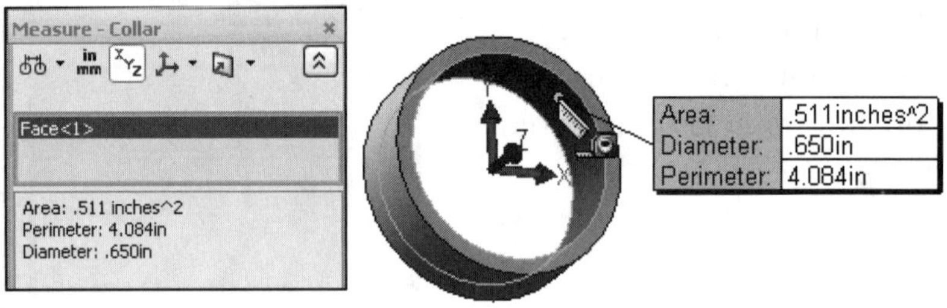

6. **Clear** all selections. **Close** the Measure dialog box. **Close** the model.

Locate the Center of mass, and Principal moments of inertia relative to the default coordinate location

The Mass Properties tool displays the mass properties of a part or assembly model, or the section properties of faces or sketches.

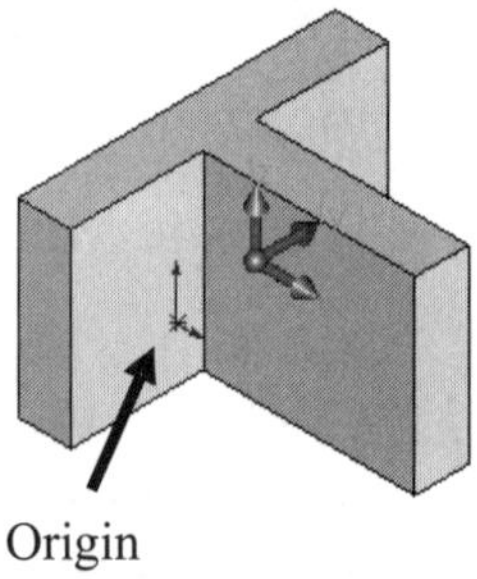

☼ The results are displayed in the Mass Properties dialog box. The principal axes and Center of mass are displayed graphically on the model in the Graphics window.

Origin

On the CSWA exam, you may see the word centroid. What is a centroid? Let's view a simple example of a triangle. In the plane of any triangle ABC, let:

- D = midpoint of side BC
- E = midpoint of side CA
- F = midpoint of side AB

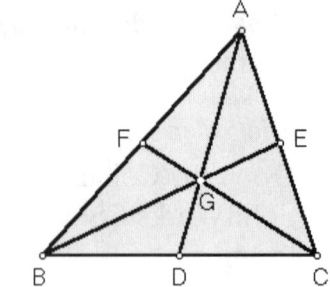

From the sketch, the lines AD, BE, CF come together to a single point. This point is called the centroid of the triangle ABC. There are many other points that are called triangle centers, but unlike most of them, "centroid" works on arbitrary shapes.

☼ Centroid is sometimes referred to as the Center of mass or Center of gravity. The Center of mass, (center of gravity) of a solid is similar to the Centroid of a Solid. However, calculating the Centroid involves only the geometrical shape of the solid. The center of gravity is equal to the Centroid if the body is homogenous, "constant density".

Basic Theory and Drawing Theory

Tutorial: Mass properties 2-1

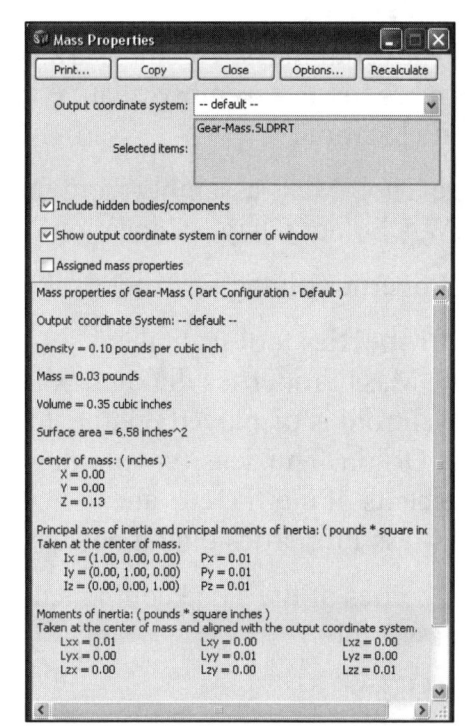

Use the Mass Properties tool to calculate the density, mass, volume, surface area, and Center of mass for a part relative to the default coordinate location.

1. **Open** the Gear-Mass part from the SolidWorks CSWA Folder\Chapter2 location. 6061 Alloy is the applied material.

Activate the Mass Properties dialog box.

2. Click the **Mass Properties** tool from the Evaluate tab in the CommandManager. The Mass Properties dialog box is displayed. View the various Mass Properties of the Gear-Mass part.

3. **Close** the Mass Properties dialog box.

4. **Close** the model.

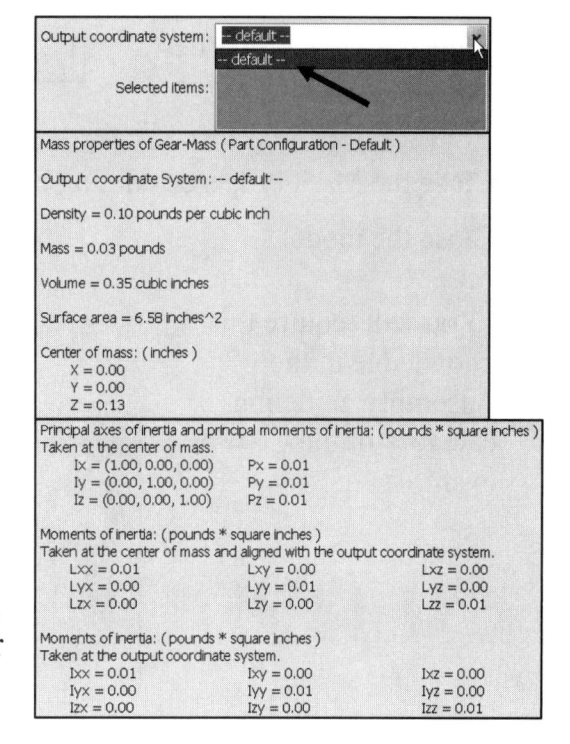

☼ By default, the Centroid is relative to the part or assembly default coordinate location.

View the Center of mass based on the default part Origin location. The density, mass, volume, surface area, Center of mass, Principal axes of inertia and principal moments of inertia, and Moments of inertia are displayed.

☼ To evaluate components or solid bodies in an assembly or multi-body part documents, click the **component** or **body**, and click **Recalculate**. If no component or solid body is selected, the mass properties for the entire assembly or multi-body part are reported.

Page 2 - 26

Basic Theory and Drawing Theory

Tutorial: Mass properties 2-2

Calculate the density, overall mass, volume, surface area, and Center of mass for an assembly relative to the default coordinate location.

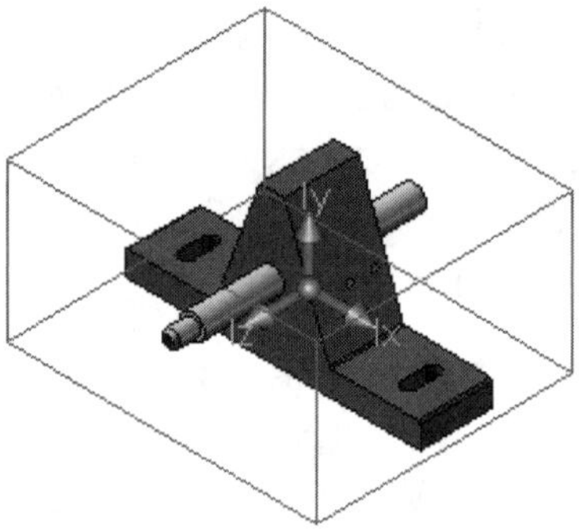

1. **Open** the Wedge-Tube-Mass assembly from the SolidWorks CSWA Folder\Chapter2 location.

Activate the Mass Properties dialog box.

2. Click the **Mass Properties** tool from the Evaluate tab. The Mass Properties dialog box is displayed. The Centroid is displayed relative to default assembly Origin. The density, volume, surface area, moments of inertia, etc. are displayed in the Mass Properties dialog box.

Obtain the Mass Properties of the Tube component in the Wedge-Tube-Mass assembly.

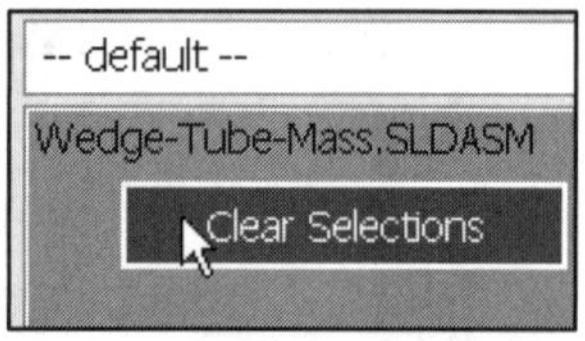

3. **Clear** all selected items.

4. Click **Tube** from the FeatureManager. Tube-1 is displayed in the Selected items box.

5. Click **Recalculate** from the Mass Properties dialog box.

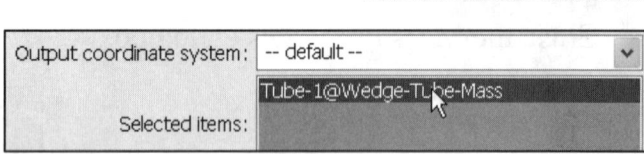

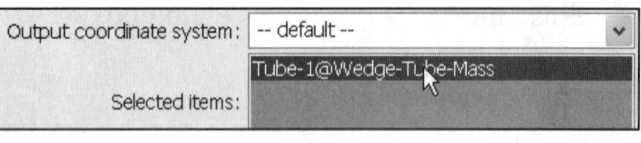

6. **View** the Tube properties. The Centroid location is relative to Tube part Origin.

7. **Close** the Mass Properties dialog box.

8. **Close** the model.

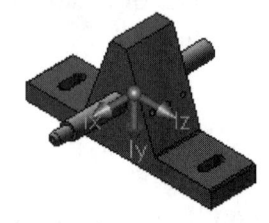

💡 You will require this knowledge in the Assembly modeling category of the CSWA exam.

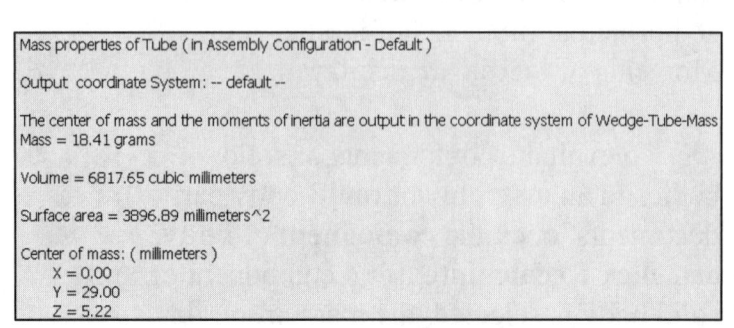

```
Mass properties of Tube ( in Assembly Configuration - Default )

Output coordinate System: -- default --

The center of mass and the moments of inertia are output in the coordinate system of Wedge-Tube-Mass
Mass = 18.41 grams

Volume = 6817.65 cubic millimeters

Surface area = 3896.89 millimeters^2

Center of mass: ( millimeters )
    X = 0.00
    Y = 29.00
    Z = 5.22
```

Identify the function and elements of a part and assembly FeatureManager design tree

You will need to understand and identify the icons, colors, suffix, and symbols for the following in the FeatureManager:

- Sketch states
- Component status and properties
- Display Pane status
- Reference Configurations

Sketch states

Sketches are generally defined in one of the following states:

☼ Color indicates the state of the individual Sketch entities.

1. *Under Defined.* Inadequate definition of the sketch, (blue). The FeatureManager displays a minus (-) symbol before the Sketch name.

2. *Fully Defined.* Complete information, (black). The FeatureManager displays no symbol before the Sketch name.

3. *Over Defined.* Duplicate dimensions and or relations, (red). The FeatureManager displays a (+) symbol before the Sketch name. The What's Wrong dialog box is displayed.

4. *Invalid Solution Found.* The sketch is solved but results in invalid geometry. Example: such as a zero length line, zero radius arc, or a self-intersecting spline, (yellow).

5. *No Solution Found.* Indicates sketch geometry that cannot be resolved, (Brown).

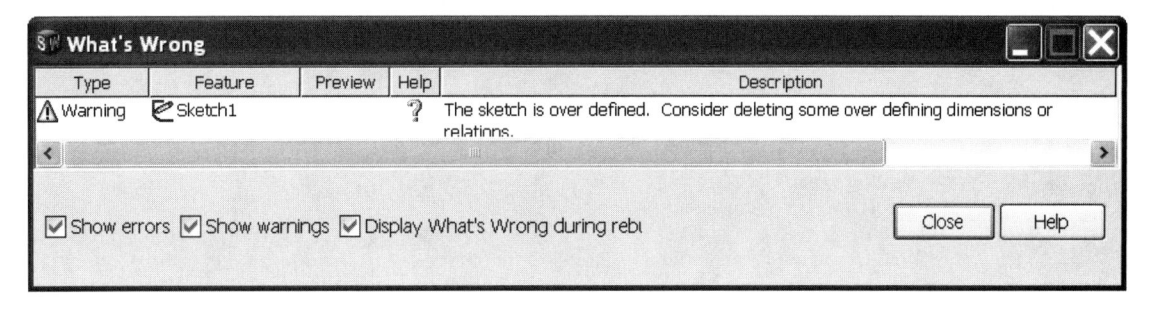

Basic Theory and Drawing Theory

☼ The SolidWorks SketchXpert is designed to assist in an Over defined state of a sketch. The SketchXpert PropertyManager is displayed as soon as you over-define a sketch. For existing over-defined sketches, click the Over Defined icon in the status bar.

With the SolidWorks software, it is not necessary to fully dimension or define sketches before you use them to create features. You should fully define sketches before you consider the part finished for manufacturing.

In the Basic Theory and Drawing Theory category, an exam question could read:

Question 1: The color of a fully defined sketch is _____?

The correct answer is Black.

Question 2: The FeatureManager displays a _____ symbol before the Sketch name in an under defined sketch?

The correct answer is (-).

Question 3: What symbol does the FeatureManager display before the Sketch name in an over defined sketch?

- A: (-)
- B: (?)
- C: (+)
- D: No symbol

The correct answer is C.

Basic Theory and Drawing Theory

☼ You can split the FeatureManager design tree and either display two FeatureManager instances, or combine the FeatureManager design tree with the ConfigurationManager or PropertyManager.

Component status and properties

Entries in the FeatureManager design tree have specific definitions. Understanding syntax and states saves time when creating and modifying assemblies. Review the columns of the MGPTube part syntax in the FeatureManager.

Column 1: A resolved component (not in lightweight state) displays a plus ⊞ icon. The plus icon indicates that additional feature information is available. A minus ⊟ icon displays the fully expanded feature list.

Column 2: Identifies a component's (part or assembly) relationship with other components in the assembly.

	Component States:
Symbol:	**State:**
⊞ 🔧	Resolved part. A yellow part icon indicates a resolved state. A blue part icon indicates a selected, resolved part. The component is fully loaded into memory and all of its features and mates are editable.
⊞ 🔧	Lightweight part. A blue feather on the part icon indicates a lightweight state. When a component is lightweight, only a subset of its model data is loaded in memory.
⊞ 🔧	Out-of-Date Lightweight. A red feather on the part icon indicates out-of-date references. This option is not available when the Large Assembly Mode is activated.
🔧	Suppressed. A gray icon indicates the part is not resolved in the active configuration.
🔧	Hidden. A clear icon indicates the part is resolved but invisible.
🔧	Hidden Lightweight. A transparent blue feather over a transparent component icon indicates that the component is lightweight and hidden.
🔧	Hidden, Out-of-Date, Lightweight. A red feather over a clear part icon indicates the part is hidden, out-of-date, and lightweight.

Icon	Description
	Hidden Smart Component. A transparent star over a transparent icon indicates that the component is a Smart Component and hidden.
	Smart Component. A star overlay is displayed on the icon of a Smart Component.
	Rebuild. A rebuild is required for the assembly or component.
	Resolved assembly. Resolved (or unsuppressed) is the normal state for assembly components. A resolved assembly is fully loaded in memory, fully functional, and fully accessible.

Column 3: The MGPTube part is fixed (f). You can fix the position of a component so that it cannot move with respect to the assembly Origin. By default, the first part in an assembly is fixed; however, you can float it at any time.

It is recommended that at least one assembly component is either fixed, or mated to the assembly planes or Origin. This provides a frame of reference for all other mates, and helps prevent unexpected movement of components when mates are added. The Component Properties are:

Component Properties in an Assembly	
Symbol:	**Relationship:**
(-)	A floating, under defined component has a minus sign (-) before its name in the FeatureManager and requires additional information.
(+)	An over-defined component has a plus sign (+) before its name in the FeatureManager.
(f)	A fixed component has a (f) before its name in the FeatureManager. The component does not move.
None	The Base component is mated to three assembly reference planes.
(?)	A question mark (?) indicates that additional information is required on the component.

Column 4: MGPTube - Name of the part.

Column 5: The symbol <#> indicates the particular inserted instance of a component. The symbol <1> indicates the first inserted instance of a component, "MGPTube" in the assembly. If you delete a component and reinsert the same component again, the <#> symbol increments by one.

Basic Theory and Drawing Theory

Column 6: The Resolved state displays the MGPTube icon with an external reference symbol, "- >". The state of external references is displayed as follows:

- If a part or feature has an external reference, its name is followed by –>. The name of any feature with external references is also followed by –>.
- If an external reference is currently out of context, the feature name and the part name are followed by ->?
- The suffix ->* means that the reference is locked.
- The suffix ->x means that the reference is broken.

☼ There are modeling situations in which unresolved components create rebuild errors. In these situations, issue the forced rebuild command, Ctrl+Q. The Ctrl+Q command rebuilds the model and all its features. If the mates still contain rebuild errors, resolve all the components below the entry in the FeatureManager that contains the first error.

In the Basic Theory and Drawing Theory category, an exam question could read:

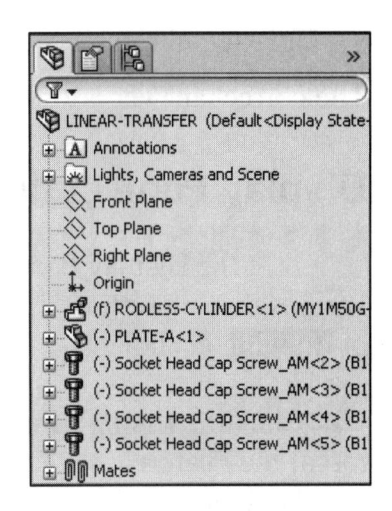

Question 1: In an assembly, the (f) symbol means?

- A: A floating component
- B: The first component
- C: A broken component
- D: A fixed component

The correct answer is D. Note: RODLESS-CYLINDER illustrated in the FeatureManager is in the Flexible state.

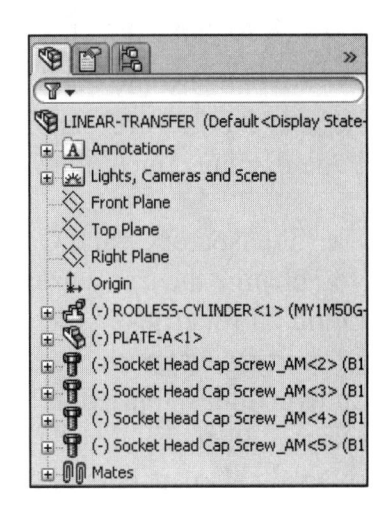

Question 2: In an assembly, the (-) symbol means?

- A: Reference is unlocked
- B: The reference is broken
- C: A floating component
- D: A fixed component

The correct answer is C.

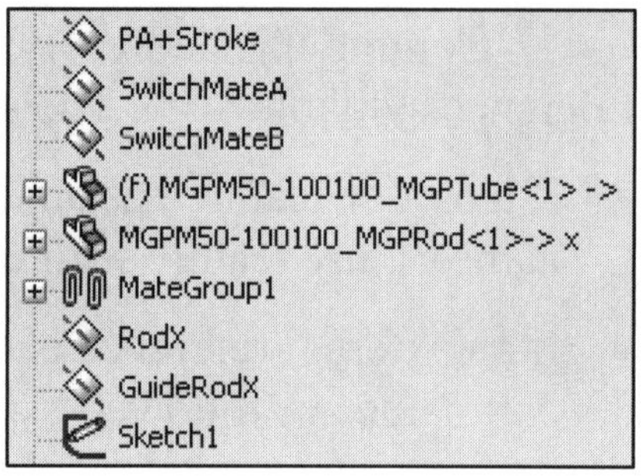

Question 3: In an assembly, the (->) symbol means?

- A: External reference
- B: Internal reference
- C: External reference locked
- D: External reference broken

The correct answer is A.

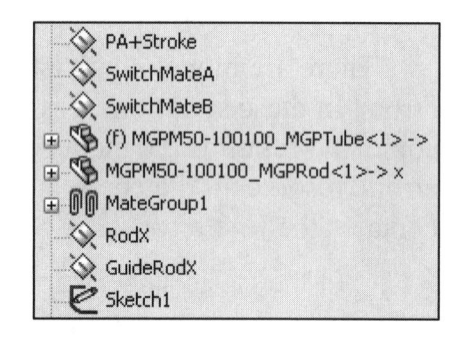

Question 4: In an assembly, the symbol after a component <#> indicates?

- A: Inserted instance of the component
- B: Inserted instance of a drawing
- C: Nothing
- D: External reference broken

The correct answer is A.

Display Pane status

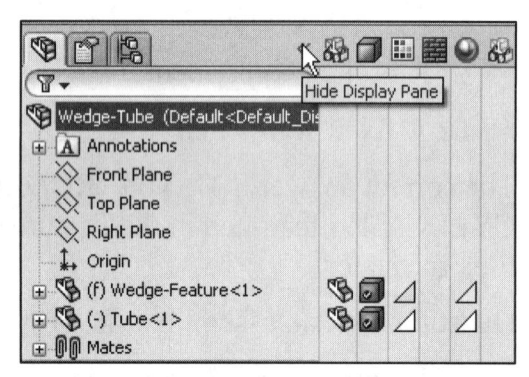

In the Display Pane, you can view various display settings in a part, assembly, or drawing document.

☼ Click the **Display Pane** icon ⟩⟩ at the top of the FeatureManager to expand the Display Pane or click the **Display Pane** icon ⟪ to close the Display Pane.

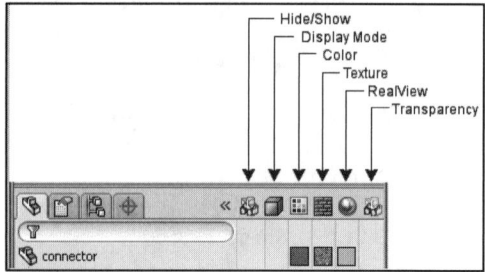

The Display Pane is displayed to the right of the FeatureManager. The Display Pane displays five columns as illustrated.

☼ Transparency and Color are linked. If you change the transparency slider under Optical Properties in the Color and Optics PropertyManager to a value other than 0.00, an override is applied in both the color and the transparency columns in the Display Pane. Likewise, if you apply transparency in the transparency column, an override also appears in the color column, because the transparency of the color has been changed to 75%.

Basic Theory and Drawing Theory

In a part, the display settings can include:

Display Setting	Icon	Description
Hide/Show		For bodies. Indicates the body is visible.
Color		For part, bodies, and features. Shows color, or blank if no color has been applied.
Texture		For part, bodies, and features. Shows texture, or blank if no texture has been applied.
RealView	Blank	No texture applied in part or assembly.
		Lower-right triangle shows the RV Appearance of the part from the part document.
Transparency		For part, bodies and features. Indicates that transparency has been applied in the Color and Optics PropertyManager, or blank if no transparency has been applied.

At this time, there are no Display Pane questions on the CSWA exam.

The Color option is not available when RealView Graphics is activated.

Reference configurations

There are many ways to create reference configurations. To specify properties for configurations: Perform one of the following task:

- In the ConfigurationManager, right-click the **part** or **assembly name** and select **Add Configuration**.

- In the ConfigurationManager, right-click a **configuration name** and select **Properties**.

- In the FeatureManager of an assembly, right-click a **component** and select **Add Configuration**. The Add Configuration PropertyManager is displayed.

Basic Theory and Drawing Theory

This provides the ability to create a configuration of the assembly component without opening the component in its own window. You must be editing the top-level assembly; the component can be at any level in the FeatureManager.

☼ At this time, there are no assembly configurations questions on the CSWA exam.

Identify the default Sketch Entities from the Sketch toolbar

You will need to understand the default Sketch Entities function and identify the Sketch Entities associated icon for the following: 3D Sketch, Smart Dimension, Line, Centerline, Rectangle, Centerpoint Arc, Tangent Arc, 3 Point Arc, Sketch Fillet, Spline, Point, and Plane.

☼ View the Sketch Entities from the default Sketch toolbar and the Consolidated drop-down toolbars.

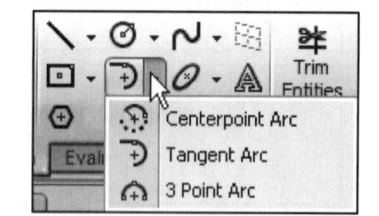

Sketching in SolidWorks is the basis for creating features. Sketch Entities provide the ability to create lines, rectangles, parallelograms, circles, etc. during the sketching process.

☼ Remember, when you create a new part or assembly, the three default planes are aligned with specific views. The plane you select for your first sketch determines the orientation of the part.

This book is targeted towards a person with a working knowledge of SolidWorks. The book will not cover each individual Sketch Entity in detail.

When do you use the following Sketch Entities?

- **3D Sketch**: To create a 3D sketch.
- **Smart Dimension**: To insert dimensions into a sketch.
- **Line**: To sketch multiple 2D lines in a sketch. The Line Sketch entity uses the Insert Line PropertyManager.

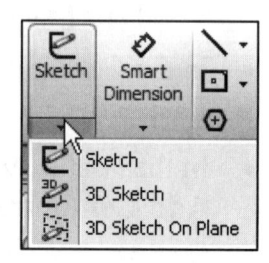

Basic Theory and Drawing Theory

- **Rectangle**: To sketch a rectangle and provides (x, y) cursor feedback coordinates on your mouse pointer. The Rectangle Sketch entity does not use a PropertyManager.

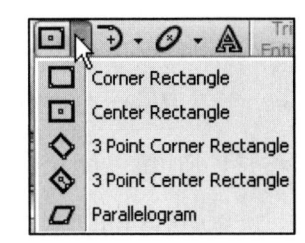

- **Circle:** To control the various properties of a perimeter-based or center-based sketched circle. The Circle and Perimeter Circle Sketch entity use the Circle PropertyManager.

- **Centerpoint Arc**: To create an arc from a centerpoint, a start point, and an end point. The Centerpoint Arc Sketch entity uses the Arc PropertyManager.

- **Tangent Arc**: To create an arc, which is tangent to the sketch entity. The Tangent Arc Sketch entity uses the Arc PropertyManager.

- **3 Point Arc**: To create an arc by specifying three points; a starting point, an endpoint, and a midpoint.

- **Sketch Fillet**: To trim the corner at the intersection of two sketch entities to create a tangent arc. The Sketch Fillet uses the Sketch Fillet PropertyManager.

- **Spline**: To create a profile that utilizes a complex curve. This complex curve is called a Spline, (Non-uniform Rational B-Spline or NURB). Create a spline with control points.

- **Centerline**: To use centerlines to create symmetrical sketch elements and revolved features, or as construction geometry. The Centerline sketch entity uses the Insert Line PropertyManager.

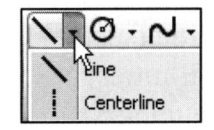

- **Point**: To insert points in your sketches and drawings. To modify the properties of a point, select the point in an active sketch, and edit the properties in the Point PropertyManager.

In the Basic Theory and Drawing Theory category, an exam question could read:

Question 1: Identify the following Sketch Entities icon .

- A: Center Arc tool
- B: Tangent Arc tool
- C: 3 Point Arc tool
- D: Point Arc tool

The correct answer is B.

Question 2: Identify the following Sketch Entities icon .

- A: Center Arc tool

- B: Tangent Arc tool
- C: 3 Point Arc tool
- D: Point Arc tool

The correct answer is C.

Identify the default Sketch tools from the Sketch toolbar

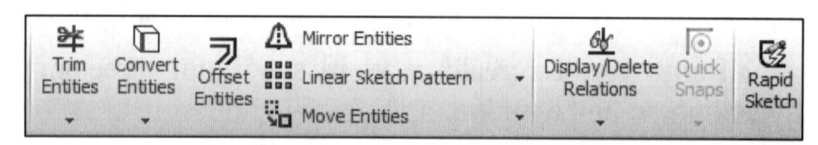

You will need to understand the default Sketch tool function and identify the Sketch tool associated icon for the following: Add Relation, Display/Delete Relations, Mirror Entities, Convert Entities, Offset Entities, and Trim Entities.

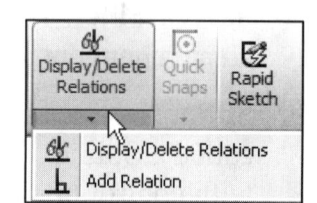

☼ At this time, the following Sketch tools are not addressed on the CSWA exam: Spline on Surface, Intersection Curve, Face Curves, Split Entities, Scale Entities, No Solve Move, Ellipse, Partial Ellipse, Parabola, Quick Snaps, Move Entities, Rapid Sketch, Instant3D, and Sketch picture.

When do you use the following Sketch tools?

- **Add Relation**: To create geometric relations such as (Tangent, Equal, Horizontal, Vertical, etc.) between sketch entities, or between sketch entities and planes, axes, edges, or vertices.

- **Display/Delete Relations**: To view and edit existing relations.

- **Mirror Entities**: To mirror a new entity, both the original and the mirrored entity, mirror some or all of the sketch entities, mirror about any type of line, and mirror about edges in a drawing, part, or assembly.

- **Convert Entities**: To create one or more curves in a sketch by projecting an edge, loop, face, curve, or external sketch contour, set of edges, or set of sketch curves onto the Sketch plane.

Basic Theory and Drawing Theory

- **Offset Entities**: To offset one or more sketch entities, selected model edges, or model faces by a specified distance. For example, you can offset sketch entities such as splines or arcs, sets of model edges, loops, etc.
- **Trim Entities**: To select the trim type based on the entities you want to trim or extend. All trim types are available with 2D sketches and 2D sketches on 3D planes.

In the Basic Theory and Drawing Theory category, an exam question could read:

Question 1: Identify the following Sketch tool icon ⫽.

- A: Trim Sketch tool
- B: Point Sketch tool
- C: Angle Sketch tool
- D: Offset Entities Sketch tool

The correct answer is D.

Question 2: Identify the following Sketch tool icon ⵣ.

- A: Trim Sketch tool
- B: Cut Sketch tool
- C: Offset Angle Sketch tool
- D: None of the above

The correct answer is A.

☼ When you create mirrored entities, SolidWorks applies a symmetric relation between each corresponding pair of sketch points (the ends of mirrored lines, the centers of arcs, etc). If you change a mirrored entity, its mirror image also changes.

SolidWorks file formats for input and export

You will need to know and understand the default and available file formats in SolidWorks for the following:

- Save As type for a part, assembly, and drawing
- Open File of different format

Basic Theory and Drawing Theory

SolidWorks provides three basic document types with their own extensions:

- A part file ends with an extension (.sldprt)
- An assembly file ends with an extension (.sldasm)
- A drawing file ends with an extension of (slddrw)

☼ When you save a new document, the Save As dialog box is displayed to enter a filename or accept the default filename.

Tutorial: File formats 2-1
View the available file formats for a part.

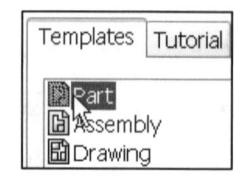

1. **Create** a New part in SolidWorks.
2. Click **Save As** from the Menu bar toolbar.
3. Click the **drop-down arrow** from the Save as type box.
4. **View** the available file formats.
5. **Close** the model.

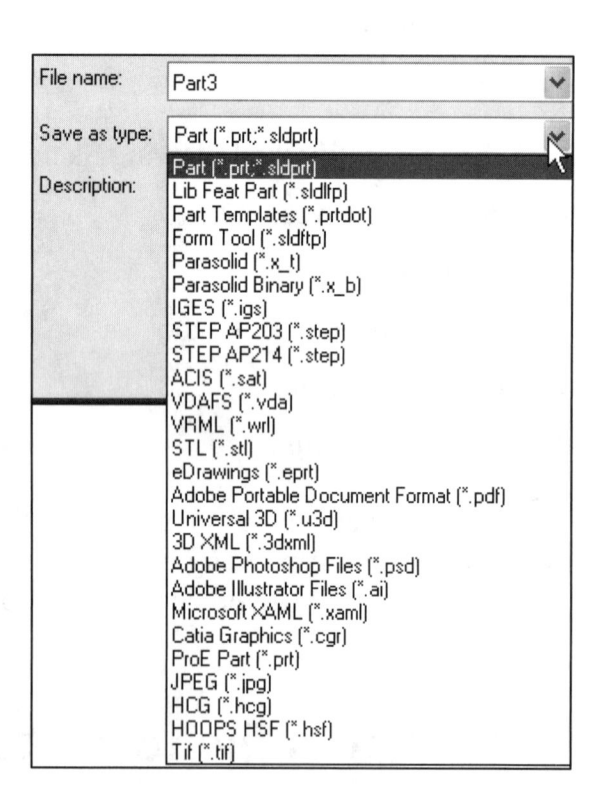

In the Basic Theory and Drawing Theory category, an exam question could read:

Question 1: What is the default part file format in SolidWorks?

- A: *.sldfop
- B: *.sldprt
- C: *.hjyy
- D: *.pprt

The correct answer is B.

Page 2 - 39

Basic Theory and Drawing Theory

Tutorial: File formats 2-2

View the available file formats for an assembly.

1. **Create** a New assembly in SolidWorks.
2. Click **Save As** from the Menu bar toolbar.
3. Click the **drop-down arrow** from the Save as type box.
4. **View** the available file formats.
5. **Close** the model.

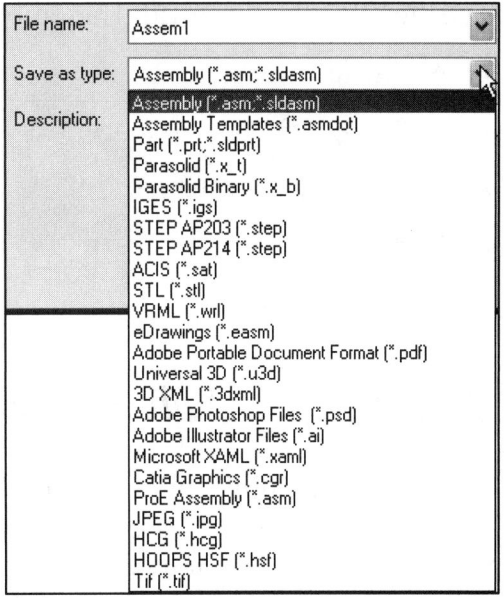

Tutorial: File formats 2-3

View the available file formats for a drawing.

1. **Create** a New drawing in SolidWorks.
2. Click **Save As** from the Menu bar toolbar. Click the **drop-down arrow** from the Save as type box.
3. **View** the available file formats.
4. **Close** the model.

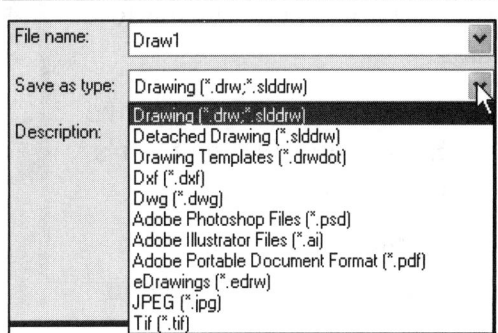

In the Basic Theory and Drawing Theory category, an exam question could read:

Question 1: Which is not a valid drawing format in SolidWorks?

- A: *.tif
- B: *.jpg
- C: * dwgg
- D: * dxf

The correct answer is C.

Question 2: Which is a valid assembly format in SolidWorks?

- A: *.tiffe
- B: *.jpgg
- C: * .assmy
- D: * .stl

The correct answer is D.

Utilize SolidWorks Help topics

Use the SolidWorks Help topics during the CSWA exam. Understand the following sections to locate needed information of various subject topics. The SolidWorks Help section is divided into three information tabs:

- **Contents** tab: Contains the SolidWorks Online User's Guide documents.
- **Index** tab: Contains additional information on key entered words.
- **Search** tab: To locate entered information.

☼ Click Help, SolidWorks Tutorials to access the online tutorials.

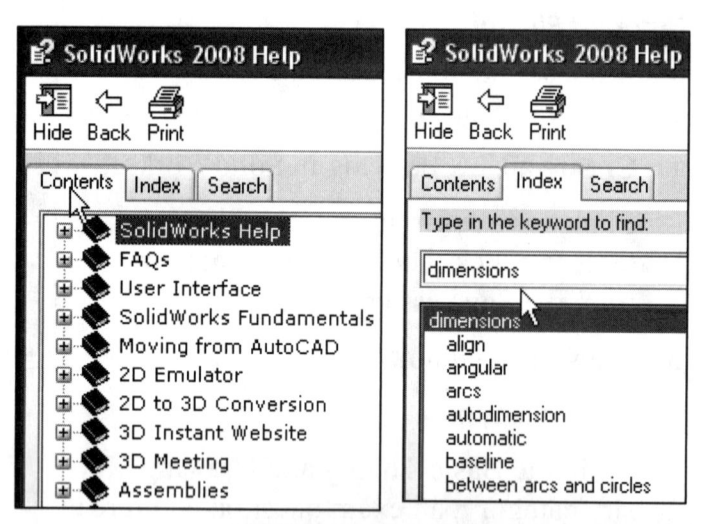

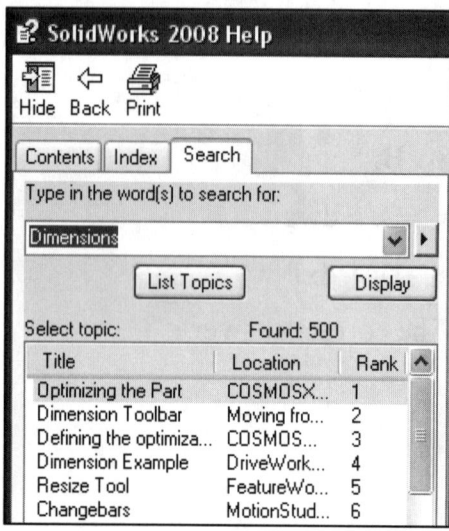

Basic Theory and Drawing Theory

Tutorial: SolidWorks help 2-1

Explore the SolidWorks Help tabs.

1. **Create** a New part in SolidWorks.
2. Click **Help, SolidWorks Help** from the Menu bar menu.
3. **View** the Online User's Guide documents under the Contents tab.
4. **Explore** the provided documents.
5. Enter **Dimensions** under the Index tab.
6. **Explore** the provided options and information.
7. Enter **Dimensions** under the Search tab.
8. Click **List Topics**.
9. **Explore** the provided options and information.
10. **Close** the model.

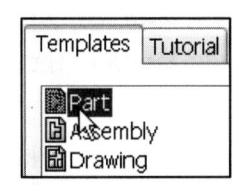

Identify the process of creating a simple drawing from a part or an assembly

You will need to recognize the process and be acquainted with the procedure for the following:

- Insert the 3 standard views: Front, Top, and Right
- Insert an Isometric named view
- Add a drawing sheet
- Insert drawing annotations
- Modify a drawing view with the scale and view mode option

☼ You are not required to create a drawing from a part or an assembly in the CSWA exam at this time. You do need the ability to understand the procedure/process to create a drawing view.

☼ Think of a drawing view as a container. Generally the contents are views of a model. When you sketch in a drawing, or insert annotations or blocks, the entities belong to the active drawing view or drawing sheet.

Tutorial: Drawing process 2-1

Understand the process to create a drawing from an existing part.

1. **Create** a New drawing. The Sheet Format/Size dialog box is displayed.

2. Select **A-Landscape**. Select **Standard sheet size**.

3. Click **OK** from the SheetFormat/Size dialog box. The Model View PropertyManager is displayed, if the Start command when creating new drawing option is check.

4. Click **Browse** from the Part/Assembly to Insert box.

5. Double click the **Gear** part from the SolidWorks CSWA Folder\Chapter2 location.

Insert views. Note: Third angle projection is used in this book.

6. Select **Multiple views** from the Number of Views box.

7. Select *Top, *Right, and *Isometric view. Note: *Front view is selected by default.

8. Click **OK** from the Model View PropertyManager. The views are displayed on Sheet1.

9. Click **Yes**.

10. **Click** inside the Isometric view boundary.

11. **Apply** Shaded With Edges to the Isometric view.

12. **Modify** the scale of the four views to 1.5:1.

13. **View** the four views in the Gear drawing.

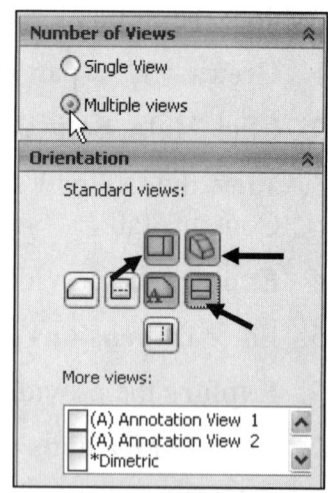

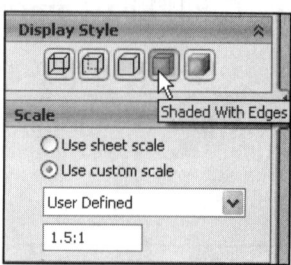

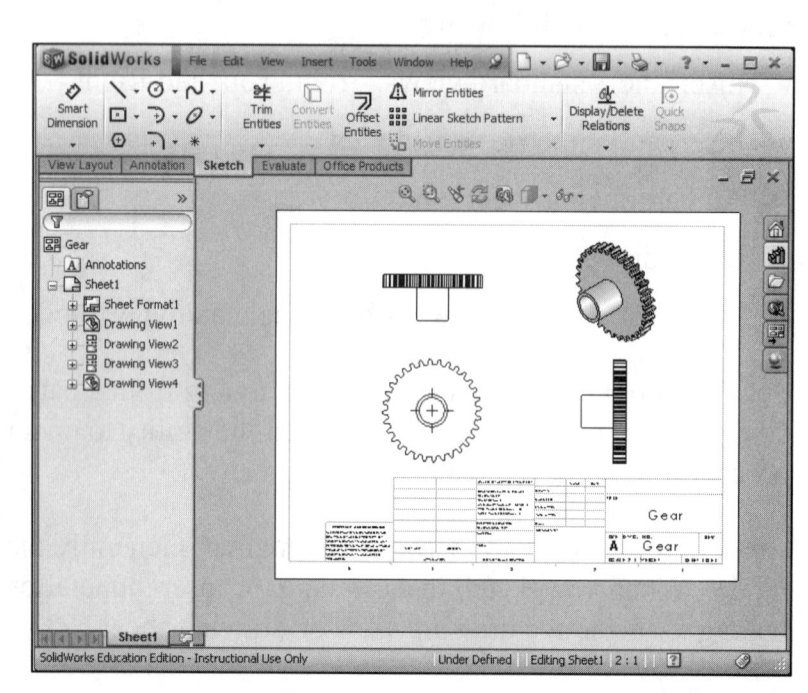

Basic Theory and Drawing Theory

14. **Insert** a Sheet in the drawing. Sheet2 is displayed.

💡 The Annotations toolbar provides tools for adding notes and symbols to a drawing, part, or assembly document. Only those annotations that are appropriate for the active document are available; the other tools are displayed in gray.

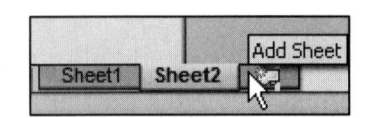

15. **Return** to Sheet1. Click the Sheet1 tab.

16. **Click** the Note tool from the Annotations toolbar.

17. **Insert** the following note as illustrated above the Front View.

18. **Save** the drawing and name it Gear. **Close** the drawing.

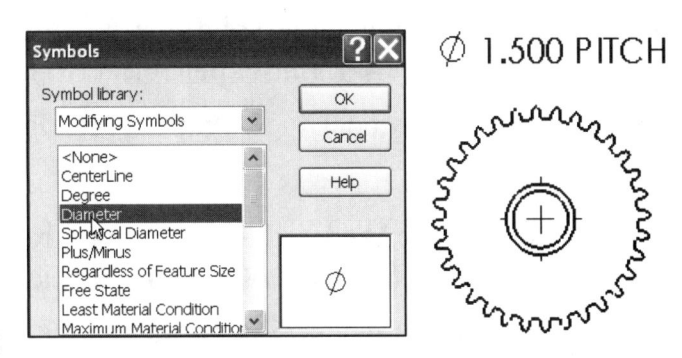

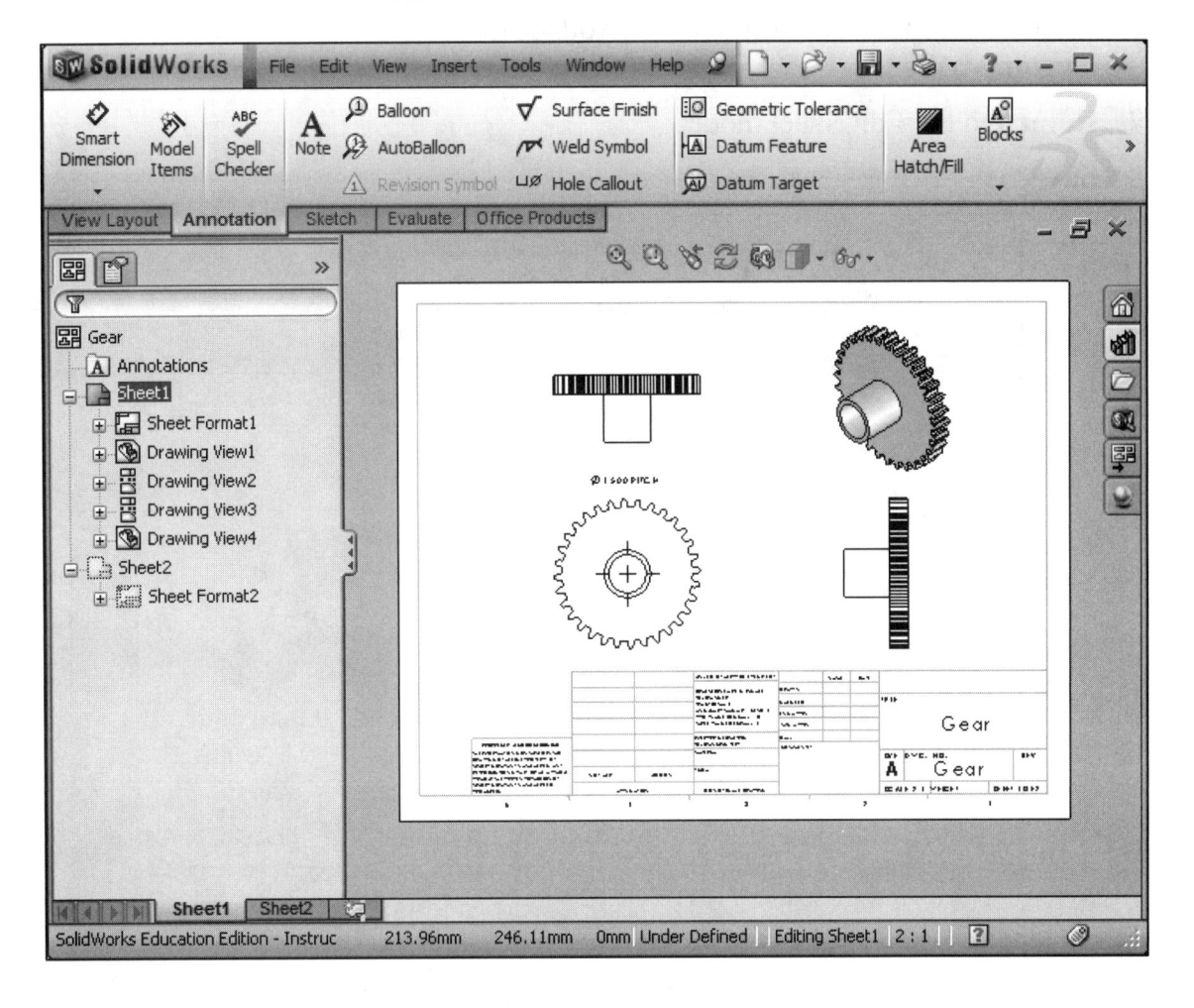

Page 2 - 44

Basic Theory and Drawing Theory

☼ You can add all types of annotations to a drawing document. You can add most types in a part or assembly document, then insert them into a drawing document. However, there are some types, such as Center Marks and Area Hatch that you can add only in a drawing document.

Tutorial: Drawing process 2-2

Understand the process to create an Isometric drawing view from an Assembly.

1. **Create** a New drawing in SolidWorks. The Sheet Format/Size dialog box is displayed. Select **A-Landscape**. Select **Standard sheet size**.

2. **Click OK**. The Model View PropertyManager is displayed.

3. **Click** Browse.

4. Double click the **Wedge-Tube** assembly from the SolidWorks CSWA Folder\Chapter2 location.

5. **Insert** an Isometric view in a Shaded mode with a 0.9:1 scale. Note: Third angle projection is used in this book.

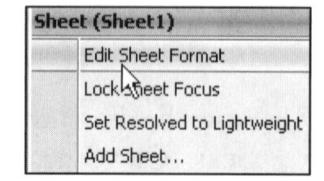

6. **Select** the Edit Sheet Format mode. Add a Title: **Customer Wedge-Tube** using 12 Font.

7. **Return** to the Edit Sheet mode.

8. **Save** the drawing. **Close** the drawing.

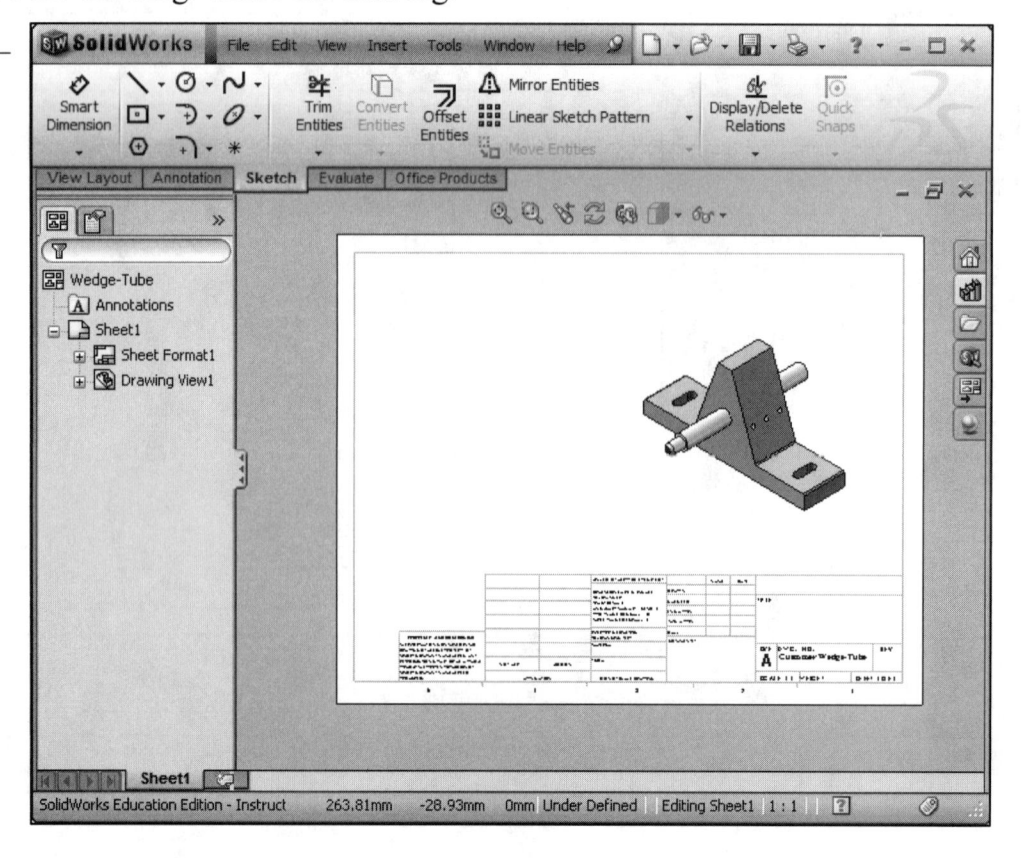

Basic Theory and Drawing Theory

Drawing section - icons

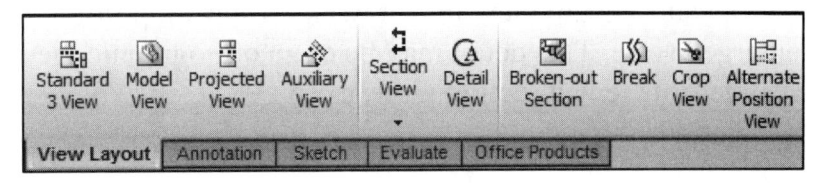

You need the ability to recognize the drawing name view type by it's icon and the procedure to create the drawing view: *Standard 3 View, Model View, Projected View, Auxiliary View, Section View, Aligned Section View, Detail View, Broken-out Section, Break, Crop View, and Alternate Position View.*

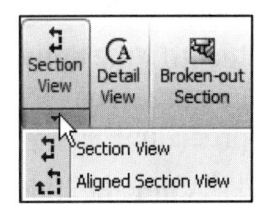

Tutorial: Drawing name view type 2-1

Identify the drawing name view type in the Drawing FeatureManager. Understand the drawing view procedure.

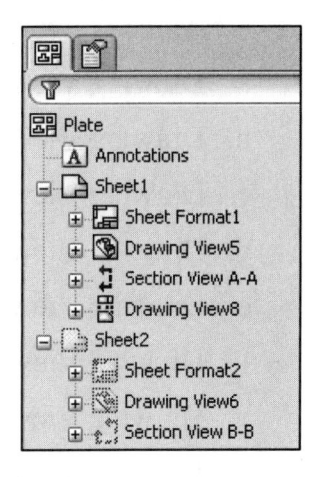

1. **Open** the Plate drawing from the SolidWorks CSWA Folder\Chapter2 location.

2. **View** Sheet1. Identify the drawing name view type that was used to create the drawing view from the FeatureManager icon. The views have been renamed.

    a. Drawing View5: Model View

    b. Section View A-A: Section View

    c. Drawing View8: Projected View

Create a Section View in a drawing by cutting the parent view with a section line. The section view can be a straight cut section or an offset section defined by a stepped section line. The section line can also include concentric arcs.

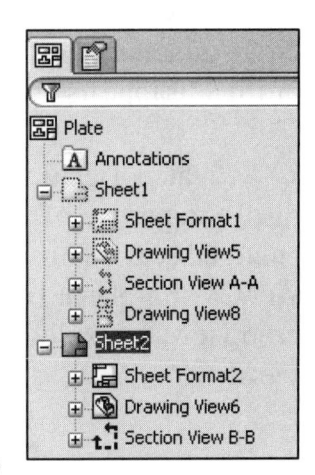

3. **View** Sheet2. Identify the drawing name view type that was used to create the drawing from the icon.

    a. Drawing View6: Model View

    b. Section View B-B: Aligned Section View

4. **Close** the drawing.

Create an Aligned section view in a drawing through a model, or portion of a model, that is aligned with a selected section line segment. The Aligned Section view is similar to a Section View, but the section line for an aligned section comprises two or more lines connected at an angle.

💡 Create a detail view in a drawing to show a portion of a view, usually at an enlarged scale. This detail may be of an orthographic view, a non-planar (isometric) view, a section view, a crop view, an exploded assembly view, or another detail view.

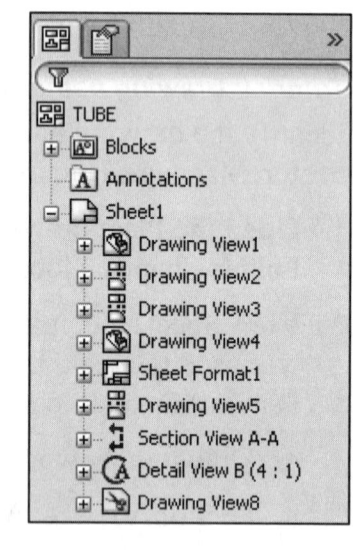

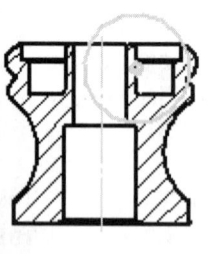

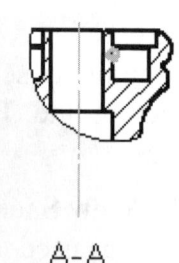

Tutorial: Drawing name view type 2-2

Identify the drawing name view type in the Drawing FeatureManager. Understand the drawing view procedure.

1. **View** the illustrated Drawing FeatureManager. Identify the drawing name view type that was used to create the drawing from the icon.

    a. Drawing View1: Model View

    b. Drawing View2: Projected View

    c. Drawing View3: Projected View

    d. Drawing View4 : Model View

    e. Drawing View5 : Projected View

    f. Drawing View6 : Section View

    g. Drawing View7: Detail View

    h. Drawing View8: Crop View

💡 Crop any drawing view except a Detail View, a view from which a Detail View has been created, or an exploded view. To create a Crop view, sketch a closed profile such as a circle or spline. The view outside the closed profile disappears as illustrated.

💡 Create a Detail view in a drawing to display a portion of a view, usually at an enlarged scale. This detail may be of an orthographic view, a non-planar (isometric) view, a section view, a crop view, an exploded assembly view, or another detail view.

In the Basic Theory and Drawing Theory section, an exam question could read:

Question 1: Identify the following Drawing View icon .

- A: Projected View
- B: Standard View
- C: Section View
- D: None of the provided

The correct answer is A.

Question 2: Identify the following Drawing View icon .

- A: Projected View
- B: Trim View
- C: Cut View
- D: Crop View

The correct answer is D.

Question 3: Identify the following Drawing View icon .

- A: Section View
- B: Broken View
- C: Break View
- D: Two Section View

The correct answer is C.

Identify the procedure to create a Named Drawing view

You need the ability to identify the procedure to create a named drawing view: *Standard 3 View, Model View, Projected View, Auxiliary View, Section View, Aligned Section View, Detail View, Broken-out Section, Break, Crop View, and Alternate Position View.*

Basic Theory and Drawing Theory

Tutorial: Drawing named procedure 2-1

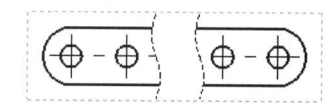

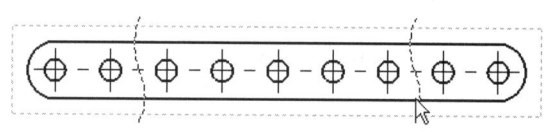

Identify the drawing name view and understand the procedure to create the name view.

1. **View** the illustrated drawing views. The top drawing view is a Break View. The Break View is created by adding a break line to a selected view.

☼ Broken views make it possible to display the drawing view in a larger scale on a smaller size drawing sheet. Reference dimensions and model dimensions associated with the broken area reflect the actual model values.

☼ In views with multiple breaks, the Break line style must be the same.

Tutorial: Drawing named procedure 2-2

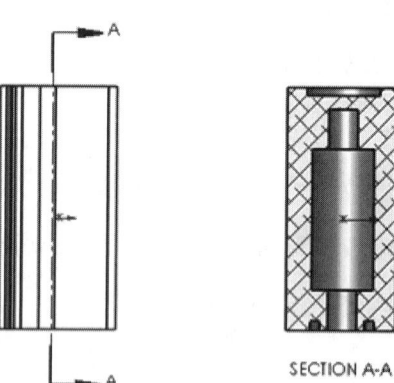

Identify the drawing name view and understand the procedure to create the name view.

1. **View** the illustrated drawing views. The right drawing view is a Section View. The Section View is created by cutting the parent view with a section line.

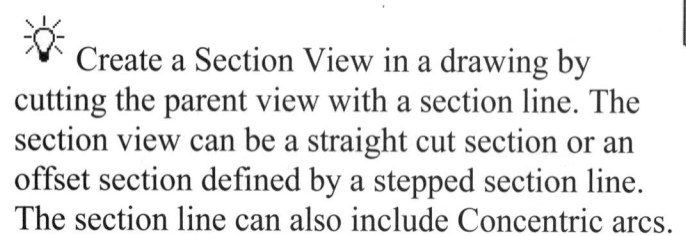

☼ Create a Section View in a drawing by cutting the parent view with a section line. The section view can be a straight cut section or an offset section defined by a stepped section line. The section line can also include Concentric arcs.

Tutorial: Drawing named procedure 2-3

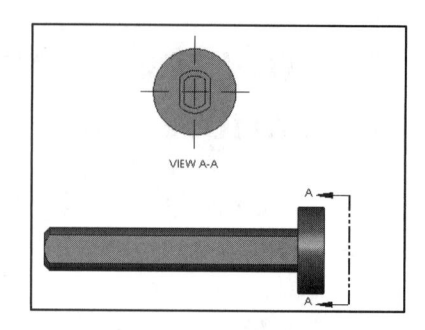

Identify the drawing name view and understand the procedure to create the name view.

1. **View** the illustrated drawing views. The Top drawing view is an Auxiliary View of the Front View. Select a reference edge to create an Auxiliary View.

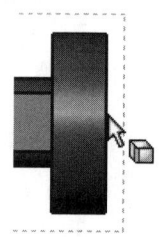

☼ An Auxiliary View is similar to a Projected View, but it is unfolded normal to a reference edge in an existing view.

Basic Theory and Drawing Theory

Tutorial: Drawing named procedure 2-4

Identify the drawing name view and understand the procedure to create the name view.

1. **View** the illustrated drawing views. The right drawing view is an Aligned Section View of the bottom view. The Section View is created by using two lines connected at an angle. Create an Aligned Section View in a drawing through a model, or portion of a model, that is aligned with a selected section line segment.

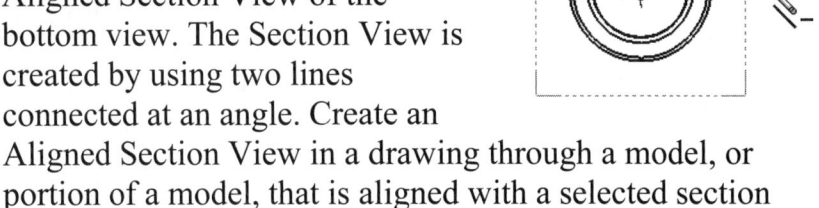

💡 The Aligned Section View is very similar to a Section View, with the exception that the section line for an aligned section comprises of two or more lines connected at an angle.

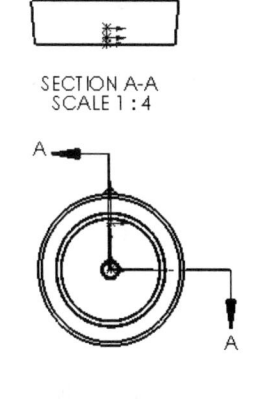

Tutorial: Drawing named procedure 2-5

Identify the drawing name view and understand the procedure to create the name view.

1. **View** the illustrated drawing views. The left drawing view is a Detail View. The Detail View is created by sketching a circle with the Circle Sketch tool. Click and drag for the location.

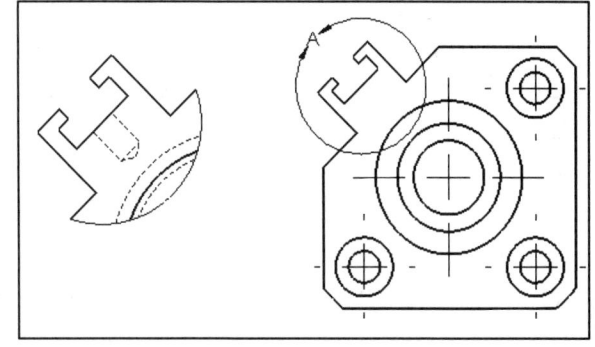

💡 The Detail View tool ⓐ provides the ability to add a detail view to display a portion of a view, usually at an enlarged scale.

💡 To create a profile other than a circle, sketch the profile before clicking the Detail View tool. Using a sketch entity tool, create a closed profile around the area to be detailed. You can add dimensions or relations to the sketch entities to position the profile precisely relative to the model.

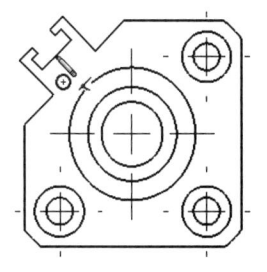

Page 2 - 50

Basic Theory and Drawing Theory

Tutorial: Drawing named procedure 2-6

Identify the drawing name view and understand the procedure to create the name view.

1. **View** the illustrated drawing views. The right drawing view is a Broken-out Section View. The Broken-out Section View is part of an existing drawing view, not a separate view. Create the Broken-out Section View with a closed profile, usually by using the Spline Sketch tool. Material is removed to a specified depth to expose inner details.

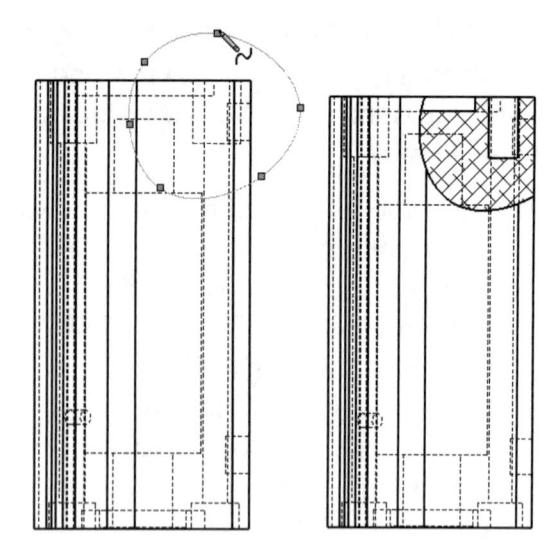

Tutorial: Drawing named procedure 2-7

Identify the drawing name view and understand the procedure to create the name view.

1. **View** the illustrated drawing view. The top drawing view is a Crop View. The Crop View is created by a closed sketch profile such as a circle, or spline as illustrated.

   The Crop View provides the ability to crop an existing drawing view. You can not use the Crop tool on a Detail View, a view from which a Detail View has been created, or an exploded view.

   ☼ Use the Crop tool to save steps. Example: instead of creating a Section View and then a Detail View, then hiding the unnecessary Section View, use the Crop tool to crop the Section View directly.

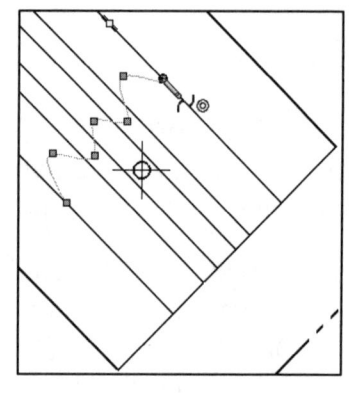

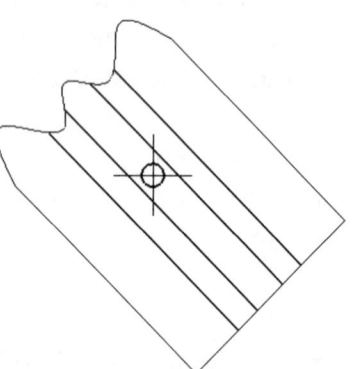

Basic Theory and Drawing Theory

Tutorial: Drawing named procedure 2-8
Identify the drawing name view and understand the procedure to create the name view.

1. **View** the illustrated drawing view. The drawing view is an Alternate Position View. The Alternate Position View tool provides the ability to superimpose an existing drawing view precisely on another. The alternate position is displayed with phantom lines.

Use the Alternate Position View is display the range of motion of an assembly. You can dimension between the primary view and the Alternate Position View. You can not use the Alternate Position View tool with Broken, Section, or Detail views.

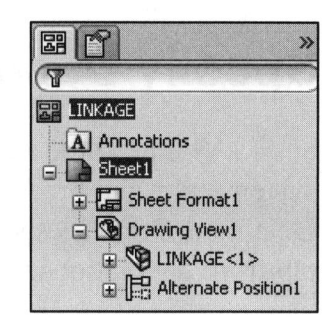

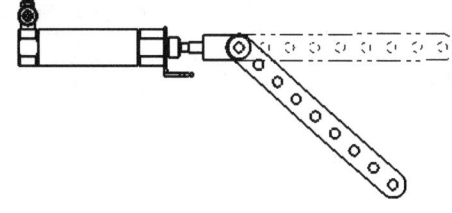

Engineering Documentation Practices

A 2D drawing view is provided to clarify the model dimensions and details for part and assembly models given in the Advanced Part and Assembly Modeling section of the exam. The ability to interpret the 2D drawing views is required.

- Example: The provided 2D detail drawing view illustrates that eight holes are required. The hole diameters are .19. The equal spacing between the holes is .55. Note: Units are provided in the problem.

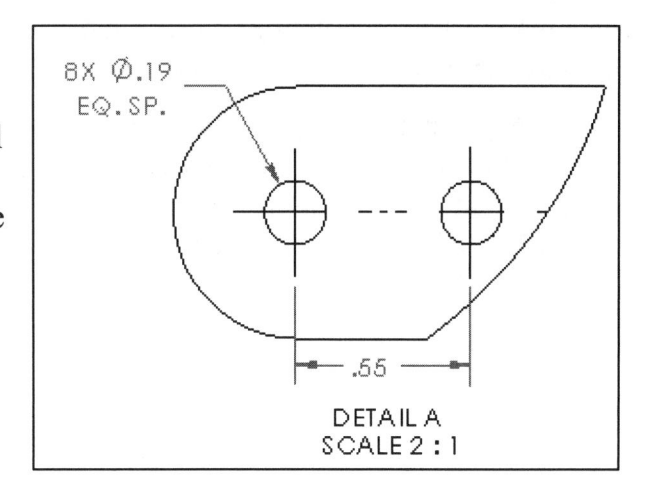

Specify Document Properties

You need the ability to identify the procedure to select system units and precision of a SolidWorks model using the Document Properties section. Access the Document Properties tab from the Options tool located in the Menu bar toolbar.

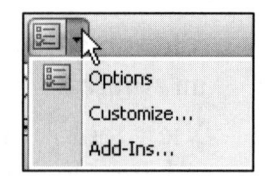

☼ In Chapter 3, you are required to build a simple part from a detailed dimensioned illustration. You are required to set the precision for the selected unit system and to set the Tolerance / Precision using the Dimension PropertyManager.

Document properties apply only to the current document. The Document Properties tab is only available when a document is open.

New documents get their document settings (such as Units, Image Quality, etc.) from the document properties of the template used to create the document.

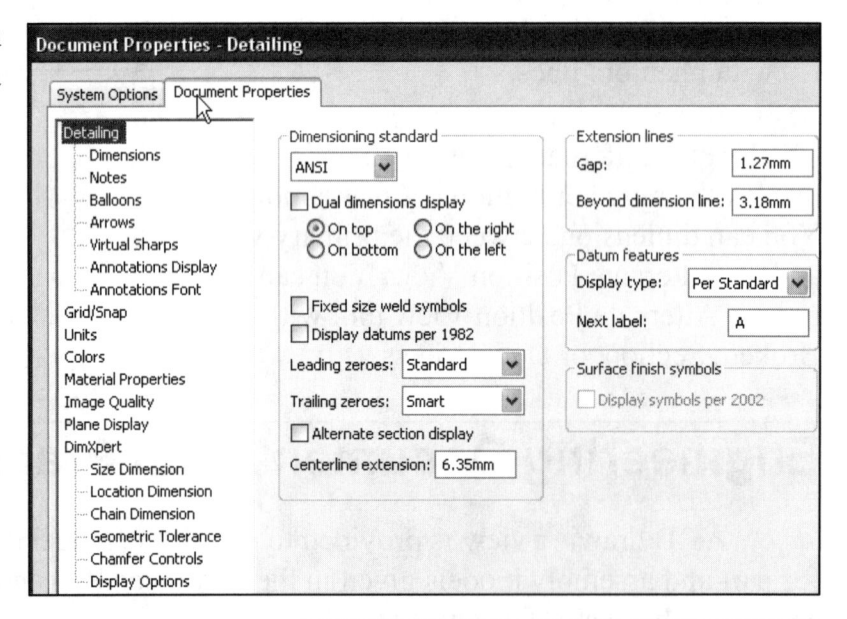

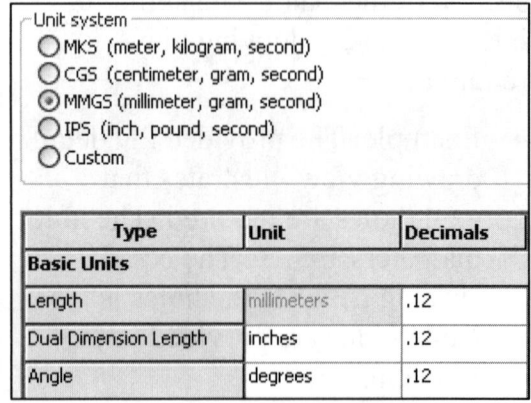

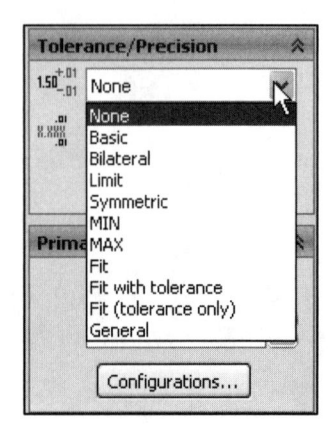

💡 Know how to set the Unit system, and Tolerance/Precision option in a document for the CSWA exam.

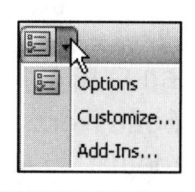

Tutorial: Document properties 2-1

Set dimensioning standard, system units, and precision.

1. **Create** a New part in SolidWorks.
2. Click **Options**, **Document Properties** tab from the Menu bar toolbar. 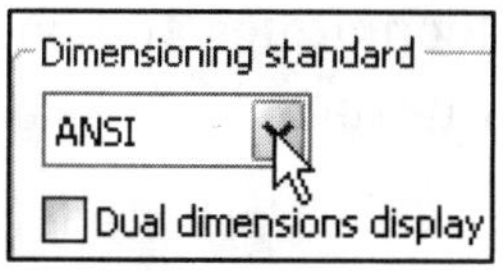
3. Select **ANSI** for Dimensioning standard.
4. Click **Units**.
5. Select **IPS** for Unit system. 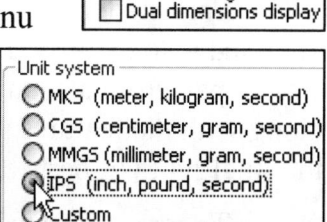
6. Select **.123** Decimal places for Length units.
7. Select **None** Decimal places for Angular units.
8. **Close** the part.

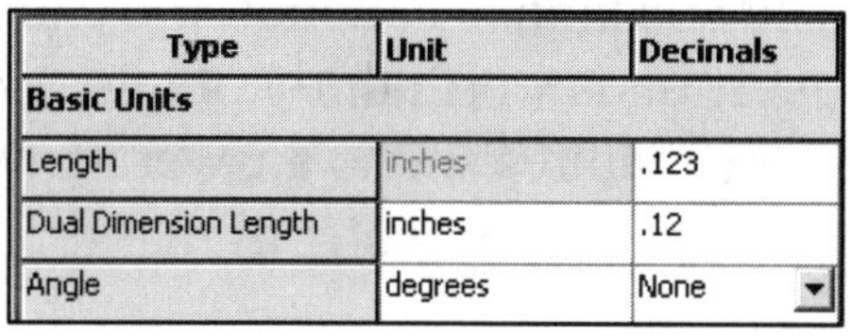

Tutorial: Document properties 2-2

Set dimensioning standard, system units, and precision.

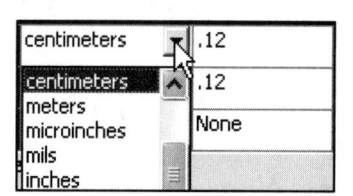

1. **Create** a New part in SolidWorks.
2. Click **Options**, **Document Properties** tab from the Menu bar menu.
3. Select **ANSI** for Dimensioning standard.
4. Select **Units**.
5. Click **Custom** for Unit system.
6. Select **centimeters** from the length units drop down box.
7. Select **.12** Decimal places for Length units.
8. Select **None** Decimal places for Angular units.
9. **Close** the part.

Summary

Basic Theory and Drawing Theory is one of the five categories on the CSWA exam. In this chapter, you covered general concepts, symbols, terminology, along with the core elements in this category.

There are two questions on the CSWA exam in this category. Each question is worth five points. The two questions are in a multiple choice single answer or fill in the blank format and requires general knowledge and understanding of the SolidWorks User Interface, FeatureManager, ConfigurationManager, Sketch toolbar, Feature toolbar, Drawing view methods, and basic 3D modeling techniques and engineering principles. Spend no more than 10 minutes on each question in this category. This is a timed exam. Manage your time.

Part Modeling is the next chapter in this book. Chapter 3 covers the knowledge to identify the part Origin, design intent, and key features to create a simple part from a detailed dimensioned illustration. The complexity of the models along with the features progressively increases throughout chapter 3 to simulate the final types of parts that could be provided on the CSWA exam.

There is one question on the CSWA exam in the Part Modeling category. The question is in a multiple choice single answer or fill in the blank format. The question is worth thirty points. You are required to create a model, with six or more features and to answer a question either on the overall mass, volume, or the location of the Center of mass relative to part Origin.

No PhotoWorks, Motion Study, Display Pane, or assembly configuration questions on the CSWA exam as this time.

Key terms

- *Aligned section view.* A drawing view that is aligned with a selected section line segment. The Aligned Section view is similar to a Section View, but the section line for an aligned section comprises two or more lines connected at an angle.

- *Alternate position view.* A drawing view in which one or more views are superimposed in phantom lines on the original view. Alternate position views are often used to show range of motion of an assembly.

- *Associativity.* Assures that changes incorporated in one document or drawing view are automatically made to all other related documents and drawing views.

- *Base feature.* The first feature of a part is called the Base feature.

- *Base sketch.* The first sketch of a part is called the Base sketch. The Base sketch is the foundation for the 3D model. Create a 2D sketch on a default plane: Front, Top, and Right in the FeatureManager design tree, or on a created plane.

Basic Theory and Drawing Theory

- *Crop view*. You can crop any drawing view except a Detail View, a view from which a Detail View has been created, or an exploded view. Create a Crop view, sketch an closed profile such as a circle or spline. The view outside the closed profile disappears.

- *ConfigurationManager*. Located on the left side of the SolidWorks Graphics window is a used to create, select, and view multiple configurations of parts and assemblies in a document.

- *Constraints*. Geometric relations such as Perpendicular, Horizontal, Parallel, Vertical, Coincident, Concentric, etc. Insert constraints to your model to incorporate design intent.

- *Coordinate system*. A system of planes used to assign Cartesian coordinates to features, parts, and assemblies. Part and assembly documents contain default coordinate systems; other coordinate systems can be defined with reference geometry. Coordinate systems can be used with measurement tools and for exporting documents to other file formats.

- *Design intent*. Design intent is the intellectual arrangements of features and dimensions of a design. Design intent governs the relationship between sketches in a feature, features in a part, and parts in an assembly.

- *Design Library*. Located in the Task Pane, the Design Library provides a central location for reusable elements such as parts, assemblies, etc.

- *Detail view*. A portion of a larger view, usually at a larger scale than the original view.

- *Drawing*. A 2D representation of a 3D part or assembly. The extension for a SolidWorks drawing file name is .SLDDRW.

- *Document*. A file containing a part, assembly, or drawing.

- *Features*. Individual shapes created by Sketch Entities tools: lines, circles, rectangles, etc. that when combined, creates the part. Features are geometry building blocks, they add or remove material, and are created from 2D or 3D sketched profiles or from edges and faces of existing geometry.

- *FeatureXpert*. Powered by SolidWorks Intelligent Feature Technology (SWIFT™), helps manages fillet and draft features; only for constant radius fillets and neutral plane drafts.

- *FeatureManager design tree*. Located on the left side of the SolidWorks Graphics window and provides an outline view of the active part, assembly, or drawing. The FeatureManager design tree and the Graphics window are dynamically linked. You can select features, sketches, drawing views, and construction geometry in either pane.

Basic Theory and Drawing Theory

- *Fly-out FeatureManager design tree.* Allows you to view and select items in the PropertyManager and the FeatureManager design tree at the same time.
- *Mass Properties tool.* Displays the mass properties of a part or assembly model, or the section properties of faces or sketches.
- *Measure tool.* Measures distance, angle, radius, and size of and between lines, points, surfaces, and planes in sketches, 3D models, assemblies, or drawings.
- *Model view.* A drawing view of a part or assembly
- *Parameter.* A value used to define a sketch or feature, often a dimension or instance.
- *Predefined view.* A drawing view in which the view position, orientation, and so on can be specified before a model is inserted. You can save drawing documents with predefined views as templates.
- *Rollback.* Suppresses all items below the rollback bar.
- *Section view.* A drawing view in which you cut the parent view with a section line. The section view can be a straight cut section or an offset section defined by a stepped section line. The section line can also include concentric arcs.
- *Sketch Status.* Sketch states include the following: Over Defined, Under Defined, Fully Defined, Driven, Dangling, Invalid, and Not Solved.
- *Standard 3 views.* The three orthographic views; front, right, and top that are often the basis of a drawing.

Check your understanding

1: Identify the following Feature icon.
- A: Extruded Cut feature
- B: Mirror feature
- C: Fillet feature
- D: Chamfer feature

2: Identify the following Feature icon.
- A: Extruded Boss/Base feature
- B: Round Pattern feature
- C: Circular Pattern feature
- D: Chamfer feature

3. Identify the following Feature icon.
- A: Hole Wizard feature
- B: Simple Hole feature
- C: Circular Pattern feature
- D: Top feature

4. Identify the following Feature icon.
- A: Shell feature
- B: Loft feature
- C: Round feature
- D: Rib feature

5. Identify the following Feature icon.
- A: Shell feature
- B: Loft feature
- C: Round feature
- D: Revolved Cut feature

6. Identify the number of instances in the illustrated model.
- A: 7
- B: 5
- C: 9
- D: None

7. Identify the Sketch plane for the Extrude1 feature.
- A: Top Plane
- B: Front Plane
- C: Right Plane
- D: Left Plane

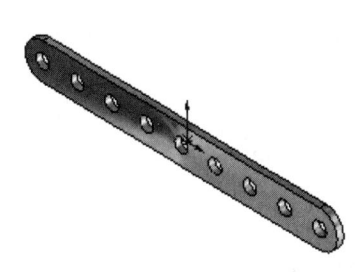

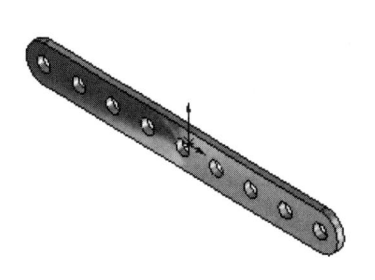

8. Identify the following Sketch Entities icon .
- A: Center Arc tool
- B: Tangent Arc tool
- C: 3 Point Arc tool
- D: Point Arc tool

9. Identify the following Sketch Entities icon .
- A: Center Arc tool
- B: Tangent Arc tool
- C: 3 Point Arc tool
- D: Point Arc tool

10. Identify the following Sketch Entities icon .
- A: Centerpoint Arc tool
- B: Circle tool
- C: 3 Point Arc tool
- D: Singlepoint Arc tool

11. A fully defined sketch is displayed in what color?
- A: Blue
- B: Black
- C: Red
- D: None of the listed

12. What symbol does the FeatureManager display before the Sketch name in an under defined sketch?
- A: (-)
- B: (?)
- C: (+)
- D: No symbol

Basic Theory and Drawing Theory

13. Which is not a valid drawing format in SolidWorks?
- A: *.tif
- B: *.jpg
- C: * .dwgg
- D: * .dxf

14. Which is a valid assembly format in SolidWorks?
- A: *.tiffe
- B: *.jpgg
- C: * .assmy
- D: * .stl

15. Identify the following Drawing View icon.
- A: Projected View
- B: Trim View
- C: Cut View
- D: Crop View

16. Identify the following Drawing View icon.
- A: Section View
- B: Broken View
- C: Break View
- D: Aligned Section View

17. Identify the following Drawing View icon.
- A: Projected View
- B: Standard 3 View
- C: Break View
- D: Aligned Section View

Basic Theory and Drawing Theory

18. Identify the illustrated Drawing view. 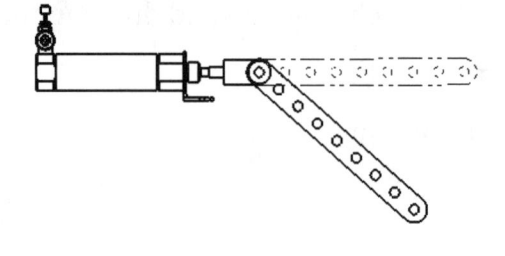
- A: Projected View
- B: Alternative Position View
- C: Extended View
- D: Aligned Section View

19. Identify the illustrated Drawing view. 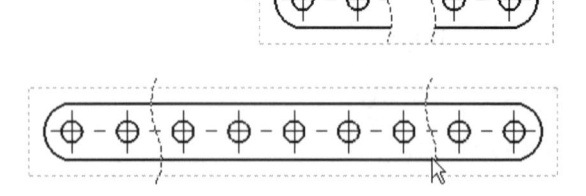
- A: Crop View
- B: Break View
- C: Broken-out Section View
- D: Aligned Section View

20. Identify the illustrated Drawing view. 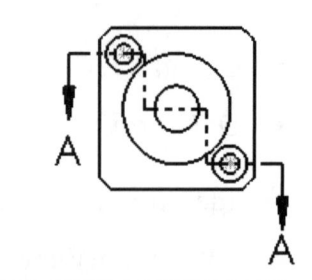
- A: Section View
- B: Crop View
- C: Broken-out Section View
- D: Aligned Section View

21. Identify the view procedure. To create the following view, you need to insert a: 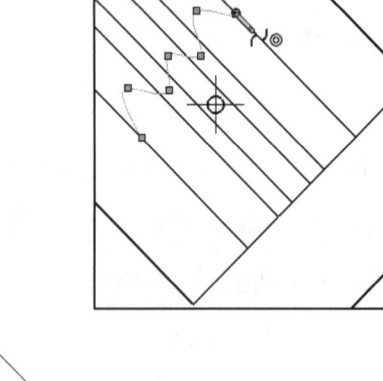
- A: Rectangle Sketch tool
- B: Closed Profile: Spline
- C: Open Profile: Circle
- D: None of the above

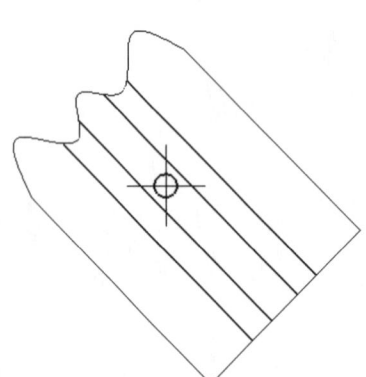

22. Identify the view procedure. To create the following view, you need to insert a:

- A: Open Spline
- B: Closed Spline
- C: 3 Point Arc
- D: None of the above

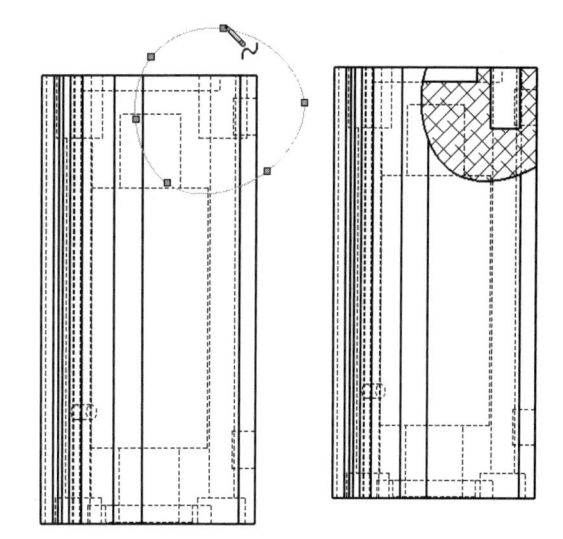

23. Identify the illustrated view type.

- A: Crop view
- B: Section view
- C: Projected view
- D: None of the above

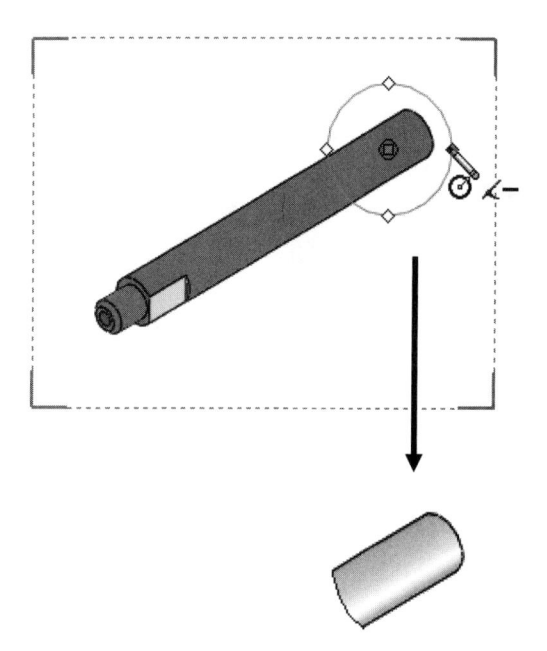

Notes:

CHAPTER 3: PART MODELING

Chapter Objective

Part Modeling is one of the five categories on the CSWA exam. This chapter covers the knowledge to identify the part Origin, design intent, and key features to build a simple part from a detailed dimensioned illustration.

There is one question on the CSWA exam in this category. The question is in a multiple choice single answer or a fill in the blank format. The question is worth thirty points. You are required to build a model, with six or more features and to answer a question either on the overall mass, volume, or the location of the Center of mass relative to the default part Origin.

☼ The complexity of the models along with the features progressively increases throughout this chapter to simulate the final types of parts that would be provided on the CSWA exam.

☼ The main difference between the Part Modeling category and the Advanced Part modeling category is the complexity of the sketches and the number of dimensions and geometric relations along with an increase in the number of features.

On the completion of the chapter, you will be able to:

- Read and understand an Engineering document:
  - Identify the Sketch plane, part Origin location, part dimensions, geometric relations, and design intent of the sketch and feature

- Build a part from a detailed dimensioned illustration using the following SolidWorks tools and features:
  - 2D & 3D Sketch tools
  - Extruded Boss/Base
  - Extruded Cut
  - Fillet
  - Mirror
  - Revolved Base
  - Chamfer

Part Modeling

- Reference geometry
- Plane
- Axis
- Calculate the overall mass and volume of the created part
- Locate the Center of mass for the created part relative to the Origin

Read and understand an Engineering document

What is an Engineering document? In a SolidWorks application, each part, assembly, and drawing is referred to as a document, and each document is displayed in the Graphics window.

💡 An asterisk (*) beside the document name in the title bar indicates that the document has changed since it was last saved.

Tutorial: Simple part 3-1

1. Open the **Line1** part from the SolidWorks CSWA Folder\Chapter3 location. The model displays all edges on perpendicular planes.

2. Double-click the **Extrude1** feature from the FeatureManager. Extrude1 is the Base feature. Sketch1 is the Base sketch and is fully defined. View the displayed dimensions.

💡 With SolidWorks, it is not necessary to fully dimension or define sketches before you use them to create features. However, you should fully define sketches before you consider a machined part complete.

View the selected Sketch plane for the part.

3. Right-click **Sketch 1** from the FeatureManager.

4. Click **Exit Sketch Plane**. The selected Sketch plane is the Front Plane. The Front Plane was chosen in order to efficiently sketch a single profile.

5. Click **OK**.

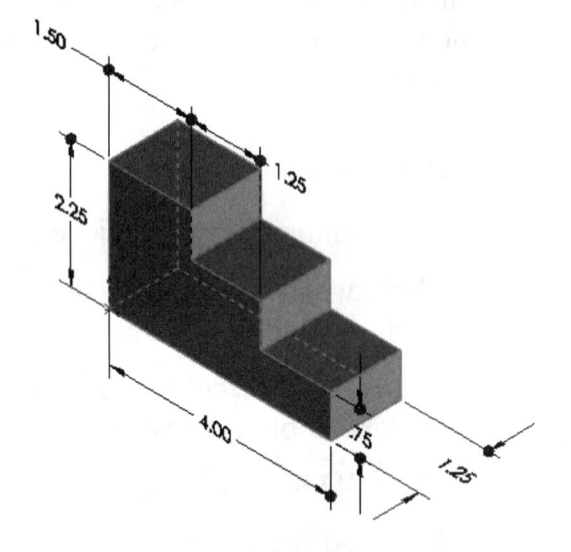

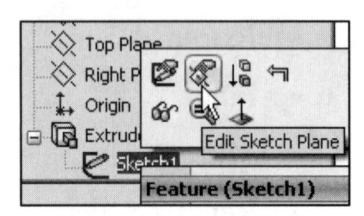

Part Modeling

6. **View** the part Origin location in the Graphics window. The part Origin is located in the front left corner of the model. Whenever you select a plane or a face and open a sketch, an Origin is created in alignment with the selected plane or face.

☼ The model origin is displayed in blue and represents the (X, Y, Z) coordinate system of the model. When a sketch is active, the sketch Origin is displayed in red and represents the (X, Y, Z) coordinate system of the sketch.

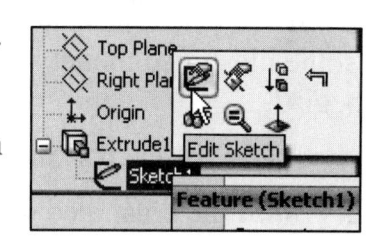

7. **Edit** Sketch1 in the FeatureManager. Sketch1 was created using the Sketch Line tool.

8. **View** the displayed geometric relations and dimensions in the Graphics window.

9. **Identify** the design intent of Sketch1. The design intent was to utilize an Equal and Vertical geometric relation for a quick design change. Equal geometric relations were used between segments 1 and 2.

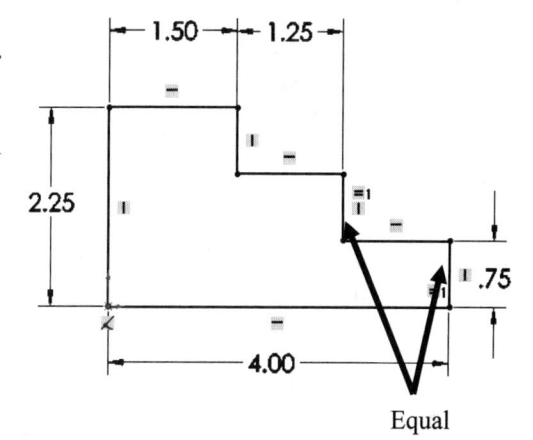

☼ Add geometric relations before dimensions in a sketch.

10. **Edit** Extrude1 in the FeatureManager. Blind is the default End Condition in Direction 1. Depth = 1.25in.

11. **View** the direction of the Extrude1 feature. The design intent of the feature is to extrude the part in a single direction. The part Origin remains in the front left corner of the model.

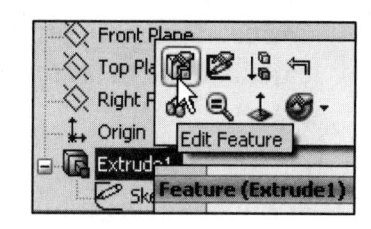

12. **Close** the model.

☼ Know and understand the location of the part Origin at all times during the exam.

☼ There are numerous ways to create the models in this chapter. The goal is to display different design intents and techniques.

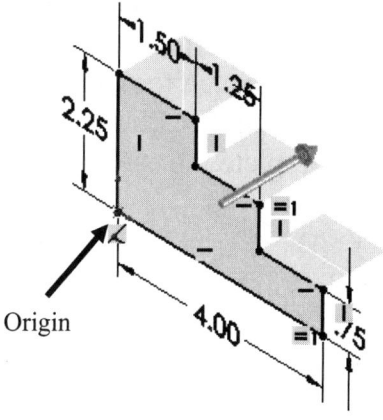

Page 3 - 3

Part Modeling

Build a simple part from a detailed dimensioned illustration

Tutorial: Simple part 3-2

Build this part. Calculate the overall mass of the illustrated model. Where do you begin?

1. **Create** a New part in SolidWorks.

2. **Build** the illustrated dimensioned model. The illustrated model displays all edges on perpendicular planes. The exam provides dimensions, (direct and indirect) material, and document property information.

   Given:
   A = 3.50
   B = .70
   Material: 1060 Alloy
   Density = 0.0975 lb/in^3
   Units: IPS
   Decimal places = 2

   Think about the steps that you would take to build the illustrated model. Read the provided information carefully for each model. Units are represented in the IPS system, (inch, pound, second).

   ☼ A = 3.50in, B = .70in

3. **Set** the document properties for the model. The correct document properties are required to calculate the mass for the part. Decimal places = 2.

4. Create **Sketch1**. Sketch1 is the Base sketch. Sketch1 a fully defined sketch based on the illustrated model. Identify the Sketch plane, (Front Plane). Select the Line Sketch tool. Insert the needed geometric relations, and dimensions to complete the sketch.

   ☼ There are numerous ways to create the models in this chapter.

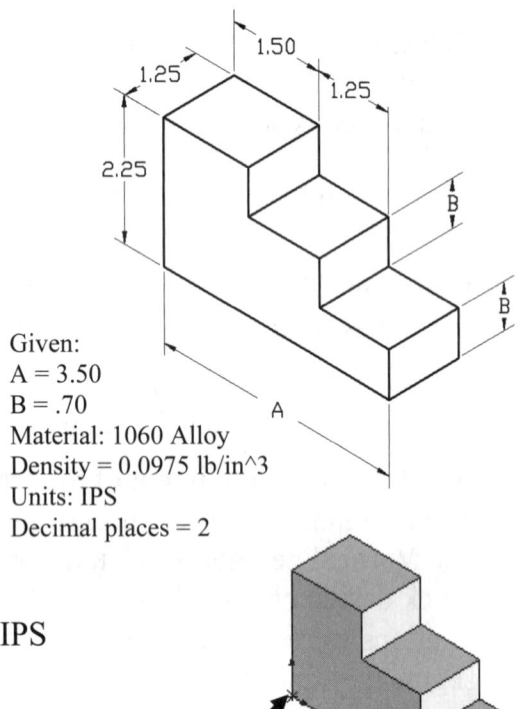

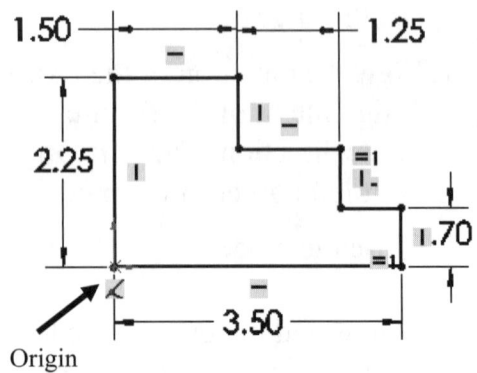

   ☼ ANSI is the dimensioning standard that is used in this book.

5. Create the **Extrude1** feature. Extrude1 is the Base feature. Blind is the default End Condition in Direction 1. Depth = 1.25in. Select the extrude direction to maintain the part Origin in the front left corner of the model.

6. **Assign** 1060 Alloy material to the part. Material is required to calculate the overall mass of the part.

💡 The Material icon is displayed in the Part FeatureManager regardless of whether a material is applied. Right-click the **icon** to view a list of the ten most recently used materials.

7. **Calculate** the overall mass of the part.

💡 Click the **Mass Properties** tool from the Evaluate tab in the CommandManager. The Mass Properties dialog box is displayed.

Questions on the exam are in a multiple choice single answer or fill in the blank format. In this category, Part Modeling an exam question could read: Build this model. Calculate the overall mass of the part.

- A: .59 pounds
- B: .69 pounds
- C: .67 pounds
- D: .96 pounds

The correct answer is B.

8. **Save** the part and name it Mass 3-1.

9. **Close** the model.

💡 Use the Options button in the Mass Properties dialog box to apply custom settings to units.

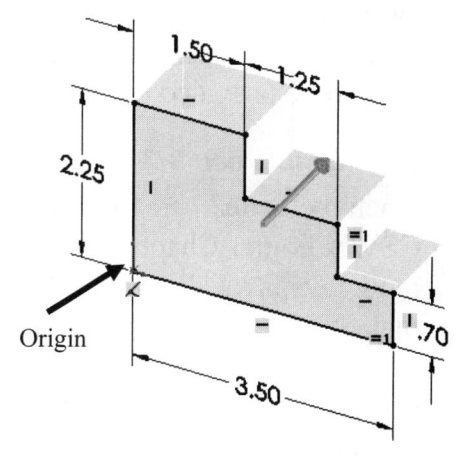

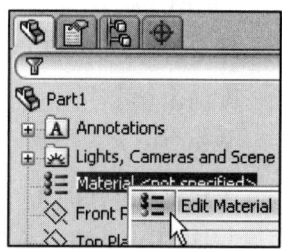

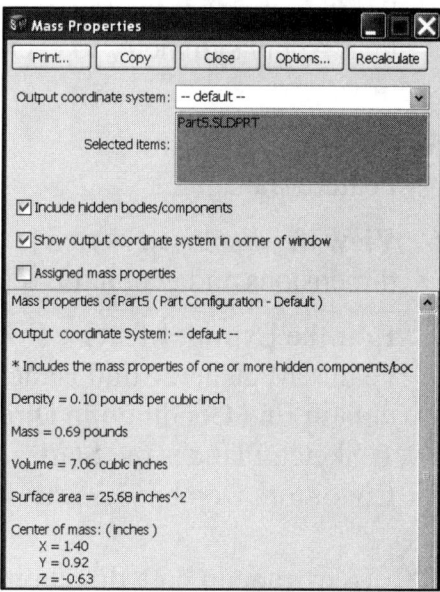

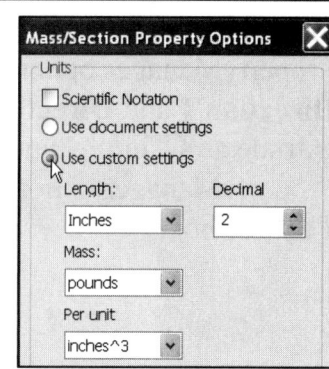

Part Modeling

💡 Remember, the complexity of the models along with the features progressively increases throughout this chapter to simulate the final types of parts that would be provided on the CSWA exam.

Tutorial: Simple part 3-3

1. Open the **Line2** part from the SolidWorks CSWA Folder\Chapter3 location. The model displays all edges on perpendicular planes. The Top Plane is the Sketch plane. The Top Plane was chosen to efficiently sketch a single profile.

2. **Double-click** the Extrude1 feature from the FeatureManager.

3. **View** the dimensions in the Graphics window.

4. **View** the part Origin location. The part Origin is located at the Mid Point, (Mid Point relation) of the left most vertical line. Apply symmetry to decrease design time.

5. **Edit** Sketch1 in the FeatureManager.

6. **View** the displayed sketch dimensions and geometric relations.

7. **Edit** the Extrude1 feature from the FeatureManager. Blind is the default End Condition in Direction 1. Sketch Plane is the Start Condition. Depth = 1.75in.

💡 To extrude in both directions from the Sketch plane, set the PropertyManager options in **Direction 1** and **Direction 2**. To extrude as a Thin feature, set the PropertyManager options in the **Thin Feature** box.

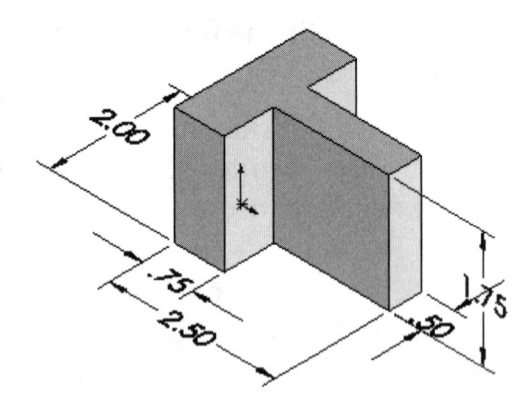

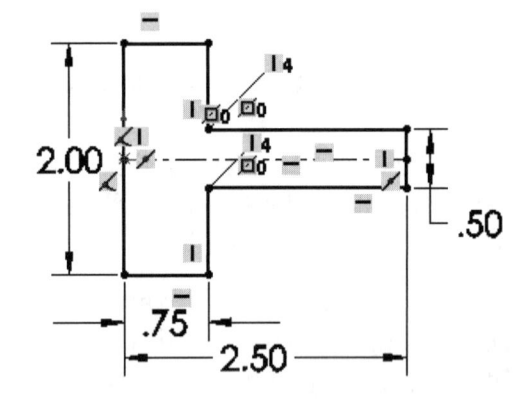

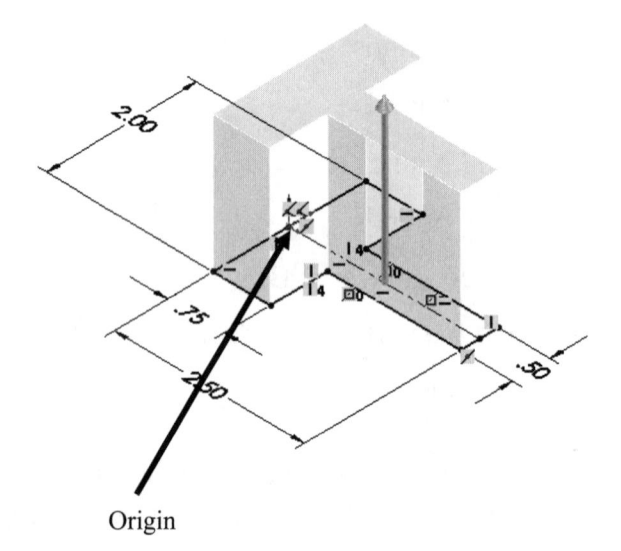

Origin

Part Modeling

💡 At this time, sheet metal parts are not on the CSWA exam.

Tutorial: Simple part 3-4

Build this model. Calculate the overall mass of the part with the provided information.

1. **Create** a New part in SolidWorks.

2. **Build** the illustrated dimensioned model. The model displays all edges on perpendicular planes. Think about the required steps to build this model.

💡 A = 3.00in, and B = .75in

3. **Set** the document properties for the model. From the provided information, the units are in the IPS unit system and the Length unit precision is two decimal places.

4. Create **Sketch1**. Sketch1 is the Base sketch. Sketch1 is the profile for the Extrude1 feature. Identify the Sketch plane, part Origin location, and the required geometric relations and dimensions for the sketch. The fully defined sketch could look like the supplied illustration. Apply construction geometry. Use the Mirror Sketch tool. Apply symmetry to decrease design time.

Given:
A = 3.00
B = .75
Material: Copper
Density = 0.321 lb/in^3
Units: IPS
Decimal places = 2

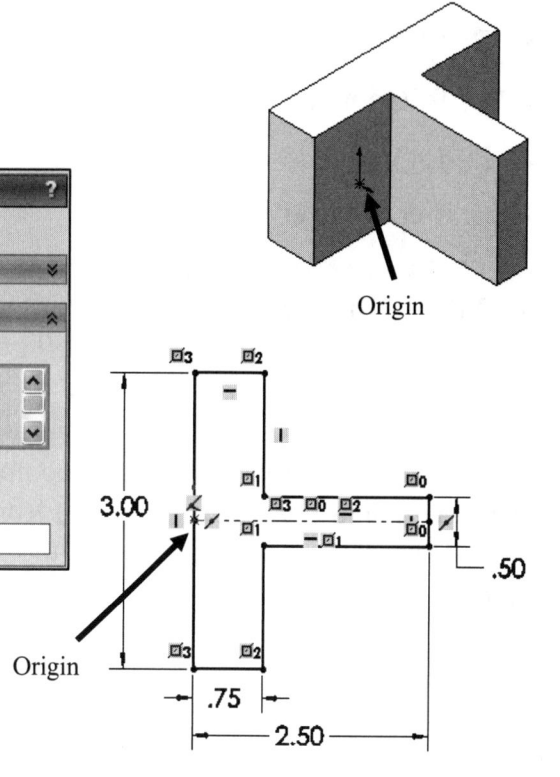

💡 When you create a new part or assembly, the three default planes are aligned with specific views. The plane you select for the Base sketch determines the orientation of the part.

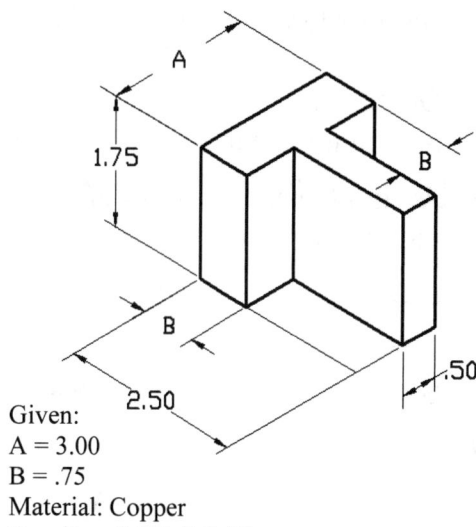

Page 3 - 7

Part Modeling

5. Create the **Extrude1** feature. Extrude1 is the Base feature. Sketch Plane is the default Start Condition. Blind is the default End Condition. Depth = 1.75in. Select the extrude direction to maintain the part Origin location.

6. **Assign** Copper material to the part. Material is required to calculate the overall mass of the part.

7. **Calculate** the overall mass of the part. The overall mass = 1.76 pounds.

8. **Save** the part and name it Mass 3-2.

9. **Close** the model.

In this category, Part Modeling an exam question could read: Build this model. What is the overall mass of the part in pounds?

- A: 1.76 pounds
- B: 1.99 pounds
- C: 1.56 pounds
- D: 1.43 pounds

The correct answer is A.

As an exercise, modify the Mass 3-2 part using the MMGS (millimeter, gram, second) unit system. The overall mass of the part = 797.59 grams. Save the part and name it Mass 3-2-MMGS.

☼ Modify the Unit system of a document, Click **Options**, **Document Properties** tab from the Menu bar toolbar. Click **Units**, and select the desired **Unit system**.

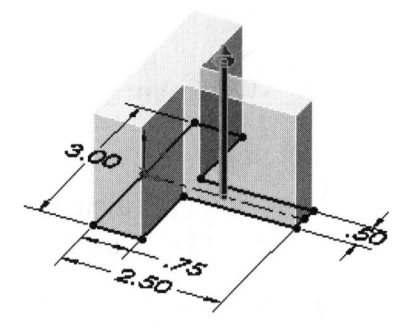

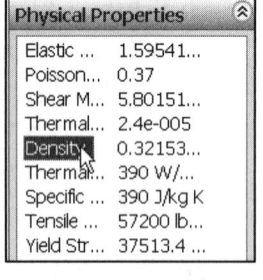

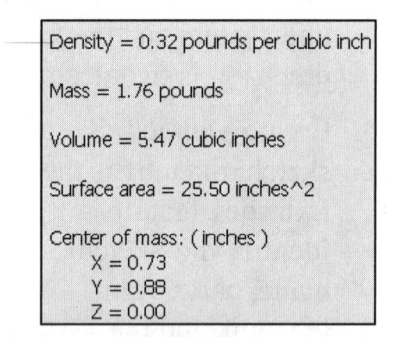

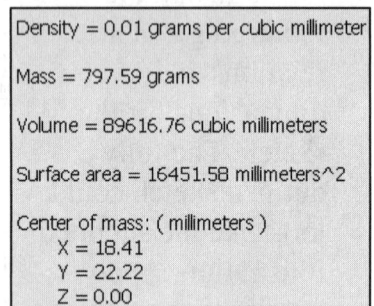

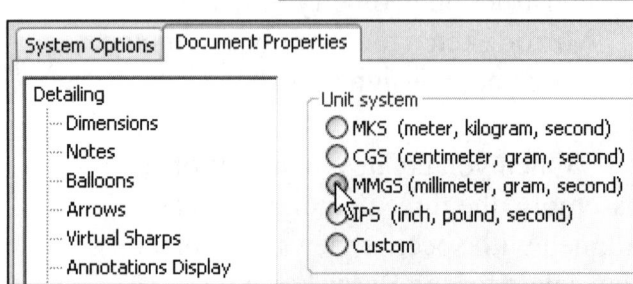

Page 3 - 8

Part Modeling

Tutorial: Volume / Center of mass 3-1

Build this model. Calculate the volume of the part and locate the Center of mass with the provided information.

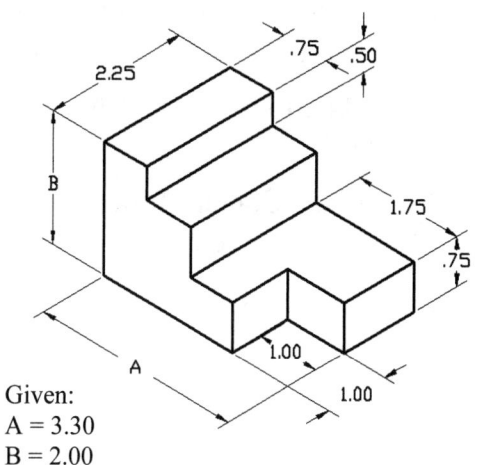

1. **Create** a New part in SolidWorks.

2. **Build** the illustrated dimensioned model. The model displays all edges on perpendicular planes. Think about the steps to build the model. Insert two features: Extruded Base and Extruded-Cut. The part Origin is located in the front left corner of the model.

3. **Set** the document properties for the model.

Given:
A = 3.30
B = 2.00
Material: 2014 Alloy
Density = .101 lb/in^3
Units: IPS
Decimal places = 2

4. Create **Sketch1**. Select the Front Plane as the Sketch plane. Sketch1 is the Base sketch. Sketch1 is the profile for the Extrude1 feature. Insert the required geometric relations and dimensions.

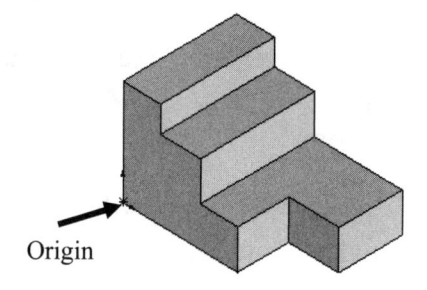

5. Create the **Extruded Base** feature. Extrude1 is the Base feature. Blind is the default End Condition in Direction 1. Depth = 2.25in. Identify the extrude direction to maintain the location of the Origin.

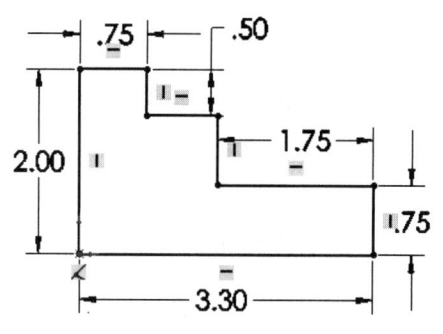

6. Create **Sketch2**. Select the Top right face as the Sketch plane. Sketch a square. Sketch2 is the profile for the Extruded Cut feature. Insert the required geometric relations and dimensions.

7. Create the **Extruded Cut** feature. Select Through All for End Condition in Direction 1.

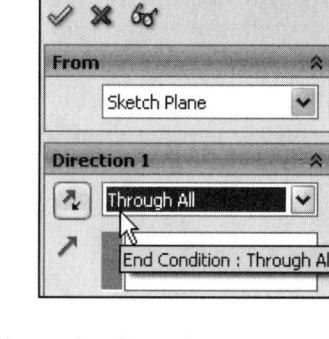

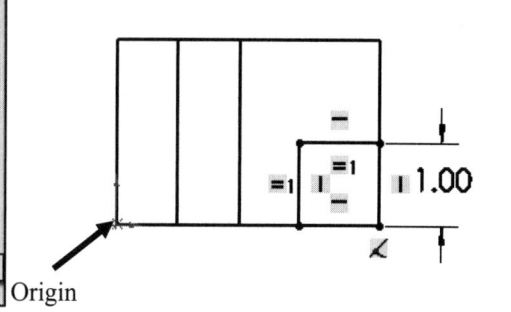

💡 Remember, there are numerous ways to create the models in this and other chapters.

Page 3 - 9

Part Modeling

8. **Assign** 2014 Alloy material to the part. Material is required to locate the Center of mass.

9. **Calculate** the volume. The volume = 8.28 cubic inches.

10. **Locate** the Center of mass. The location of the Center of mass is derived from the part Origin.

    - X: 1.14 inches
    - Y: 0.75 inches
    - Z: -1.18 inches

11. **Save** the part and name it Volume-Center of mass 3-1.

12. **Close** the model.

💡 The principal axes and Center of mass are displayed graphically on the model in the Graphics window.

```
Mass = 0.84 pounds
Volume = 8.28 cubic inches
Surface area = 29.88 inches^2
Center of mass: ( inches )
    X = 1.14
    Y = 0.75
    Z = -1.18
```

Tutorial: Volume / Center of mass 3-2

Build this model. Calculate the volume of the part and locate the Center of mass with the provided information.

1. **Create** a New part in SolidWorks.

2. **Build** the illustrated dimensioned model. The model displays all edges on perpendicular planes. Think about the steps that are required to build this model. Remember, it was stated earlier in the book, "There are numerous ways to build the models in this chapter. The goal is to display different design intents and techniques".

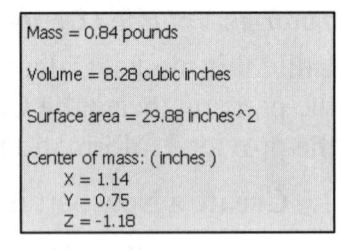

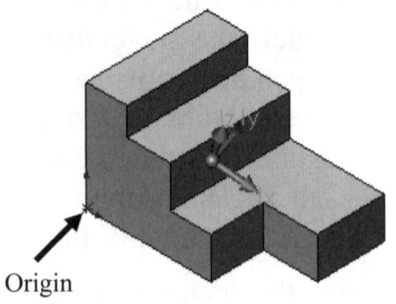

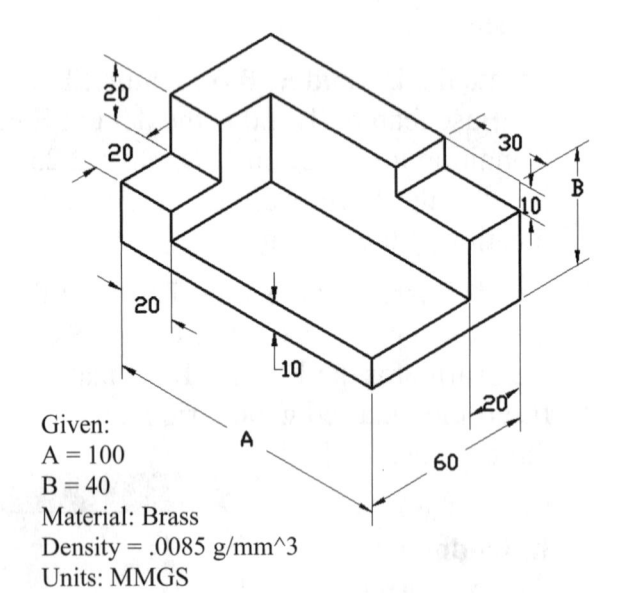

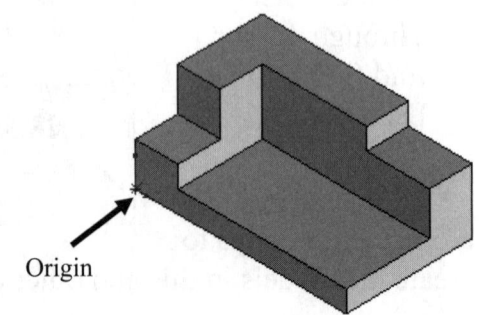

Given:
A = 100
B = 40
Material: Brass
Density = .0085 g/mm^3
Units: MMGS

Page 3 - 10

Part Modeling

☼ The CSWA is a three hour timed exam. Work efficiently.

View the provided Part FeatureManagers. Both FeatureManagers create the same illustrated model. In Option1, there are four sketches and four features (Extruded Base, three Extruded Cuts) that are used to build the model.

In Option2, there are three sketches and three features (Extruded Boss/ Base) that are used to build the model. Which FeatureManager is better? In a timed exam, optimize your time and use the least amount of features through mirror, pattern, symmetry, etc.

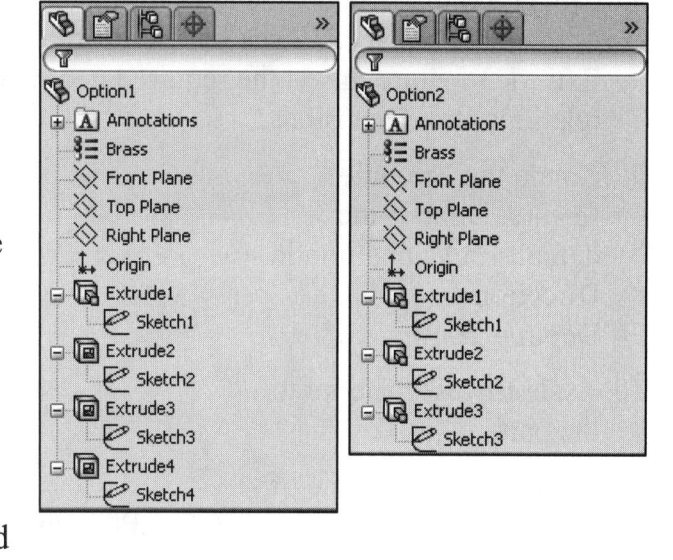

☼ Use centerlines to create symmetrical sketch elements and revolved features, or as construction geometry.

Create the model using the Option2 Part FeatureManager.

3. **Set** the document properties for the model.

4. Create **Sketch1**. Select the Top Plane as the Sketch plane. Sketch a rectangle. Insert the required dimensions.

5. Create the **Extruded Base** feature. Extrude1 is the Base feature. Blind is the default End Condition in Direction 1. Depth = 10mm.

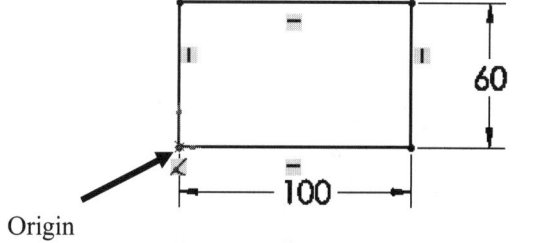

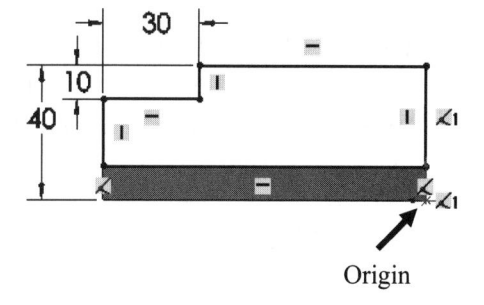

6. Create **Sketch2**. Select the back face of Extrude1. Select Normal To view. Sketch2 is the profile for the Extrude2 feature. Insert the required geometric relations and dimensions as illustrated.

7. Create the **Extrude2** feature. Blind is the default End Condition in Direction 1. Depth = 20mm. Note the direction of the extrude feature, towards the front of the model.

Page 3 - 11

Part Modeling

8. Create **Sketch3**. Select the left face of Extrude1 as the Sketch plane. Sketch3 is the profile for the Extrude3 feature. Insert the required geometric relations and dimensions.

9. Create the **Extrude3** feature. Blind is the default End Condition in Direction 1.
Depth = 20mm.

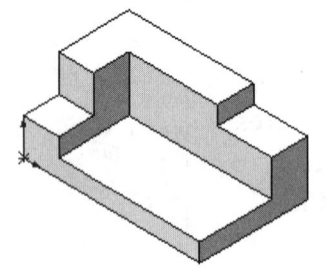

10. **Assign** Brass material to the part.

11. **Calculate** the volume of the model. The volume = 130,000.00 cubic millimeters.

12. **Locate** the Center of mass. The location of the Center of mass is derived from the part Origin.

- X: 43.36 millimeters
- Y: 15.00 millimeters
- Z: -37.69 millimeters

13. **Save** the part and name it Volume-Center of mass 3-2.

14. **Calculate** the volume of the model using the IPS unit system. The volume = 7.93 cubic inches.

15. **Locate** the Center of mass using the IPS unit system. The location of the Center of mass is derived from the part Origin.

- X: 1.71 inches
- Y: 0.59 inches
- Z: -1.48 inches

16. **Save** the part and name it Volume-Center of mass 3-2-IPS.

17. **Close** the model.

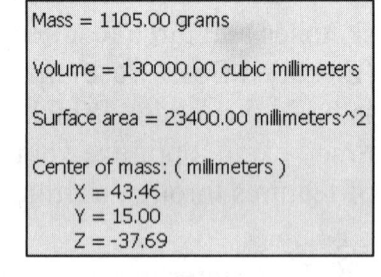

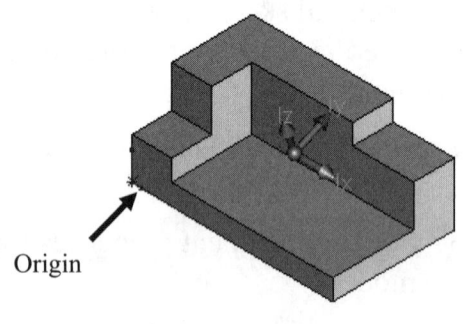

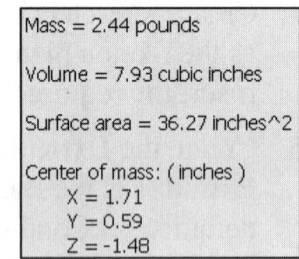

Page 3 - 12

Part Modeling

Tutorial: Mass-Volume 3-3

Build this model. Calculate the overall mass of the illustrated model with the provided information.

1. **Create** a New part in SolidWorks.

2. **Build** the illustrated model. The model displays all edges on perpendicular planes. Think about the steps required to build the model. Apply the Mirror Sketch tool to the Base sketch. Insert an Extruded Base and Extruded-Cut feature.

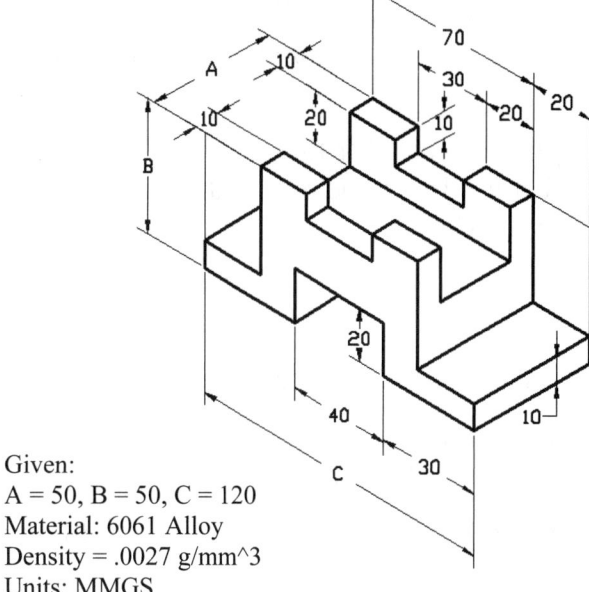

Given:
A = 50, B = 50, C = 120
Material: 6061 Alloy
Density = .0027 g/mm^3
Units: MMGS

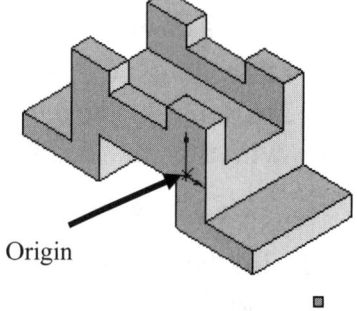

3. **Set** the document properties for the model.

💡 To activate the Mirror Sketch tool, click **Tools**, **Sketch Tools**, **Mirror** from the Menu bar menu. The Mirror PropertyManager is displayed.

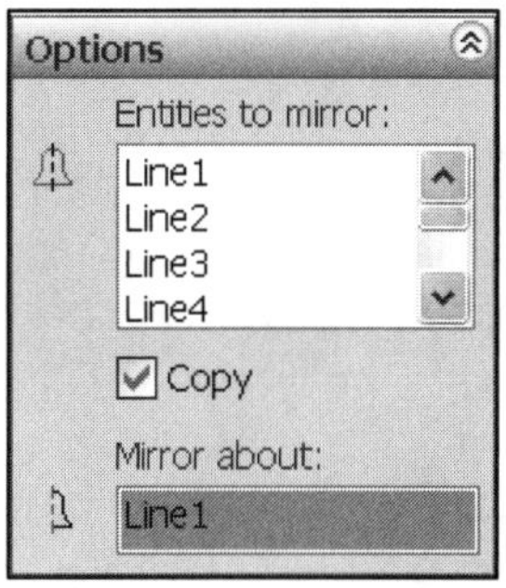

4. Create **Sketch1**. Select the Front Plane as the Sketch plane. Apply the Mirror Sketch tool. Select the construction geometry to mirror about as illustrated. Select the Entities to mirror. Insert the required geometric relations and dimensions.

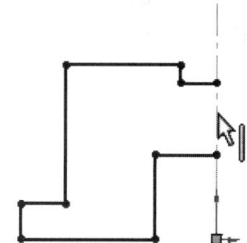

💡 Construction geometry is ignored when the sketch is used to create a feature. Construction geometry uses the same line style as centerlines.

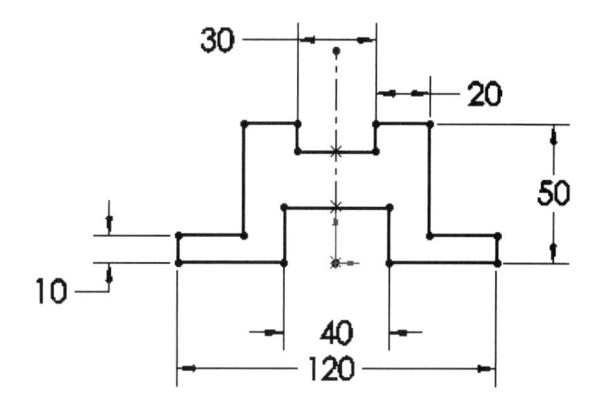

Page 3 - 13

Part Modeling

5. Create the **Extrude1** feature. Extrude1 is the Base feature. Apply the Mid Plane End Condition in Direction 1 for symmetry. Depth = 50mm.

6. Create **Sketch2**. Select the right face for the Sketch plane. Sketch2 is the profile for the Extruded Cut feature. Insert the required geometric relations and dimensions. Apply construction geometry.

7. Create the **Extruded Cut** feature. Through All is the selected End Condition in Direction 1.

8. **Assign** 6061 Alloy material to the part.

9. **Calculate** the overall mass. The overall mass = 302.40 grams.

10. **Save** the part and name it Mass-Volume 3-3.

11. **Close** the model.

```
Mass = 302.40 grams
Volume = 112000.00 cubic millimeters
Surface area = 26200.00 millimeters^2
Center of mass: ( millimeters )
    X = 0.00
    Y = 19.20
    Z = 0.00
```

Tutorial: Mass-Volume 3-4

Build this model. Calculate the volume of the part and locate the Center of mass with the provided information.

1. **Create** a New part in SolidWorks.

2. **Build** the illustrated model. The model displays all edges on perpendicular planes.

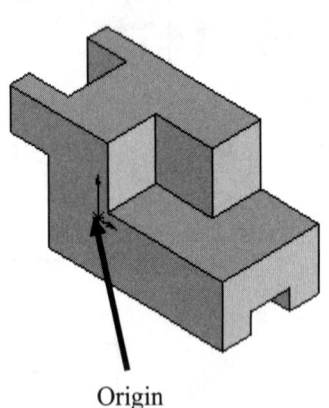

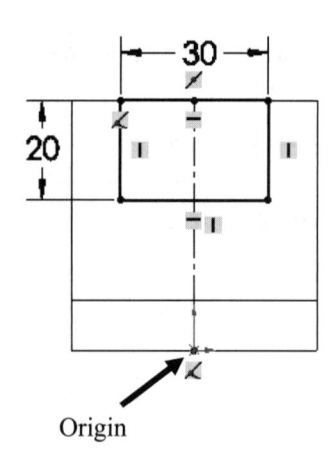

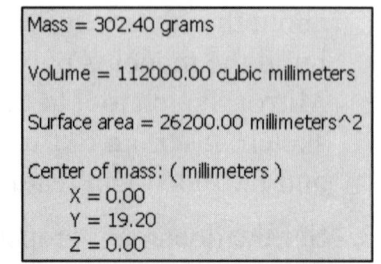

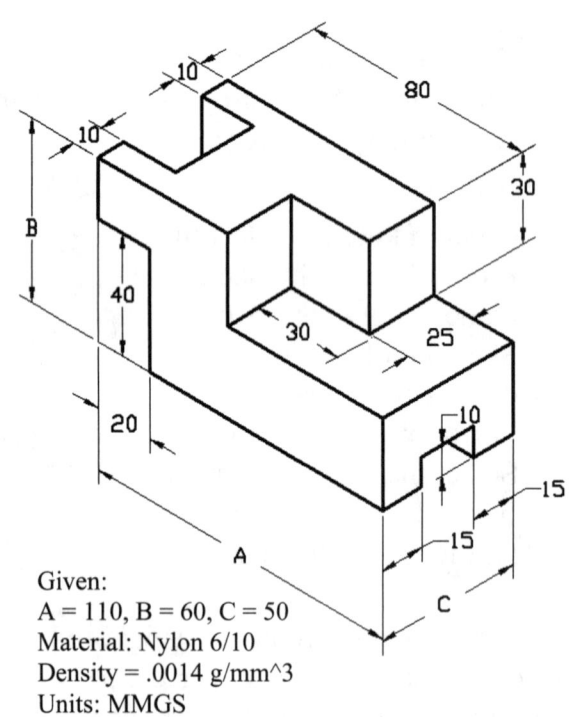

Given:
A = 110, B = 60, C = 50
Material: Nylon 6/10
Density = .0014 g/mm^3
Units: MMGS

Origin

Page 3 - 14

Part Modeling

View the provided Part FeatureManagers. Both FeatureManagers create the same model. In Option4, there are three sketches and three features that are used to build the model.

In Option3, there are four sketches and four features that are used to build the model. Which FeatureManager is better? In a timed exam, optimize your design time and use the least amount of features. Use the Option4 FeatureManager in this tutorial. As an exercise, build the model using the Option3 FeatureManager.

3. **Set** the document properties for the model.

4. Create **Sketch1**. Select the Right Plane as the Sketch plane. Sketch1 is the Base sketch. Apply the Mirror Sketch tool. Select the construction geometry to mirror about as illustrated. Select the Entities to mirror. Insert the required geometric relations and dimensions.

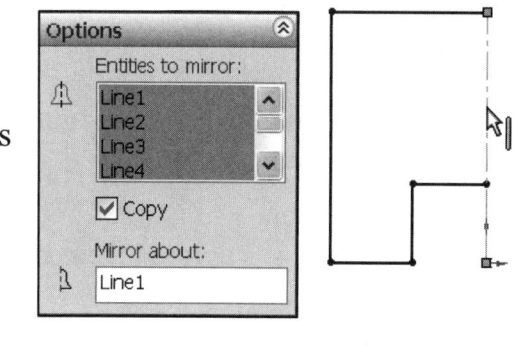

5. Create the **Extrude1** feature. Extrude1 is the Base feature. Blind is the default End Condition in Direction 1. Depth = (A - 20mm) = 90mm. Note the direction of Extrude1.

6. Create **Sketch2**. Select the Top face of Extrude1 for the Sketch plane. Sketch2 is the profile for the Extrude2 feature. Insert the required geometric relations and dimensions.

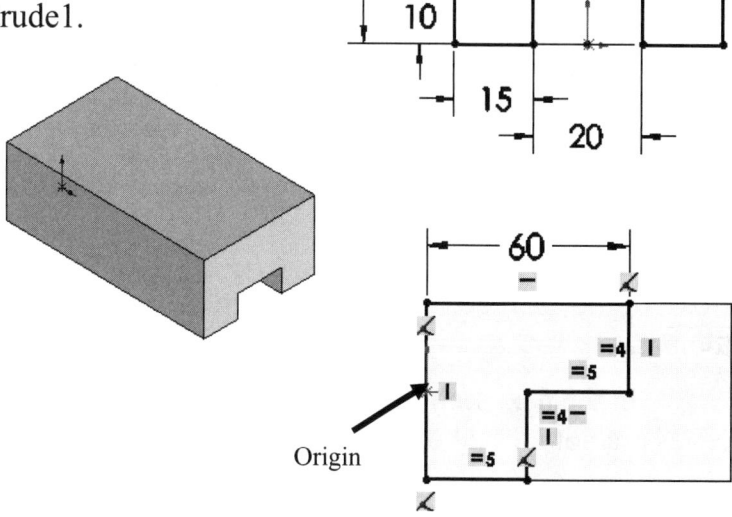

Page 3 - 15

Part Modeling

7. Create the **Extrude2** feature. Blind is the default End Condition in Direction 1. Depth = 30mm.

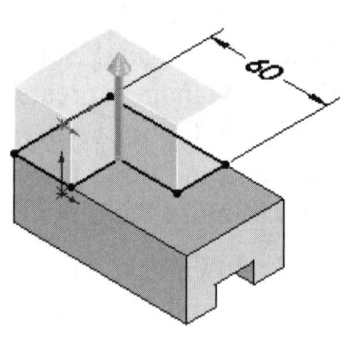

8. Create **Sketch3**. Select the left face of Extrude1 for the Sketch plane. Apply symmetry. Insert the required geometric relations and dimensions. Use construction reference geometry.

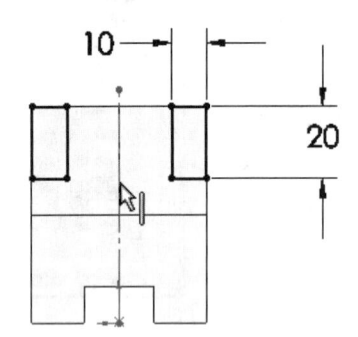

☼ The 20mm dimension for Sketch3 was calculated by: (B - 40mm) = 20mm.

9. Create the **Extrude3** feature. Blind is the default End Condition in Direction 1. Depth = 20mm. Note the direction of Extrude3.

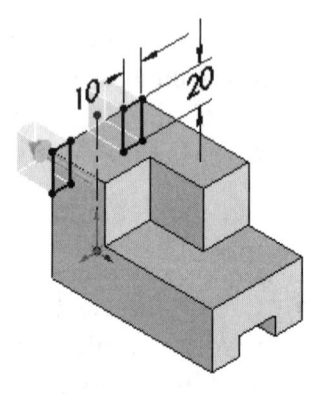

10. **Assign** Nylon 6/10 material to the part.

11. **Calculate** the volume. The volume = 192,500.00 cubic millimeters.

12. **Locate** the Center of mass. The location of the Center of mass is derived from the part Origin.

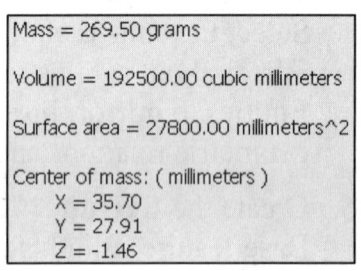

   - X: 35.70 millimeters
   - Y: 27.91 millimeters
   - Z: -1.46 millimeters

13. **Save** the part and name it Mass-Volume 3-4.

14. **Close** the model.

    In the previous section, all of the models that you created displayed all edges on perpendicular planes and used the Extruded Base, Extruded Boss, or the Extruded Cut feature from the Features toolbar.

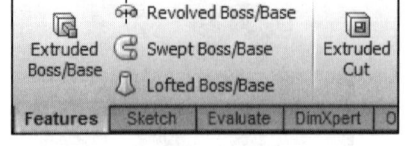

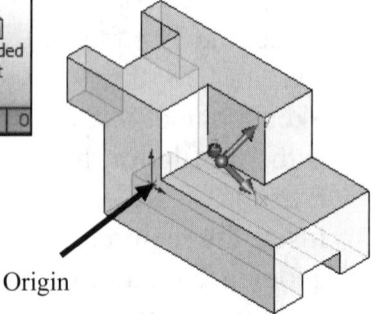

    In the next section, build models where all edges are not located on perpendicular planes.

Mass = 269.50 grams
Volume = 192500.00 cubic millimeters
Surface area = 27800.00 millimeters^2
Center of mass: (millimeters)
  X = 35.70
  Y = 27.91
  Z = -1.46

Origin

Page 3 - 16

Part Modeling

First, let's review a simple 2D Sketch for a Extruded Cut feature.

Tutorial: Simple Cut 3-1

1. **Create** a New part in SolidWorks.

2. **Build** the illustrated model. Start with a 60mm x 60mm x 100mm block. System units = MMGS. Decimal places = 2. Note the location of the part Origin.

3. **Set** the document properties for the model.

4. Create **Sketch1**. Select the Front Plane as the Sketch plane. Sketch a square as illustrated. Insert the required dimension. The part Origin is located in the bottom left corner of the model.

5. Create the **Extruded Base** feature. Apply the Mid Plane End Condition in Direction 1. Depth = 100mm.

6. Create **Sketch2**. Select the front face as the Sketch plane. Apply the Line Sketch tool. Sketch a diagonal line. Select the front right vertical midpoint as illustrated.

7. Create the **Extruded Cut** feature. Through All for End Condition in Direction 1 and Direction 2 is selected by default.

8. **Save** the part and name it Simple-Cut 3-1. View the FeatureManager.

9. **Close** the model.

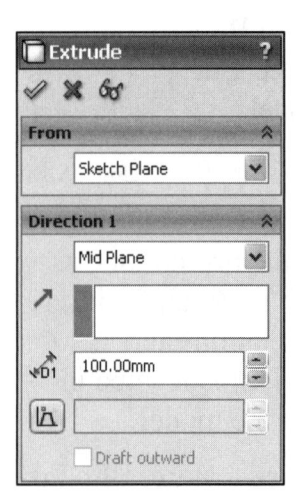

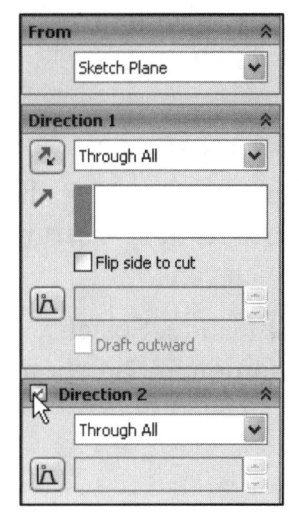

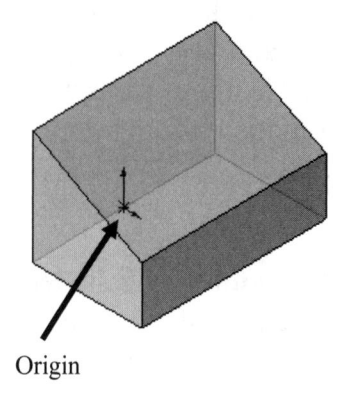

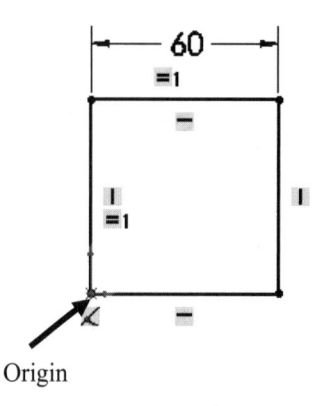

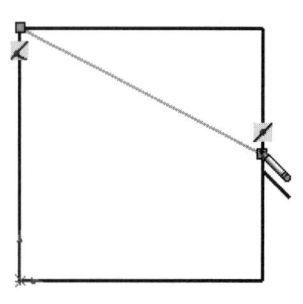

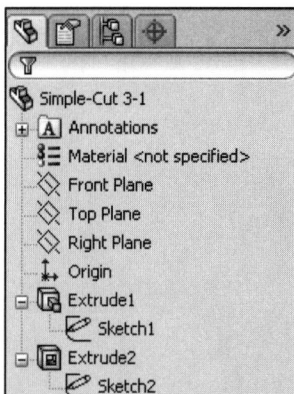

Page 3 - 17

Part Modeling

Tutorial: Mass-Volume 3-5

Build this model. Calculate the overall mass of the part and locate the Center of mass with the provided information.

1. **Create** a New part in SolidWorks.

2. **Build** the illustrated model. All edges of the model are not located on perpendicular planes. Insert an Extruded Base feature and three Extruded Cut features to build the model.

   Given:
   A = 110, B = 60, C = 50
   Material: Plain Carbon Steel
   Density = .0078 g/mm^3
   Units: MMGS

3. **Set** the document properties for the model.

4. Create **Sketch1**. Select the Front Plane as the Sketch plane. Sketch a rectangle. Insert the required geometric relations and dimensions. The part Origin is located in the left bottom corner of the model.

5. Create the **Extruded Base** feature. Blind is the default End Condition in Direction 1. Depth = 50mm. Note the direction of Extrude1.

6. Create **Sketch2**. Select the top face of Extrude1 for the Sketch plane. Sketch2 is the profile for the Extruded Cut feature. Insert the required relations and dimensions.

7. Create the **Extruded Cut** feature. Select Through All for End Condition in Direction 1.

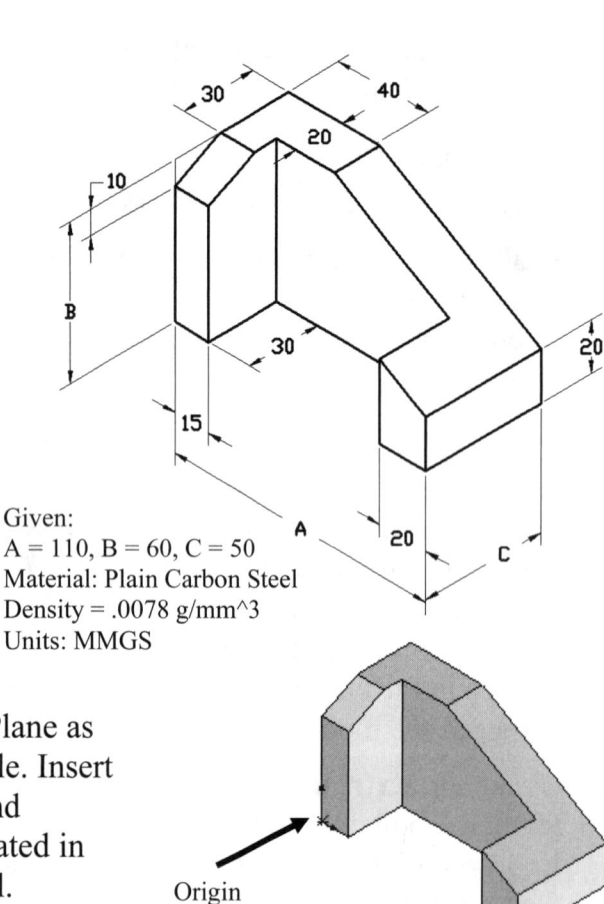

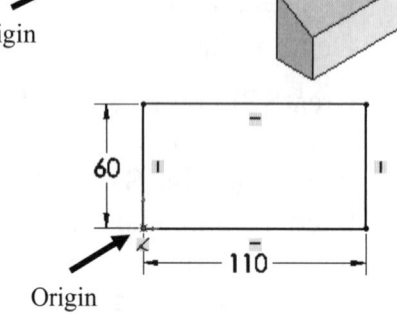

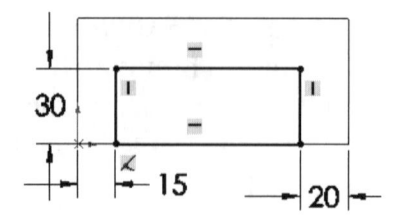

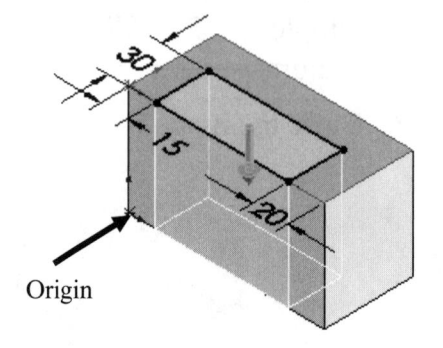

Page 3 - 18

8. Create **Sketch3**. Select the back face of Extrude1 as the Sketch plane. Sketch a diagonal line. Insert the required geometric relations and dimensions.

9. Create the **Extruded Cut** feature. Through All for End Condition in Direction 1 and Direction 2 is selected by default. Note the direction of the Extrude feature.

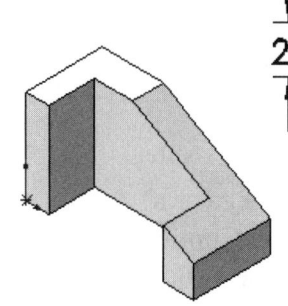

10. Create **Sketch4**. Select the left face of Extrude1 as the Sketch plane. Sketch a diagonal line. Insert the required geometric relations and dimensions.

11. Create the **Extruded Cut** feature. Through All for End Condition in Direction 1 and Direction 2 is selected by default.

12. **Assign** Plain Carbon Steel material to the part.

13. **Calculate** the overall mass. The overall mass = 1130.44 grams.

14. **Locate** the Center of mass. The location of the Center of mass is derived from the part Origin.

- X: 45.24 millimeters
- Y: 24.70 millimeters
- Z: -33.03 millimeters

In this category, Part Modeling an exam question could read: Build this model. Locate the Center of mass with respect to the part Origin.

- A: X = 45.24 millimeters, Y = 24.70 millimeters, Z = -33.03 millimeters
- B: X = 54.24 millimeters, Y = 42.70 millimeters, Z = 33.03 millimeters
- C: X = 49.24 millimeters, Y = -37.70 millimeters, Z = 38.03 millimeters
- D: X = 44.44 millimeters, Y = -24.70 millimeters, Z = -39.03 millimeters

The correct answer is A.

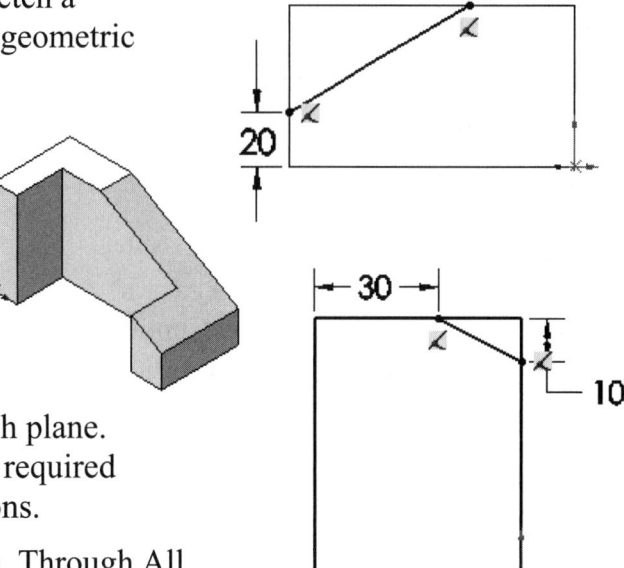

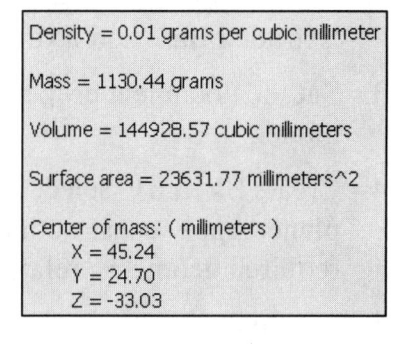

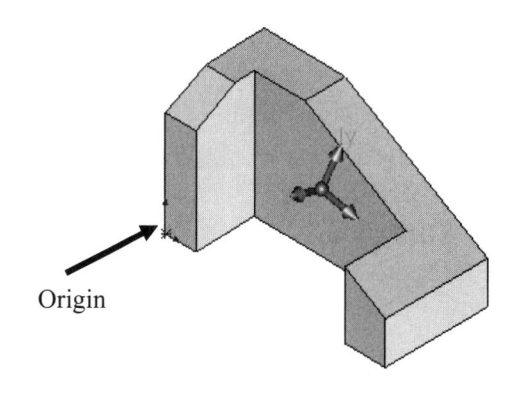

Part Modeling

💡 The principal axes and Center of mass are displayed graphically on the model in the Graphics window.

15. **Save** the part and name it Mass-Volume 3-5.

16. **Close** the model.

Tutorial: Mass-Volume 3-6

Build this model. Calculate the overall mass of the part and locate the Center of mass with the provided information.

1. **Create** a New part in SolidWorks.

2. **Build** the illustrated model. All edges of the model are not located on perpendicular planes. Think about the steps required to build the model. Insert two features: Extruded Base and Extruded Cut.

Given:
A = 3.00, B = 1.00
Material: 6061 Alloy
Density = .097 lb/in^3
Units: IPS
Decimal places = 2

3. **Set** the document properties for the model.

4. Create **Sketch1**. Select the Right Plane as the Sketch plane. Apply construction geometry. Insert the required geometric relations and dimensions.

5. Create the **Extruded Base** feature. Extrude1 is the Base feature. Apply symmetry. Select Mid Plane as the End Condition in Direction 1. Depth = 3.00in.

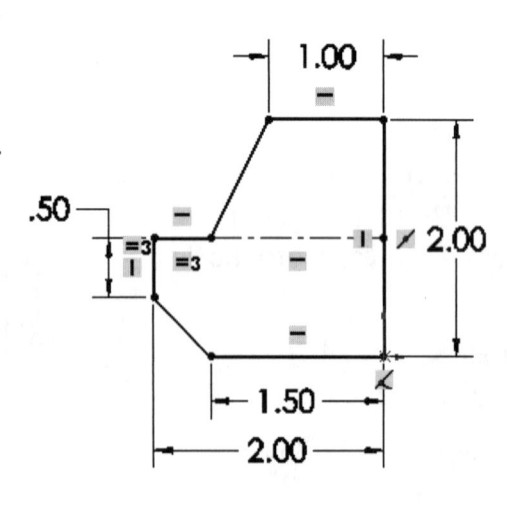

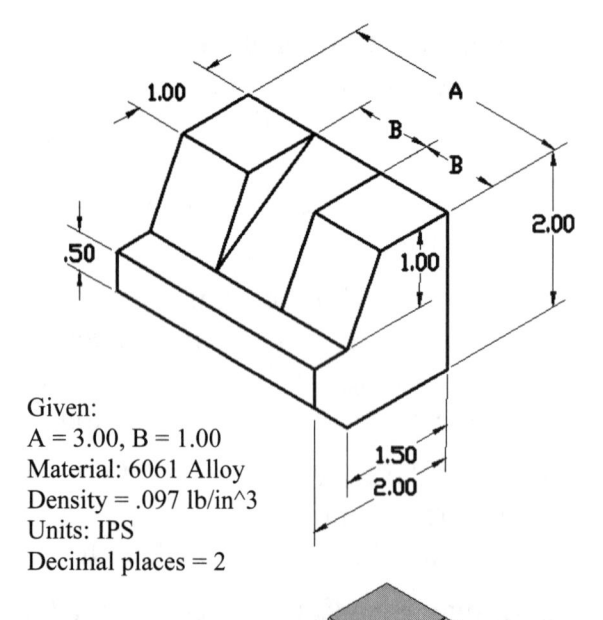

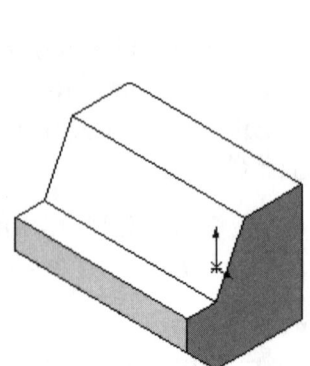

Page 3 - 20

6. Create **Sketch2**. Select the Right Plane as the Sketch plane. Select the Line Sketch tool. Insert the required geometric relations. Sketch2 is the profile for the Extruded Cut feature.

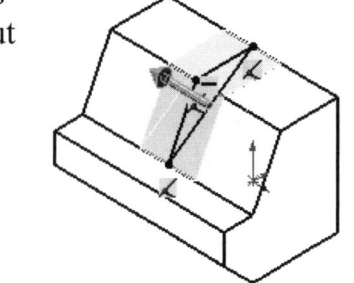

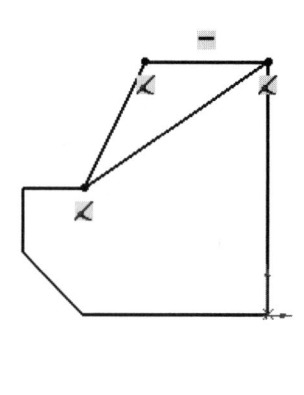

7. Create the **Extruded Cut** feature. Apply symmetry. Select Mid Plane as the End Condition in Direction 1. Depth = 1.00in.

8. **Assign** 6061 Alloy material to the part.

9. **Calculate** overall mass. The overall mass = 0.87 pounds.

10. **Locate** the Center of mass. The location of the Center of mass is derived from the part Origin.

- X: 0.00 inches
- Y: 0.86 inches
- Z: 0.82 inches

```
Mass = 0.87 pounds
Volume = 8.88 cubic inches
Surface area = 28.91 inches^2
Center of mass: (inches)
   X = 0.00
   Y = 0.86
   Z = 0.82
```

In this category, Part Modeling an exam question could read: Build this model. Locate the Center of mass with respect to the part Origin.

- A: X = 0.10 inches, Y = -0.86 inches, Z = -0.82 inches
- B: X = 0.00 inches, Y = 0.86 inches, Z = 0.82 inches
- C: X = 0.15 inches, Y = -0.96 inches, Z = -0.02 inches
- D: X = 1.00 inches, Y = -0.89 inches, Z = -1.82 inches

The correct answer is B.

11. **Save** the part and name it Mass-Volume 3-6.

12. **Close** the model.

As an exercise, modify the Mass-Volume 3-6 part using the MMGS unit system. Assign Nickel as the material. Calculate the overall mass. The overall mass of the part = 1236.20 grams. Save the part and name it Mass-Volume 3-6-MMGS.

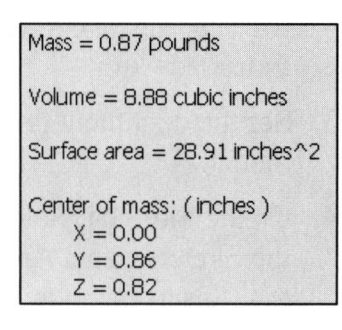

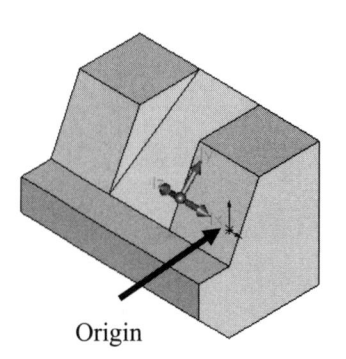

Part Modeling

Tutorial: Mass-Volume 3-7

Build this model. Calculate the overall mass of the part and locate the Center of mass with the provided information.

1. **Create** a New part in SolidWorks.

2. **Build** the illustrated model. All edges of the model are not located on perpendicular planes. Think about the steps required to build the model. Insert two features: Extruded Base and Extruded Cut.

3. **Set** the document properties for the model.

4. Create **Sketch1**. Select the Right Plane as the Sketch plane. Apply the Line Sketch tool. Insert the required geometric relations and dimension. The location of the Origin is in the left lower corner of the sketch.

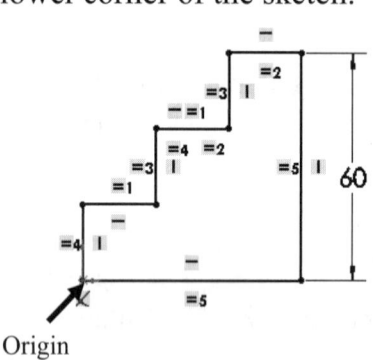

Given:
A = 110, B = 60, C = 60
Material: Ductile Iron
Density = .0079 g/mm^3
Units: MMGS

5. Create the **Extruded Base** feature. Extrude1 is the Base feature. Blind is the default End Condition in Direction 1. Depth = 110mm.

6. Create **Sketch2**. Select the Front Plane as the Sketch plane. Sketch a diagonal line. Complete the sketch. Sketch2 is the profile for the Extruded Cut feature.

7. Create the **Extruded Cut** feature. Through All for End Condition in Direction 1 and Direction 2 is selected by default.

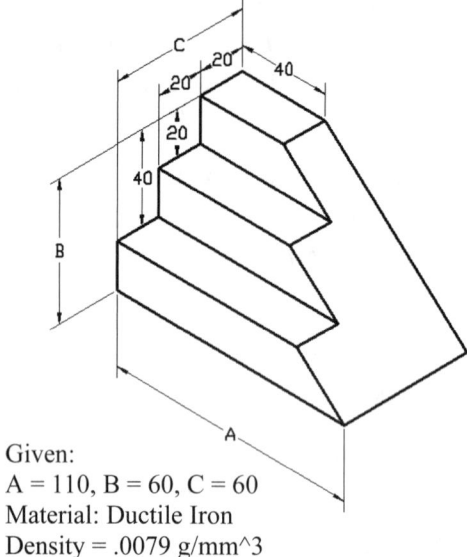

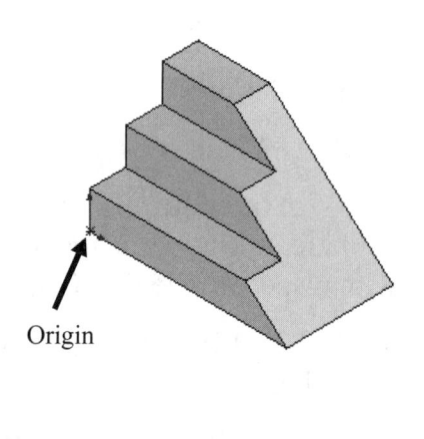

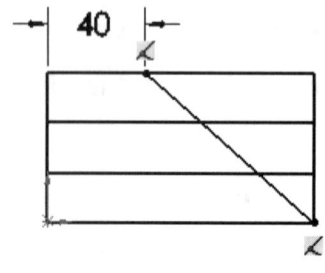

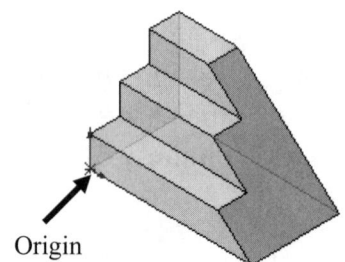

Page 3 - 22

Part Modeling

8. **Assign** Ductile Iron material to the part.
9. **Calculate** overall mass. The mass = 1569.47 grams.

```
Mass = 1569.47 grams
Volume = 198666.67 cubic millimeters
Surface area = 25487.82 millimeters^2
Center of mass: ( millimeters )
    X = 43.49
    Y = 19.73
    Z = -35.10
```

10. **Locate** the Center of mass. The location of the Center of mass is derived from the part Origin.

- X: 43.49 millimeters
- Y: 19.73 millimeters
- Z: -35.10 millimeters

11. **Save** the part and name it Mass-Volume 3-7.
12. **Close** the model.

In this category, Part Modeling an exam question could read: Build this model. Locate the Center of mass with respect to the part Origin.

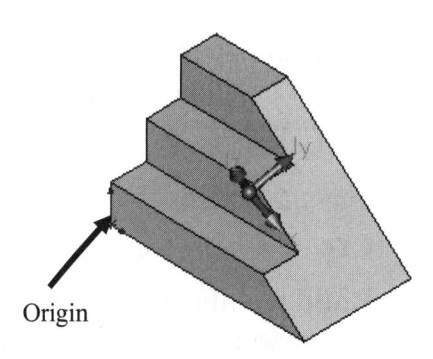

- A: X = -43.99 millimeters, Y = 29.73 millimeters, Z = -38.10 millimeters
- B: X = -44.49 millimeters, Y = -19.73 millimeters, Z = 35.10 millimeters
- C: X = 43.49 millimeters, Y = 19.73 millimeters, Z = -35.10 millimeters
- D: X = -1.00 millimeters, Y = 49.73 millimeters, Z = -35.10 millimeters

The correct answer is C.

As an exercise, locate the Center of mass using the IPS unit system, and re-assign copper material. Re-calculate the Center of mass location, with respect to the part Origin. Save the part and name it Mass-Volume 3-7-IPS.

- X: 1.71 inches
- Y: 0.78 inches
- Z: -1.38 inches

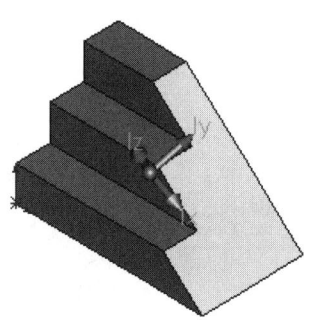

```
Mass = 3.90 pounds
Volume = 12.12 cubic inches
Surface area = 39.51 inches^2
Center of mass: ( inches )
    X = 1.71
    Y = 0.78
    Z = -1.38
```

Part Modeling

2D vs. 3D Sketching

Up to this point, the models that you created in this chapter started with a 2D sketch. Sketches are the foundation for creating features. SolidWorks provides the ability to create either 2D or 3D sketches. A 2D sketch is limited to a flat 2D Sketch plane. A 3D sketch can include 3D elements.

💡 As you create a 3D sketch, the entities in the sketch exist in 3D space. They are not related to a specific Sketch plane as they are in a 2D sketch.

You may need to apply a 3D sketch in the CSWA exam. Below is an example of a 3D sketch to create a Cut-Extrude feature.

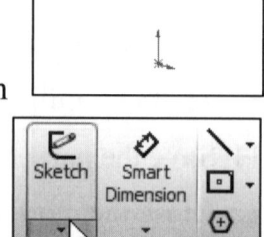

💡 The complexity of the models increases through-out this chapter to simulate the types of models that can be provided on the CSWA exam.

Tutorial 3DSketch 3-1

1. **Create** a New part in SolidWorks.

2. **Build** the illustrated model. Insert two features: Extruded Base and Extruded Cut. Apply the 3D Sketch tool to create the Extruded Cut feature. System units = MMGS. Decimal places = 2.

3. **Set** the document properties for the model.

4. Create **Sketch1**. Select the Front Plane as the Sketch plane. Sketch a rectangle. The part Origin is located in the bottom left corner of the sketch. Insert the illustrated geometric relations and dimensions.

5. Create the **Extruded Base** feature. Apply symmetry. Select the Mid Plane End Condition in Direction 1. Depth = 100.00mm.

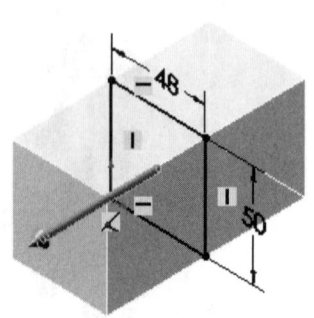

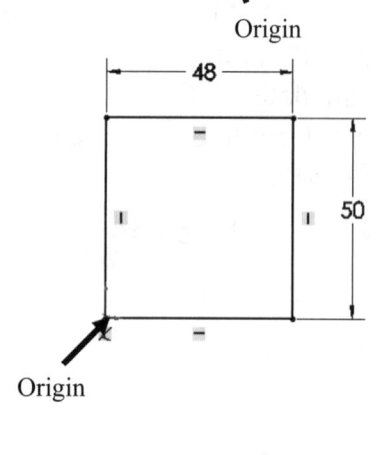

Page 3 - 24

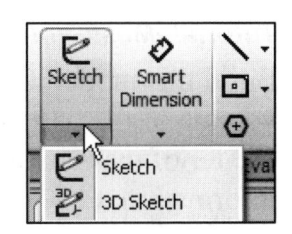

Part Modeling

💡 Click **3D Sketch** from the Sketch toolbar. Select the proper Sketch tool.

6. Create **3DSketch1**. Use the Line Sketch tool. 3DSketch1 is a four point sketch as illustrated. 3DSketch1 is the profile for Extruded Cut feature.

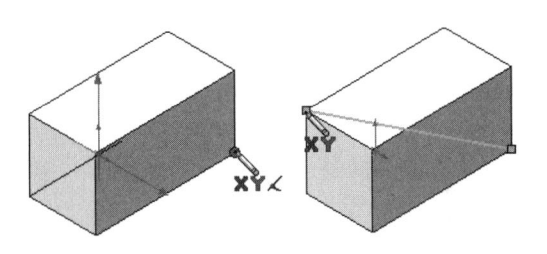

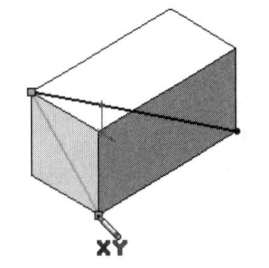

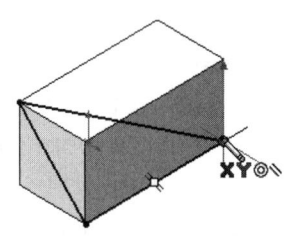

7. Create the **Extruded Cut** feature. Select the front right vertical edge as illustrated to remove the material. Edge<1> is displayed in the Direction of Extrusion box.

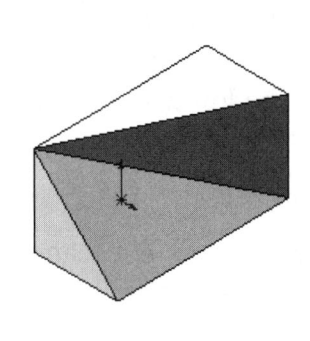

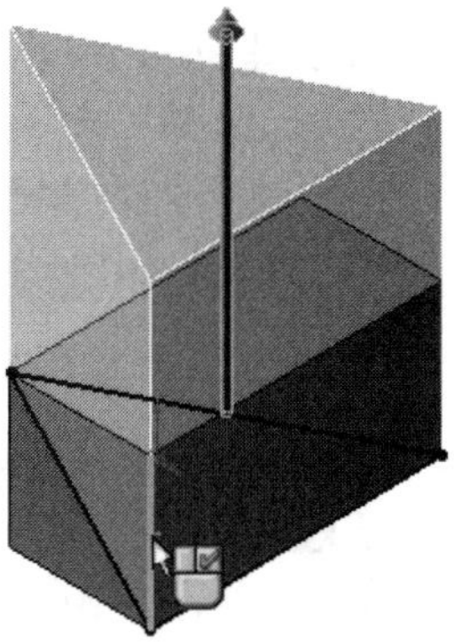

8. **Save** the part and name it 3DSketch 3-1.

9. **Close** the model.

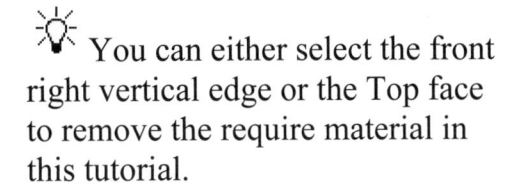

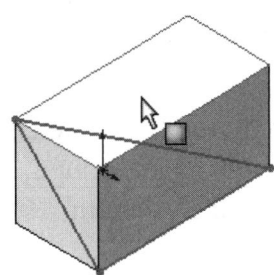

💡 You can either select the front right vertical edge or the Top face to remove the require material in this tutorial.

You can use any of the following tools to create 3D sketches: Lines, All Circle tools, All Rectangle tools, All Arcs tools, Splines, and Points.

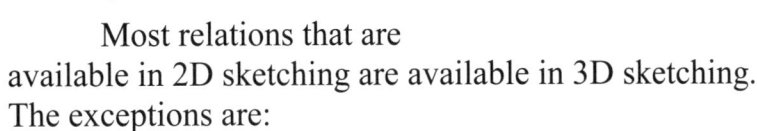

Most relations that are available in 2D sketching are available in 3D sketching. The exceptions are:

- Symmetry
- Patterns
- Offset

Page 3 - 25

Part Modeling

Tutorial: Mass-Volume 3-8

Build this model. Calculate the volume of the part and locate the Center of mass with the provided information.

1. **Create** a New part in SolidWorks.

2. **Build** the illustrated model. All edges of the model are not located on perpendicular planes. Insert two features: Extruded Base, and Extruded Cut. Apply a closed four point 3D sketch as the profile for the Extruded Cut feature. The part Origin is located in the lower left front corner of the model.

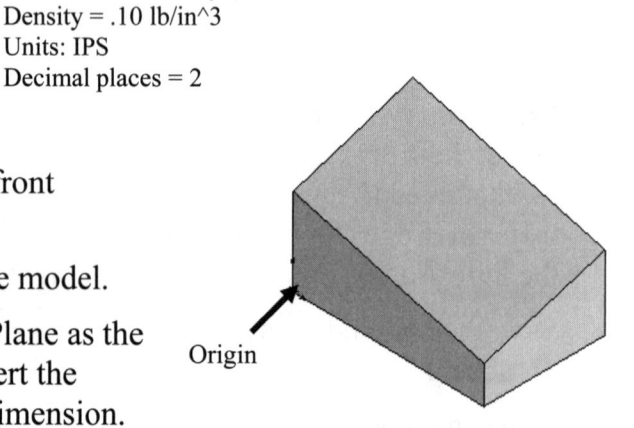

Given:
A = .75, B = 2.50
Material: 2014 Alloy
Density = .10 lb/in^3
Units: IPS
Decimal places = 2

3. **Set** the document properties for the model.

4. Create **Sketch1**. Select the Right Plane as the Sketch plane. Sketch a square. Insert the required geometric relations and dimension.

5. Create the **Extruded Base** feature. Extrude1 is the Base feature. Blind is the default End Condition in Direction 1. Depth = 4.00in.

6. Create **3DSketch1**. Apply the Line Sketch tool. Create a closed five point 3D sketch as illustrated. 3DSketch1 is the profile for the Extruded Cut feature. Insert the required dimensions.

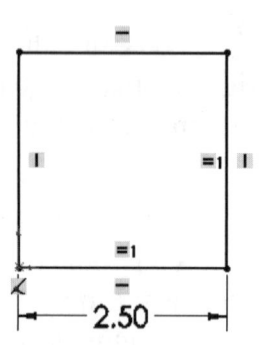

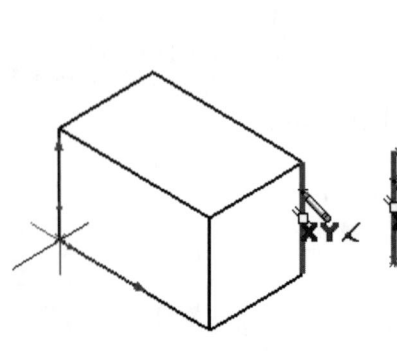

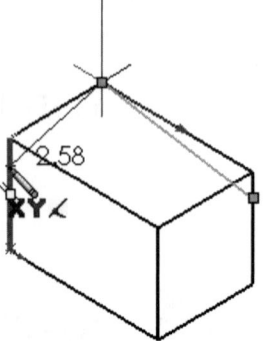

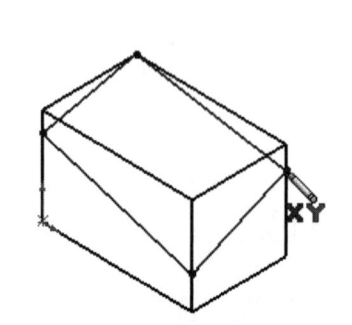

7. Create the **Extruded Cut** feature. Select the front right vertical edge as illustrated. Select Through All for End Condition in Direction 1. Note the direction of the extrude.

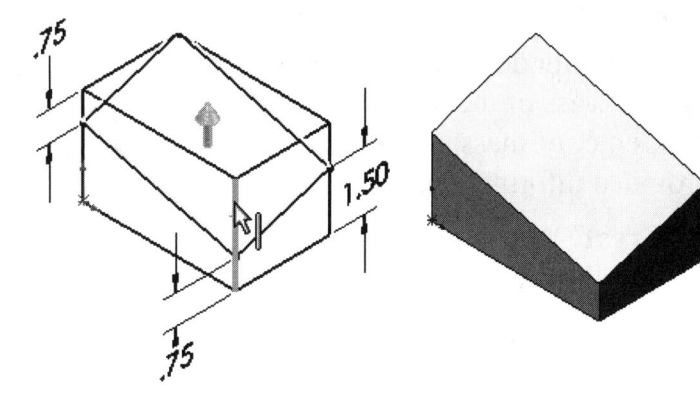

8. **Assign** the defined material to the part.

13. **Calculate** the volume. The volume = 16.25 cubic inches.

14. **Locate** the Center of mass. The location of the Center of mass is derived from the part Origin.

- X = 1.76 inches
- Y = 0.85 inches
- Z = -1.35 inches

In this category, Part Modeling an exam question could read: Build this model. What is the volume of the part?

- A: 18.88 cubic inches
- B: 19.55 cubic inches
- C: 17.99 cubic inches
- D: 16.25 cubic inches

The correct answer is D.

View the triad location of the Center of mass for the part.

13. **Save** the part and name it Mass-Volume 3-8.

14. **Close** the model.

As an exercise, calculate the overall mass of the part using the MMGS unit system, and re-assign Nickel as the material. The overall mass of the part = 2263.46 grams. Save the part and name it Mass-Volume 3-8-MMGS.

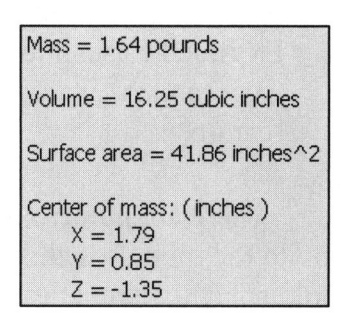

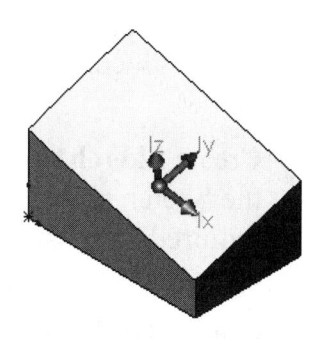

Part Modeling

Tutorial: Mass-Volume 3-9

Build this model. Calculate the overall mass of the part and locate the Center of mass with the provided information.

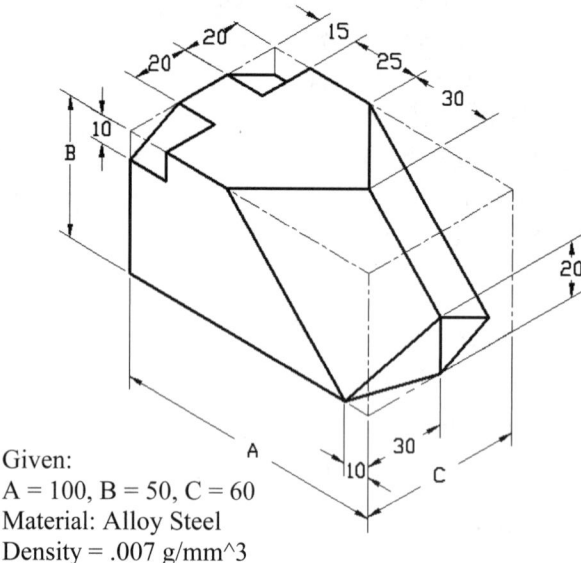

1. **Create** a New part in SolidWorks.

2. **Build** the illustrated model. Insert five sketches and five features to build the model: Extruded Base, three Extruded Cut features, and a Mirror feature.

Given:
A = 100, B = 50, C = 60
Material: Alloy Steel
Density = .007 g/mm^3
Units: MMGS

There are numerous ways to build the models in this chapter. The goal is to display different design intents and techniques.

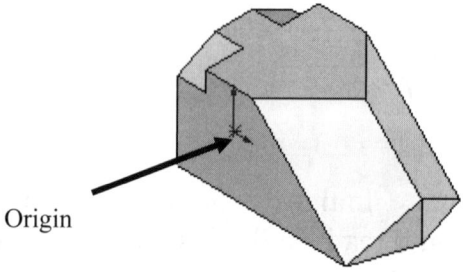

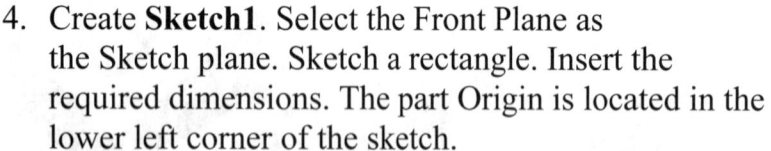

3. **Set** the document properties for the model.

Origin

4. Create **Sketch1**. Select the Front Plane as the Sketch plane. Sketch a rectangle. Insert the required dimensions. The part Origin is located in the lower left corner of the sketch.

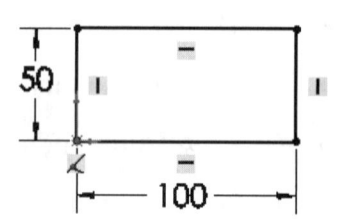

5. Create the **Extruded Base** feature. Apply symmetry. Select the Mid Plane End Condition for Direction 1. Depth = 60mm.

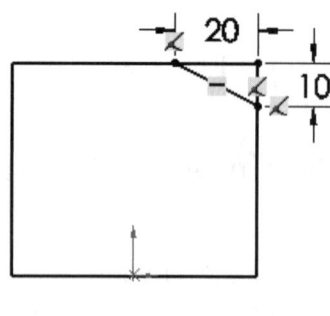

6. Create **Sketch2**. Select the left face of Extrude1 as the Sketch plane. Insert the required geometric relations and dimensions.

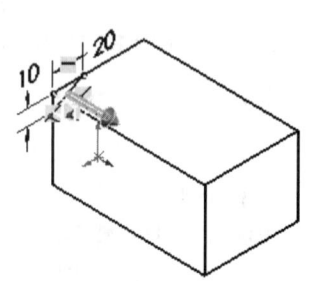

7. Create the **Extruded Cut** feature. Blind is the default End Condition in Direction 1. Depth = 15mm. Note the direction of the extrude feature.

Page 3 - 28

8. Create **Sketch3**. Select the bottom face of Extrude1 for the Sketch plane. Insert the required geometric relations and dimension.

9. Create the second **Extruded Cut** feature. Blind is the default End Condition in Direction 1. Depth = 20mm.

10. Create **Sketch4**. Select Front Plane as the Sketch plane. Sketch a diagonal line. Sketch4 is the direction of extrusion for the third Extruded Cut feature. Insert the required dimension.

11. Create **Sketch5**. Select the top face of Extrude1 as the Sketch plane. Sketch5 is the sketch profile for the third Extruded Cut feature. Apply construction geometry. Insert the required geometric relations and dimensions.

12. Create the third **Extruded Cut** feature. Select Through All for End Condition in Direction 1. Select Sketch4 in the Graphics window for Direction of Extrusion. Line1@Sketch4 is displayed in the Extrude PropertyManager.

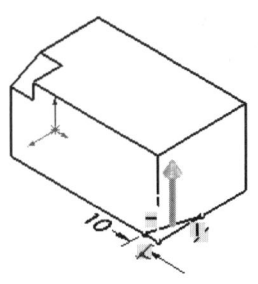

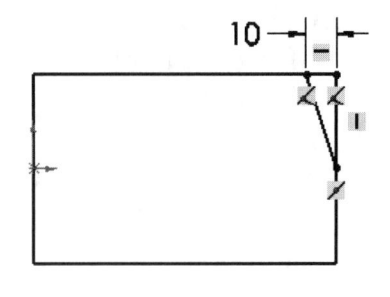

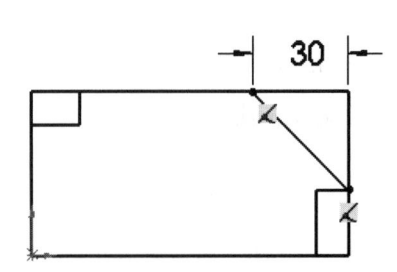

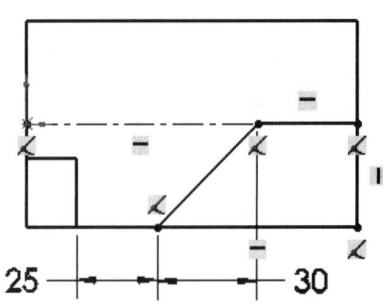

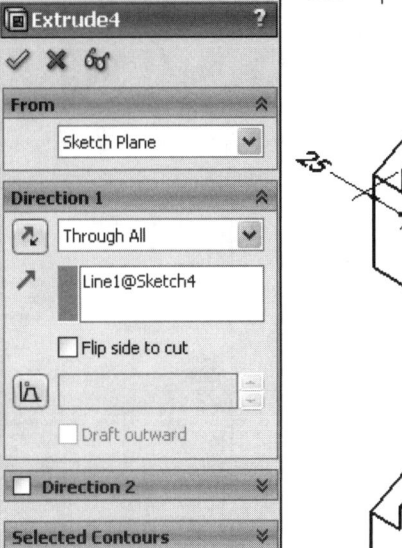

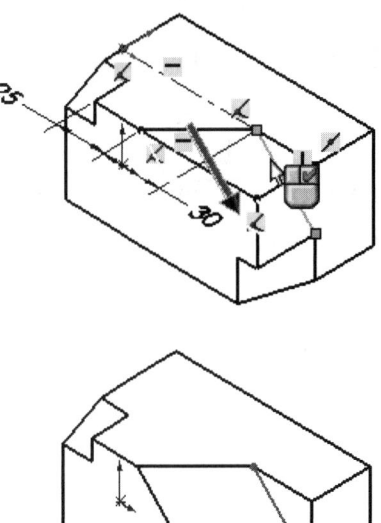

Part Modeling

13. Create the **Mirror** feature. Mirror the three Extruded Cut features about the Front Plane. Use the fly-out FeatureManager.

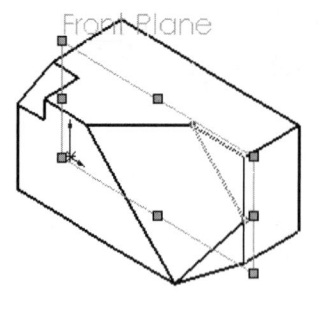

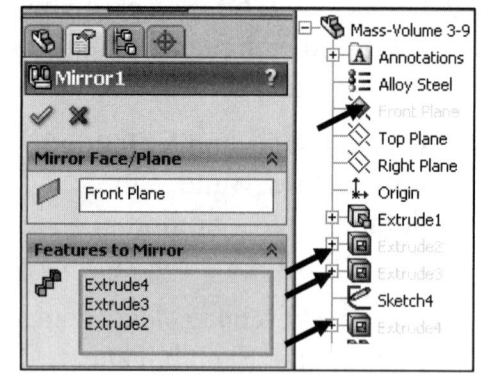

14. **Assign** Alloy Steel material to the part.

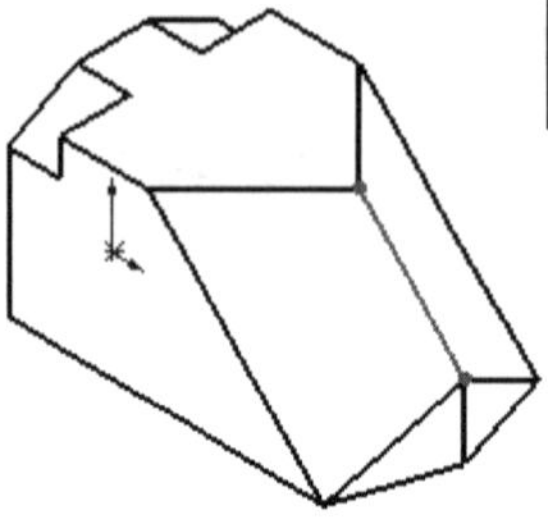

15. **Calculate** the overall mass. The overall mass = 1794.10 grams.

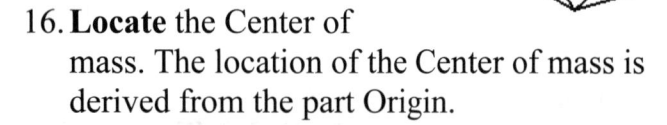

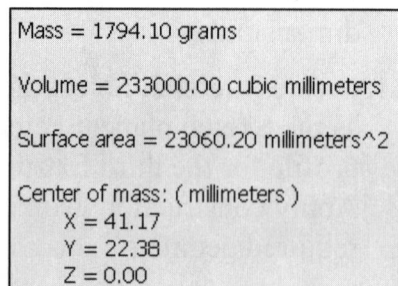

16. **Locate** the Center of mass. The location of the Center of mass is derived from the part Origin.

- X = 41.17 millimeters
- Y = 22.38 millimeters
- Z = 0.00 millimeters

View the triad location of the Center of mass for the part.

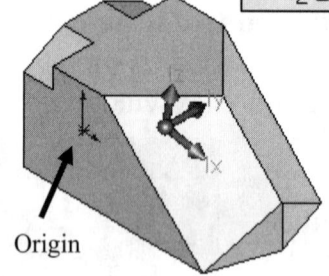

17. **Save** the part and name it Mass-Volume 3-9.

18. **Close** the model.

In Chapter 2, you addressed the procedure to set document precision through the Document Properties dialog box. In this chapter, you will address precision using the Dimension PropertyManager.

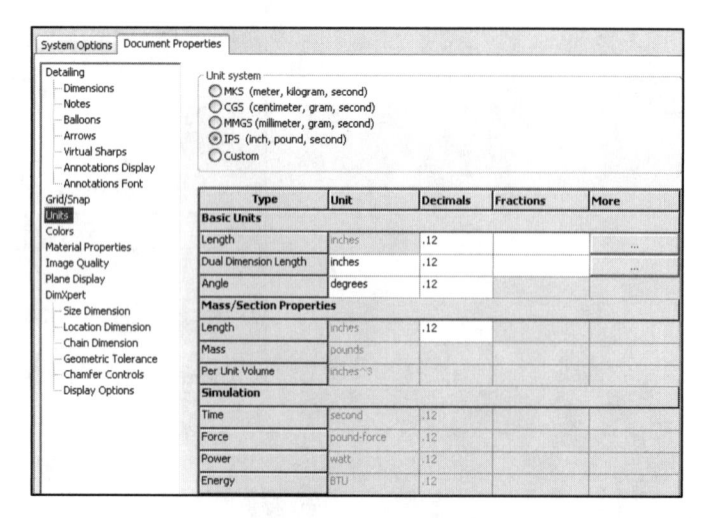

You will also address how to read and understand: Callout value, Tolerance type, and Dimension Text symbols in the Dimension PropertyManager.

Callout value

A Callout value is a value that you select in a SolidWorks document. Click a dimension in the Graphics window, the selected dimension is displayed in blue and the Dimension PropertyManager is displayed.

💡 A Callout value is available for dimensions with multiple values in the callout.

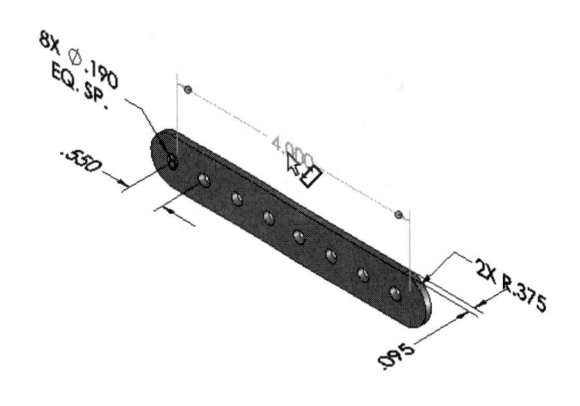

Tolerance type

A Tolerance type is selected from the available drop down list in the Dimension PropertyManager. The list is dynamic. A few examples of Tolerance type display are listed below:

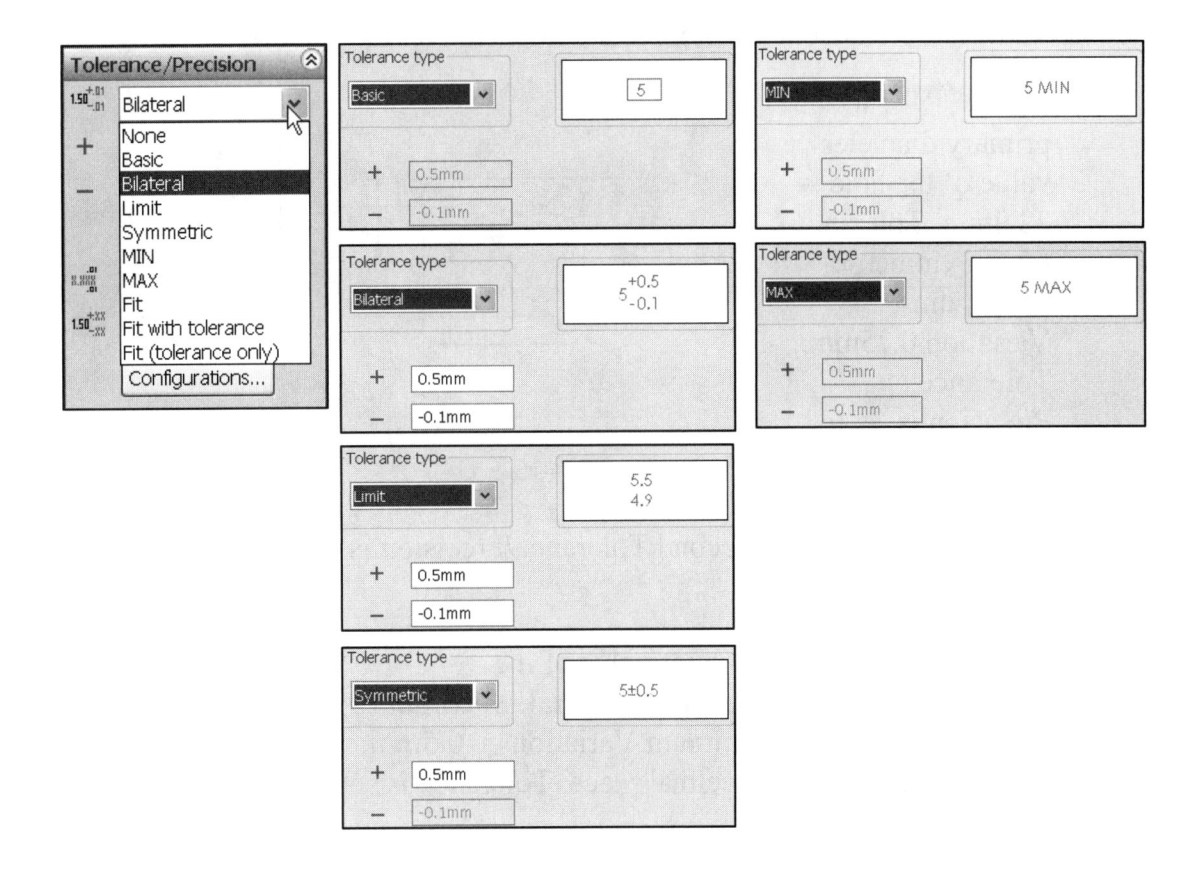

Part Modeling

Tutorial: Dimension text 3-1

1. **View** the illustrated model.
2. **Review** the Tolerance, Precision, and Dimension Text.
   a. 2X Ø.190 - Two holes with a diameter of .190. Precision is set to three decimal places.
   b. 2X R.250 - Two corners with a radius of .250. Precision is set to three decimal places.

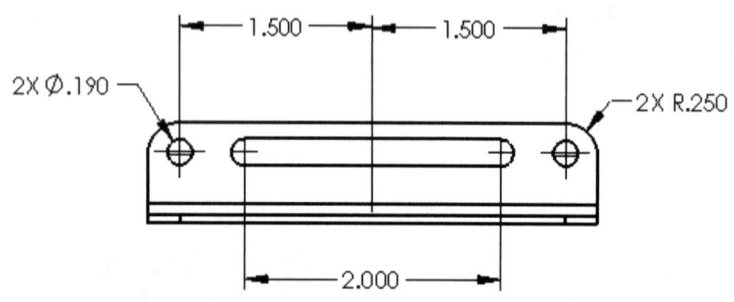

Tutorial: Dimension text 3-2

1. **View** the illustrated model.
2. **Review** the Tolerance, Precision, and Dimension text.
   a. Ø 22±0.25 - The primary diameter value of the hole = 22.0mm. Tolerance type: Symmetric. Maximum Variation 0.25mm. Tolerance / Precision is set to two decimal place.

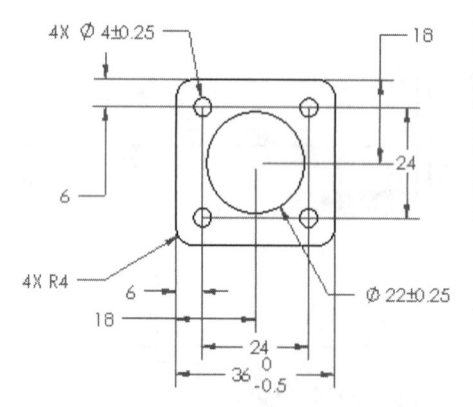

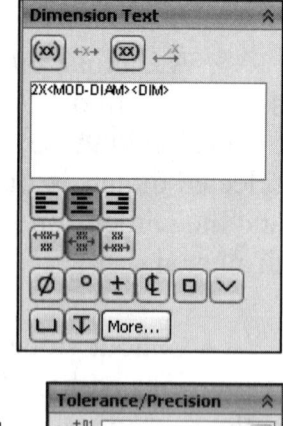

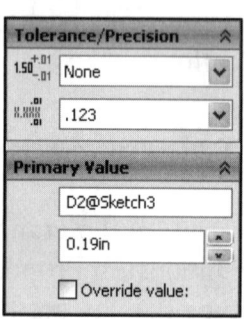

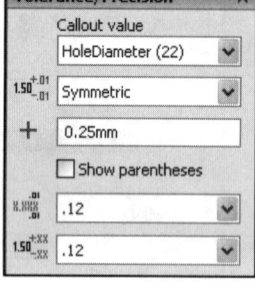

   For a Chamfer feature, a second Tolerance/Precision is available.

   b. $36^{\ 0}_{-0.5}$ - The primary diameter value of the hole = 36mm. Tolerance type: Bilateral. Maximum Variation is 0.0mm. Minimum Variation = -0.5mm. Precision is set to two decimal place. Tolerance is set to one decimal place.

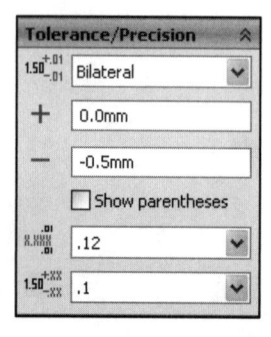

Page 3 - 32

Part Modeling

💡 Trailing zeros are removed according to the Dimension Standard.

c. **24** - The primary value = 24mm. Tolerance type: General. Tolerance / Precision is set to two decimal place.

d. **4X Ø 4±0.25** - Four holes with a primary diameter value = 4mm. Tolerance type: Symmetric. Maximum Variation = 0.25mm. Precision / Tolerance is set to two decimal place.

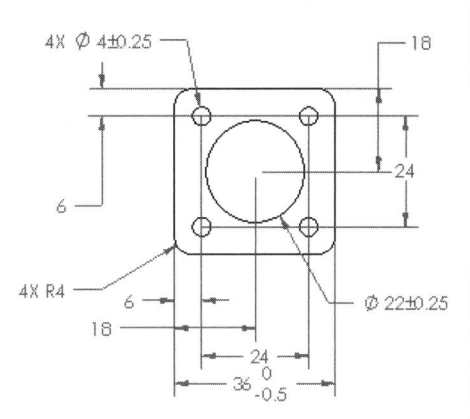

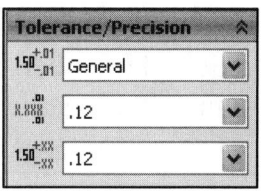

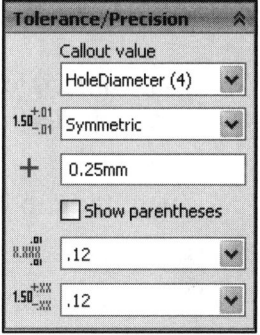

Tutorial: Dimension text 3-3

1. **View** the illustrated model.
2. **Review** the Tolerance, and Precision.

   a. **14/12** - The primary value = 12mm. Tolerance type: Limit. Maximum Variation = 2mm. Minimum Variation = 0mm. Tolerance / Precision is set to none.

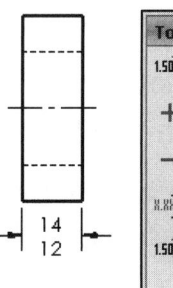

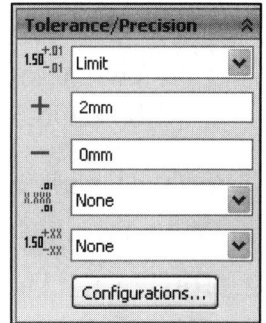

Dimension text symbols

Dimension Text symbols are displayed in the Dimension PropertyManager. The Dimension Text box provides eight commonly used symbols and a more button to access the Symbol Library. The eight displayed symbols in the Dimension Text box from left to right are: Diameter, Degree, Plus/Minus, Centerline, Square, Countersink, Counterbore, and Depth/Deep.

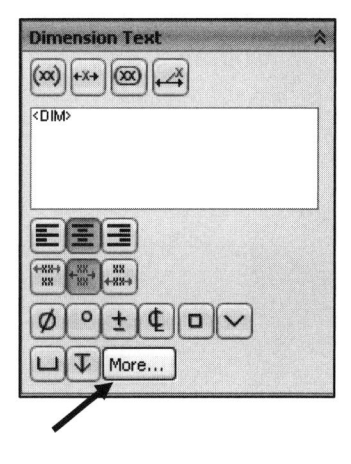

Page 3 - 33

Part Modeling

Review each symbol in the Dimension Text box and in the Symbol library. You are required to understand the meaning of these symbols in a SolidWorks document.

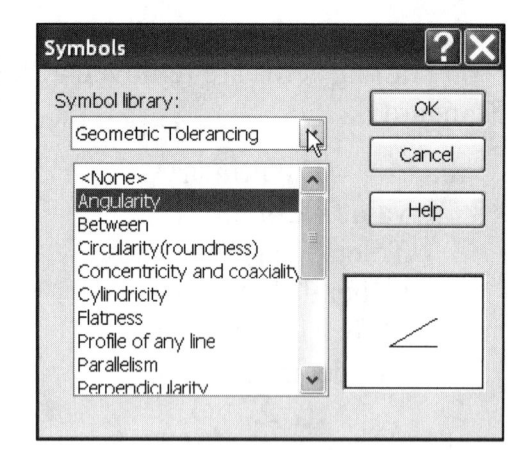

Tutorial: Dimension text symbols 3-1

1. **View** the illustrated model.

2. **Review** the Dimension Text and document symbols.

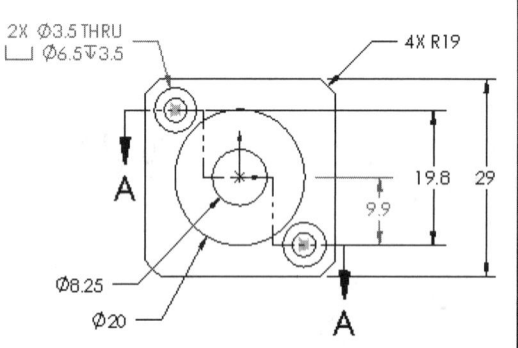

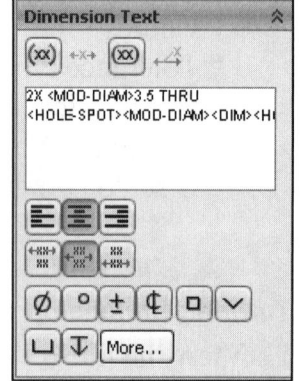

    a. 2X Ø3.5 THRU ⌴ Ø6.5▽3.5 - Two holes with a primary diameter value = 3.5mm, Cbore Ø6.5 with a depth 3.5.

Tutorial: Dimension text symbols 3-2

1. **View** the illustrated model.

2. **Review** the Dimension Text and document symbols.

    a. 2X ⌴ Ø5.5 ▽ 8.8 - Two Cbores with a primary diameter value = 5.5mm with a depth 8.8.

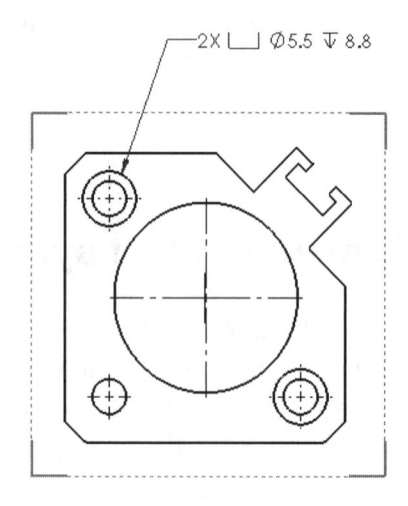

Build additional simple parts

Tutorial: Mass-Volume 3-10

Build this model. Calculate the overall mass of the part and locate the Center of mass with the provided information.

1. **Create** a New part in SolidWorks.

2. **Build** the illustrated model. Note the Depth/Deep ⇩ symbol with a 1.50 dimension associated with the hole. The hole Ø.562 has a three decimal place precision. Insert three features: Extruded Base, and two Extruded Cuts. Insert a 3D sketch for the first Extruded Cut feature.

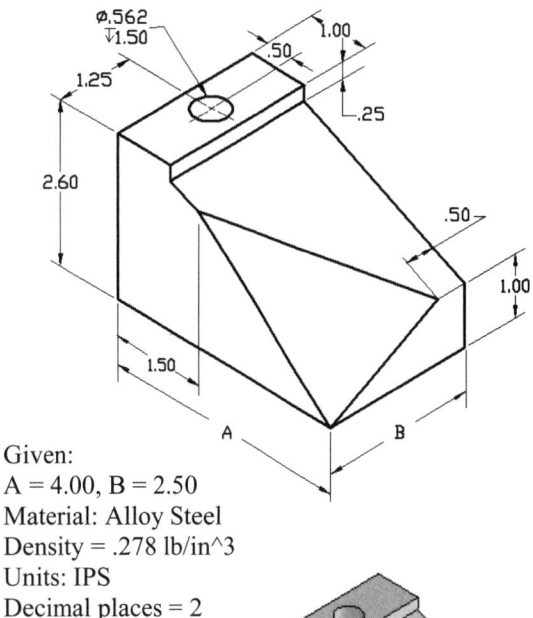

Given:
A = 4.00, B = 2.50
Material: Alloy Steel
Density = .278 lb/in^3
Units: IPS
Decimal places = 2

☼ There are numerous ways to build the models in this chapter. The goal is to display different design intents and techniques.

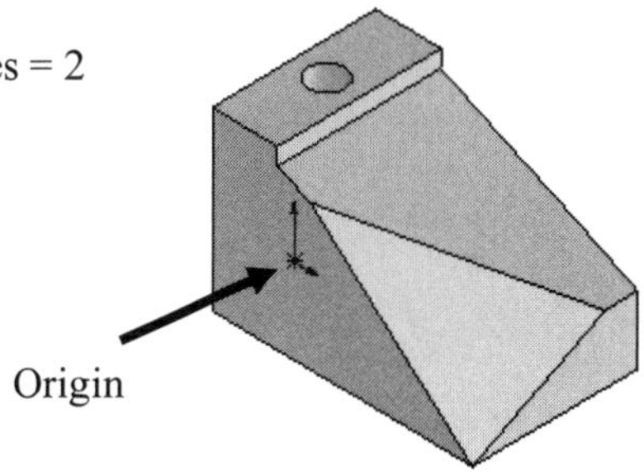

3. **Set** the document properties for the model.

4. Create **Sketch1**. Select the Front Plane as the Sketch plane. The part Origin is located in the lower left corner of the sketch. Insert the required geometric relations and dimensions.

5. Create the **Extruded Base** feature. Apply symmetry. Select the Mid Plane End Condition in Direction 1. Depth = 2.50in.

6. Create **3DSketch1**. Apply the Line Sketch tool. Create a closed four point 3D sketch. 3DSketch1 is the profile for the first Extruded Cut feature. Insert the required dimensions.

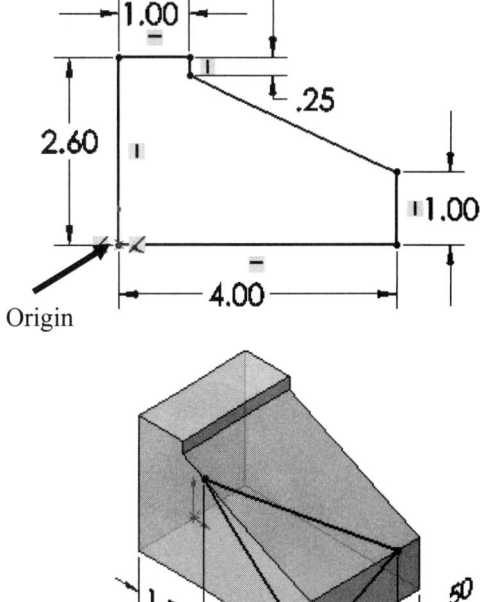

Part Modeling

7. Create the **Extruded Cut** feature. Blind is the default End Conditions. Select the top face as illustrated to be removed. Note the direction of extrude.

8. Create **Sketch2**. Select the top flat face of Extrude1. Sketch a circle. Insert the required geometric relations and dimensions. The hole diameter Ø.562 has a three decimal place precision.

9. Create the second **Extruded Cut** feature. Blind is the default End Condition. Depth = 1.50in. Note: For the exam, you do not need to insert the Depth/Deep ⊽ symbol or note.

10. **Assign** Alloy Steel material to the part.

11. **Calculate** the overall mass. The overall mass = 4.97 pounds.

12. **Locate** the Center of mass. The location of the Center of mass is derived from the part Origin.

    - X: 1.63 inches
    - Y: 1.01 inches
    - Z: -0.04 inches

    View the triad location of the Center of mass for the part.

13. **Save** the part and name it Mass-Volume 3-10.

14. **Close** the model.

    As an exercise, calculate the overall mass of the part using 6061 Alloy. Modify the A dimension from 4.00 to 4.50. Modify the hole dimension from Ø.562 to Ø.575. The overall mass of the part = 1.93 pounds. Save the part and name it Mass-Volume 3-10A.

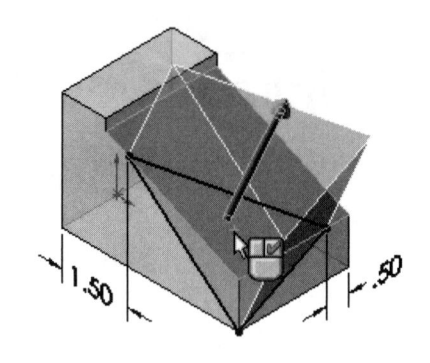

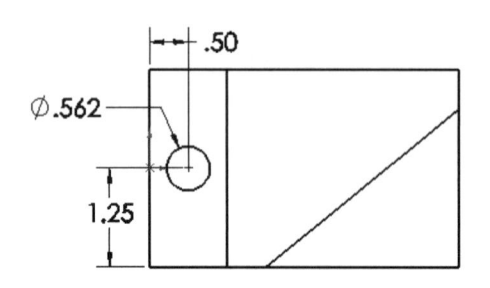

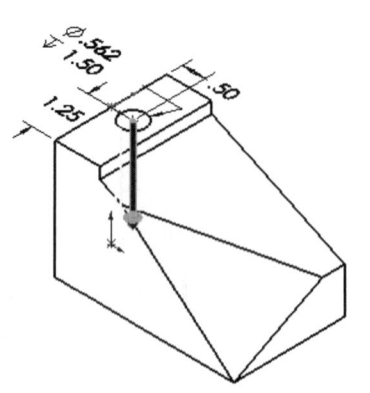

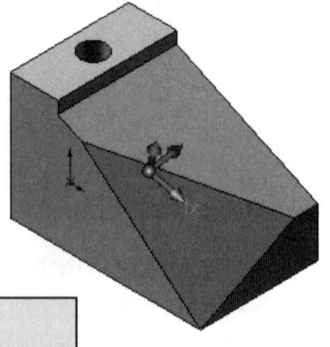

```
Mass = 1.93 pounds
Volume = 19.77 cubic inches
Surface area = 50.66 inches^2
Center of mass: ( inches )
    X = 1.83
    Y = 0.99
    Z = -0.04
```

Tutorial: Mass-Volume 3-11

Build this model. Calculate the overall mass of the part and locate the Center of mass with the provided information.

1. **Create** a New part in SolidWorks.

2. **Build** the illustrated model. Think about the required steps to build this part. Insert four features: Extruded Base, two Extruded Cuts, and a Fillet.

Given:
A = 4.00
B = R.50
Material: 6061 Alloy
Density = .0975 lb/in^3
Units: IPS
Decimal places = 2

💡 There are numerous ways to build the models in this chapter. The goal is to display different design intents and techniques.

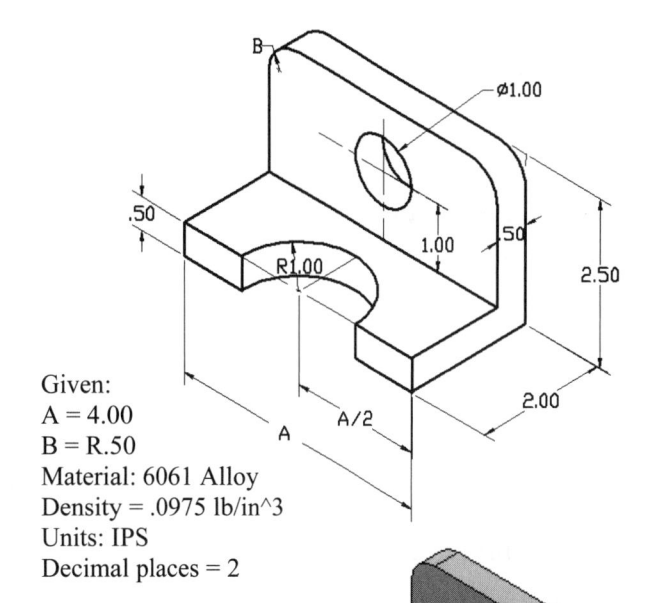

3. **Set** the document properties for the model.

4. Create **Sketch1**. Select the Right Plane as the Sketch plane. The part Origin is located in the lower left corner of the sketch. Insert the required geometric relations and dimensions.

5. Create the **Extruded Base** feature. Apply symmetry. Select the Mid Plane End Condition for Direction 1. Depth = 4.00in.

6. Create **Sketch2**. Select the top flat face of Extrude1 as the Sketch plane. Sketch a circle. The center of the circle is located at the part Origin. Insert the required dimension.

7. Create the first **Extruded Cut** feature. Select Through All for End Condition in Direction 1.

8. Create **Sketch3**. Select the front vertical face of Extrude1 as the Sketch plane. Sketch a circle. Insert the required geometric relations and dimensions.

9. Create the second **Extruded Cut** feature. Select Through All for End Condition in Direction 1.

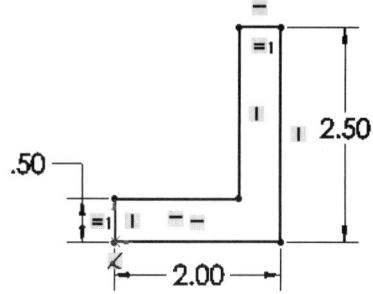

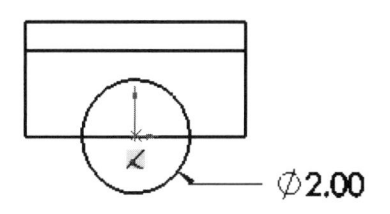

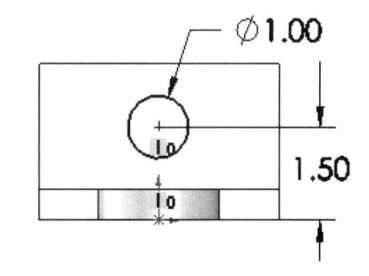

Part Modeling

10. Create the **Fillet** feature. Constant radius is selected by default. Fillet the top two edges as illustrated. Radius = .50in.

💡 A Fillet feature removes material. Selecting the correct radius value is important to obtain the correct mass and volume answer in the exam.

11. **Assign** the defined material to the part.

12. **Calculate** the overall mass. The overall mass = 0.66 pounds.

13. **Locate** the Center of mass. The location of the Center of mass is derived from the part Origin.

- X: 0.00 inches
- Y: 0.90 inches
- Z: -1.46 inches

In this category, Part Modeling an exam question could read: Build this model. Locate the Center of mass relative to the part Origin.

- A: X = -2.63 inches, Y = 4.01 inches, Z = -0.04 inches
- B: X = 4.00 inches, Y = 1.90 inches, Z = -1.64 inches
- C: X = 0.00 inches, Y = 0.90 inches, Z = -1.46 inches
- D: X = -1.69 inches, Y = 1.00 inches, Z = 0.10 inches

The correct answer is C.

14. **Save** the part and name it Mass-Volume 3-11.

15. **Close** the model.

As an exercise, calculate the overall mass of the part using the MMGS unit system, and assign 2014 Alloy material to the part. The overall mass of the part = 310.17 grams. Save the part and name it Mass-Volume 3-11-MMGS.

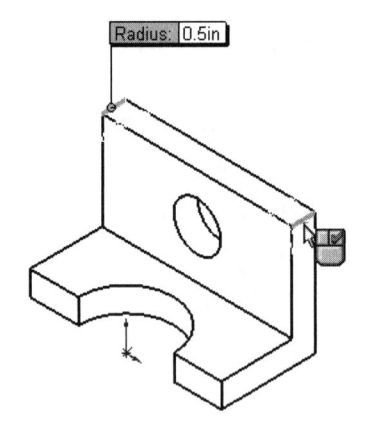

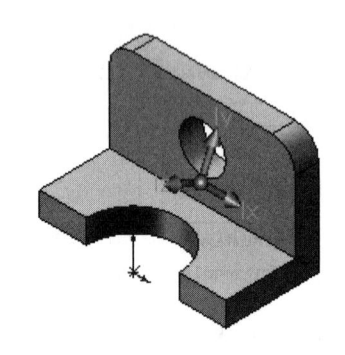

```
Mass = 310.17 grams
Volume = 110774.26 cubic millimeters
Surface area = 23865.83 millimeters^2
Center of mass: ( millimeters )
   X = 0.00
   Y = 22.83
   Z = -37.11
```

Tutorial: Mass-Volume 3-12

Build this model. Calculate the overall mass of the part and locate the Center of mass with the provided information.

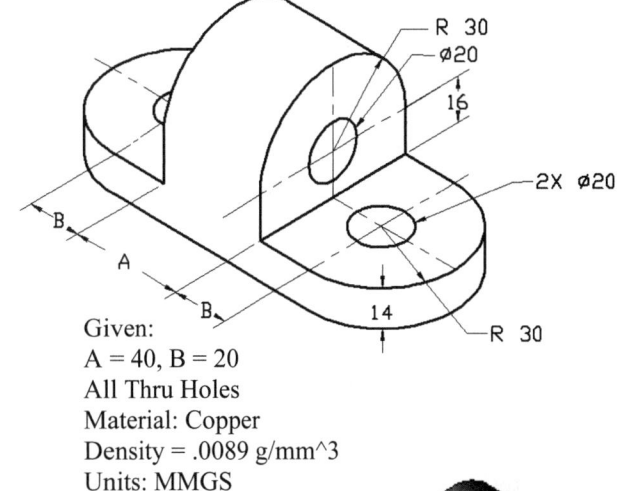

1. **Create** a New part in SolidWorks.

2. **Build** the illustrated model. Insert two features: Extruded Base, and Extruded Boss.

3. **Set** the document properties for the model.

Given:
A = 40, B = 20
All Thru Holes
Material: Copper
Density = .0089 g/mm^3
Units: MMGS

4. Create **Sketch1**. Select the Top Plane as the Sketch plane. Apply the Centerline Sketch tool. Locate the part Origin at the center of the sketch. Insert the required geometric relations and dimensions.

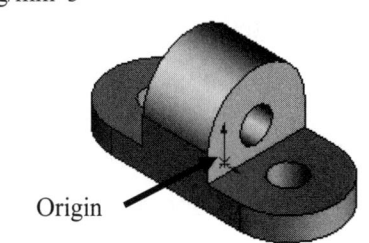

5. Create the **Extruded Base** feature. Blind is the default End Condition. Depth = 14mm.

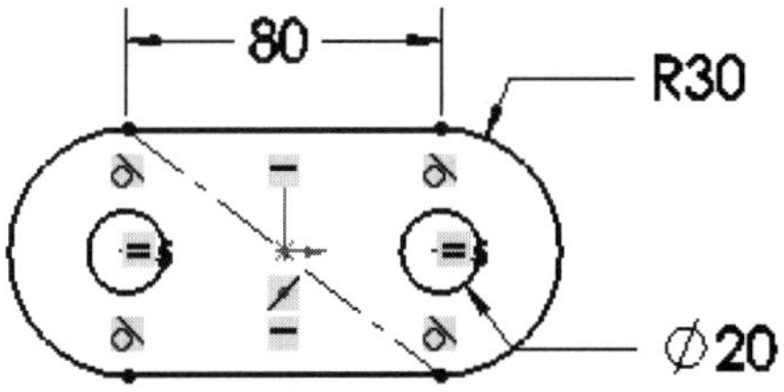

6. Create **Sketch2**. Select the Right Plane as the Sketch plane. Insert the required geometric relations and dimensions.

7. Create the **Extruded Boss** feature. Apply symmetry. Select the Mid Plane End Condition. Depth = 40mm.

8. **Assign** the defined material to the part.

9. **Calculate** the overall mass. The overall mass = 1605.29 grams.

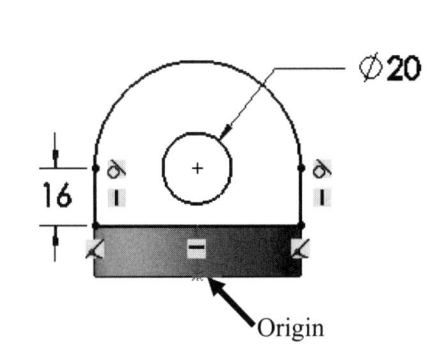

10. **Locate** the Center of mass. The location of the Center of mass is derived from the part Origin.

- X: 0.00 millimeters
- Y: 19.79 millimeters
- Z: 0.00 millimeters

Part Modeling

11. **Save** the part and name it Mass-Volume 3-12.

12. **Close** the model.

💡 There are numerous ways to build the models in this chapter. Optimize your time. The CSWA is a timed exam.

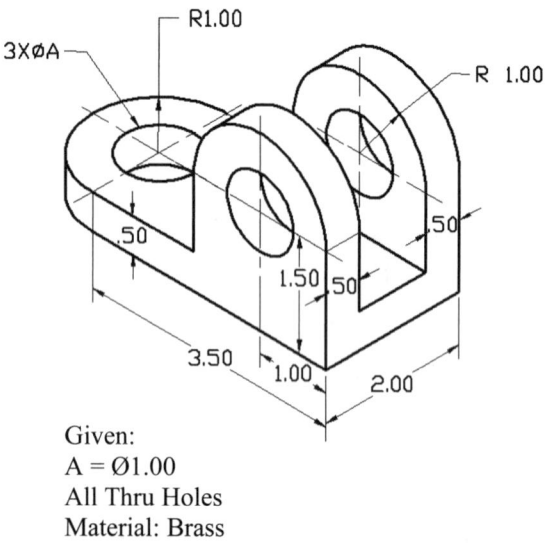

Tutorial: Mass-Volume 3-13

Build this model. Calculate the volume of the part and locate the Center of mass with the provided information.

1. **Create** a New part in SolidWorks.

2. **Build** the illustrated model. Insert three features: Extruded Base, Extruded Boss, and Mirror. Three holes are displayed with an Ø1.00in.

3. **Set** the document properties for the model.

Given:
A = Ø1.00
All Thru Holes
Material: Brass
Density = .307 lb/in^3
Units: IPS
Decimal places = 2

4. Create **Sketch1**. Select the Top Plane as the Sketch plane. Apply the Tangent Arc and Line Sketch tool. Insert the required geometric relations and dimensions. Note the location of the Origin.

5. Create the **Extruded Base** feature. Blind is the default End Condition. Depth = .50in.

6. Create **Sketch2**. Select the front vertical face of Extrude1 as the Sketch plane. Insert the required geometric relations and dimensions.

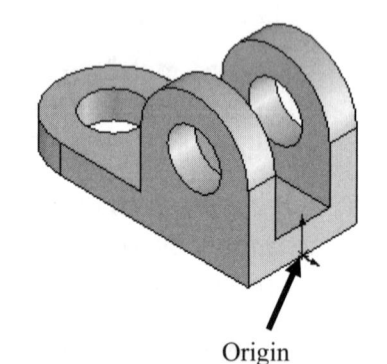

Page 3 - 40

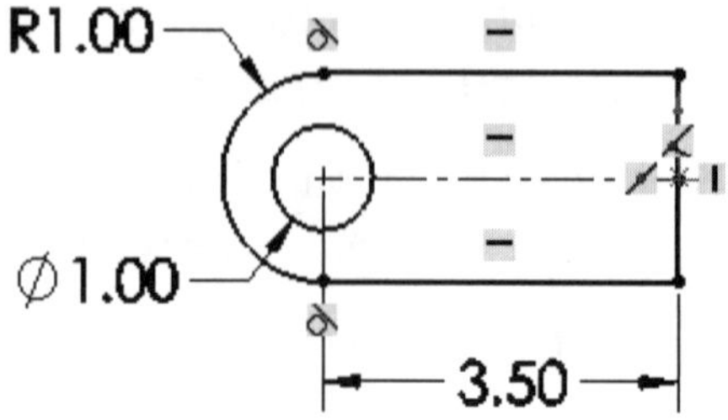

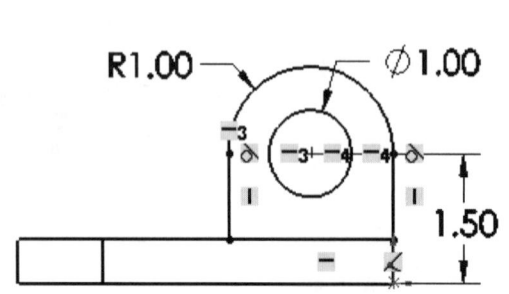

7. Create the **Extruded Boss** feature. Blind is the default End Condition in Direction 1. Depth = .50in. Note the direction of the extrude feature.

8. Create the **Mirror** feature. Apply Symmetry. Mirror Extrude2 about the Front Plane.

9. **Assign** the defined material to the part.

10. **Calculate** the volume. The volume = 6.68 cubic inches.

11. **Locate** the Center of mass. The location of the Center of mass is derived from the part Origin.

- X: -1.59 inches
- Y: 0.72 inches
- Z: 0.00 inches

In this category, Part Modeling an exam question could read: Build this model. What is the volume of the model?

- A = 6.19 cubic inches
- B = 7.79 cubic inches
- C = 7.87 cubic inches
- D = 6.68 cubic inches

The correct answer is D.

View the triad location of the Center of mass for the part.

12. **Save** the part and name it Mass-Volume 3-13.

13. **Close** the model.

As an exercise, calculate the overall mass of the part using the IPS unit system, and assign Copper material to the part. Modify the hole diameters from 1.00in to 1.125in. The overall mass of the part = 2.05 pounds. Save the part and name it Mass-Volume 3-13A.

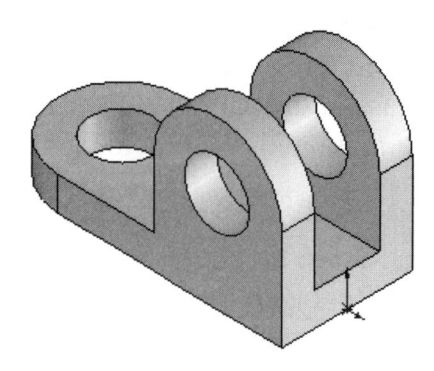

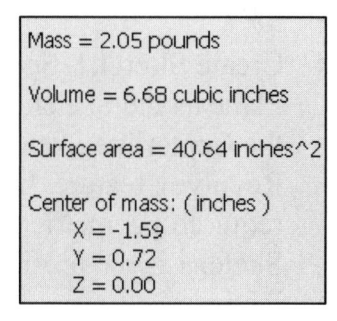

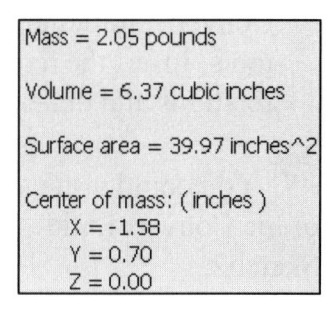

```
Mass = 2.05 pounds
Volume = 6.68 cubic inches
Surface area = 40.64 inches^2
Center of mass: ( inches )
  X = -1.59
  Y = 0.72
  Z = 0.00
```

```
Mass = 2.05 pounds
Volume = 6.37 cubic inches
Surface area = 39.97 inches^2
Center of mass: ( inches )
  X = -1.58
  Y = 0.70
  Z = 0.00
```

Part Modeling

Tutorial: Mass-Volume 3-14

Build this model. Calculate the overall mass of the part and locate the Center of mass with the provided information.

1. **Create** a New part in SolidWorks.

2. **Build** the illustrated model. Insert a Revolved Base feature and Extruded Cut feature to build this part.

3. **Set** the document properties for the model.

4. Create **Sketch1**. Select the Front Plane as the Sketch plane. Apply the Centerline Sketch tool for the Revolve1 feature. Insert the required geometric relations and dimensions. Sketch1 is the profile for the Revolve1 feature.

5. Create the **Revolved Base** feature. The default angle is 360deg. Select the centerline for the Axis of Revolution.

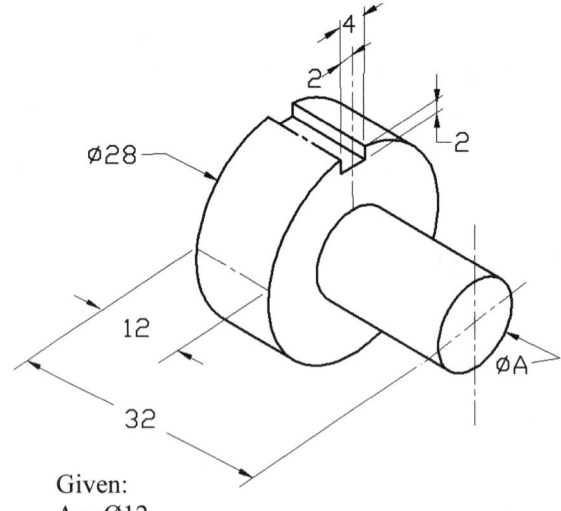

Given:
A = Ø12
Material: Cast Alloy Steel
Density = .0073 g/mm^3
Units: MMGS

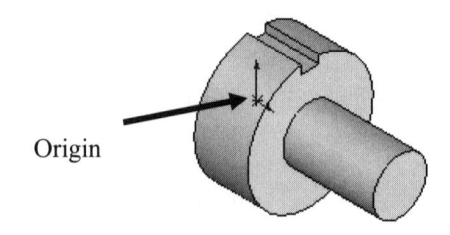

☼ A Revolve feature adds or removes material by revolving one or more profiles around a centerline.

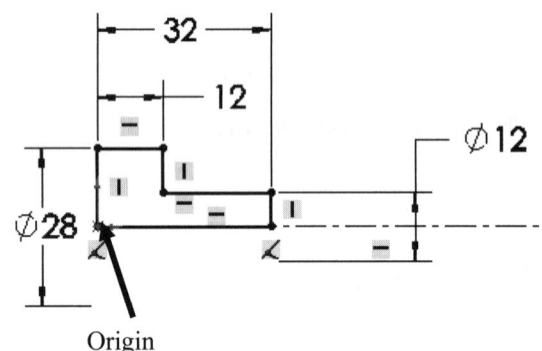

6. Create **Sketch2**. Select the right large circular face of Revolve1 as the Sketch plane. Apply reference construction geometry. Use the Convert Entities and Trim Sketch tools. Insert the required geometric relations and dimensions.

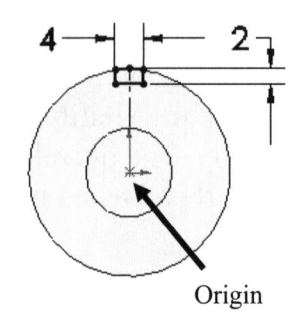

☼ You could also use the 3 Point Arc Sketch tool instead of the Convert Entities and Trim Sketch tools to create Sketch2.

7. Create the **Extruded Cut** feature. Select Through All for End Condition in Direction 1.

8. **Assign** the defined material to the part.

9. **Calculate** the overall mass. The overall mass = 69.77 grams.

10. **Locate** the Center of mass. The location of the Center of mass is derived from the part Origin.

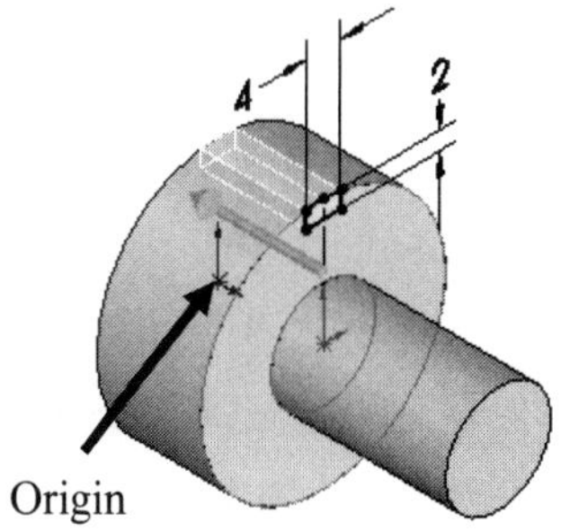

- X = 9.79 millimeters
- Y = -0.13 millimeters
- Z = 0.00 millimeters

11. **Save** the part and name it Mass-Volume 3-14.

12. **Close** the model.

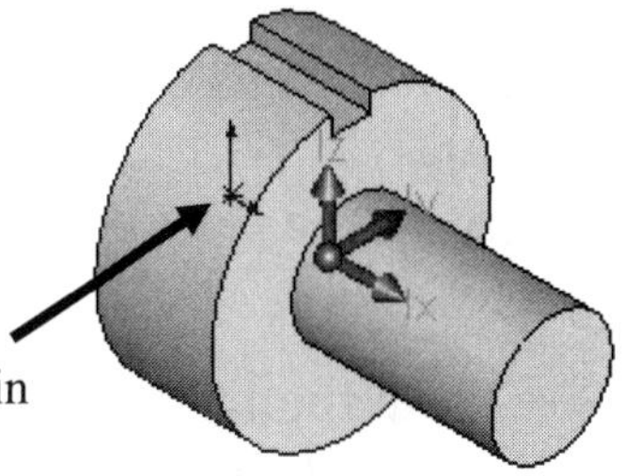

Tutorial: Mass-Volume 3-15

Build this model. Calculate the overall mass of the part and locate the Center of mass with the provided information.

1. **Create** a New part in SolidWorks.

2. **Build** the illustrated model. Insert two features: Extruded Base and Revolved Boss.

3. **Set** the document properties for the model.

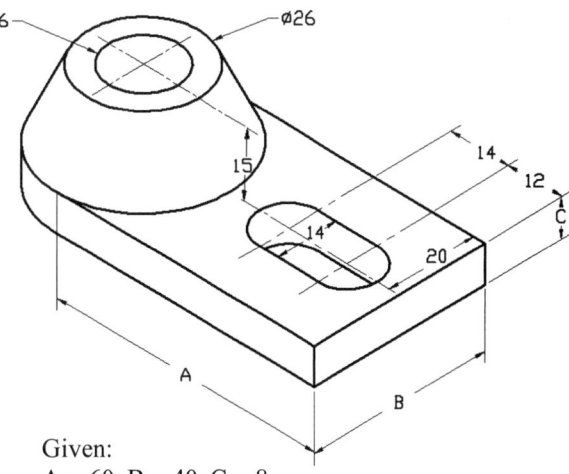

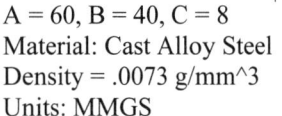

Given:
A = 60, B = 40, C = 8
Material: Cast Alloy Steel
Density = .0073 g/mm^3
Units: MMGS

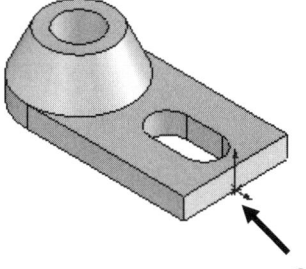

Page 3 - 43

Part Modeling

4. Create **Sketch1**. Select the Top Plane as the Sketch plane. Apply construction geometry. Apply the Tangent Arc and Line Sketch tool. Insert the required geometric relations and dimensions.

5. Create the **Extruded Base** feature. Blind is the default End Condition. Depth = 8mm.

6. Create **Sketch2**. Select the Front Plane as the Sketch plane. Apply construction geometry for the Revolved Boss feature. Insert the required geometric relations and dimension.

7. Create the **Revolved Boss** feature. The default angle is 360deg. Select the centerline for Axis of Revolution.

8. **Assign** the defined material to the part. **Calculate** the overall mass. The overall mass = 229.46 grams.

9. **Locate** the Center of mass. The location of the Center of mass is derived from the part Origin.

- X = -46.68 millimeters
- Y = 7.23 millimeters
- Z = 0.00 millimeters

In this category, Part Modeling an exam question could read: Build this model. What is the overall mass of the part?

- A: 229.46 grams
- B: 249.50 grams
- C: 240.33 grams
- D: 120.34 grams

The correct answer is A.

10. **Save** the part and name it Mass-Volume 3-15.
11. **Close** the model.

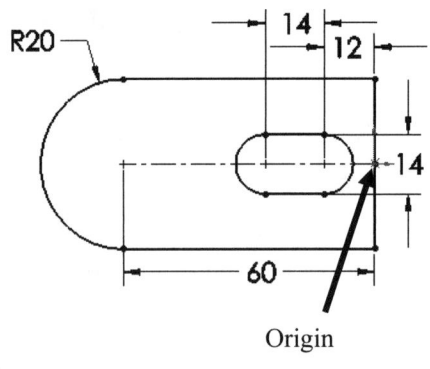

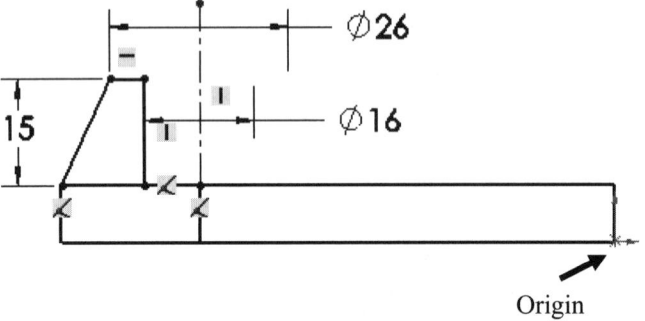

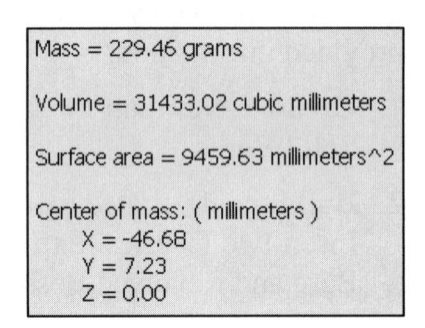

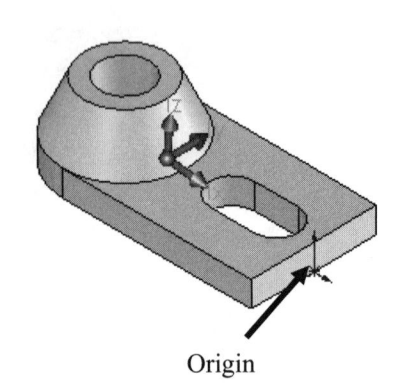

Tutorial: Mass-Volume 3-16

Build this model. Calculate the overall mass of the part and locate the Center of mass with the provided information.

1. **Create** a New part in SolidWorks.

2. **Build** the illustrated model. Insert three features: Extruded Base, Extruded Cut, and Circular Pattern. There are eight holes Ø14mm equally spaces on an Ø56mm bolt circle. The center hole = Ø22mm.

Given:
A = Ø80
Material: ABS
Density: .001 g/mm^3
Units: MMGS

3. **Set** the document properties for the model.

4. Create **Sketch1**. Select the Front Plane as the Sketch plane. Sketch two circles. The part Origin is located in the center of the sketch. Insert the required geometric relations and dimensions.

5. Create the **Extruded Base** feature. Blind is the default End Condition. Depth = 20mm.

6. Create **Sketch2**. Select the front face as the Sketch plane. Apply construction geometry to locate the seed feature for the Circular Pattern. Insert the required geometric relations and dimensions.

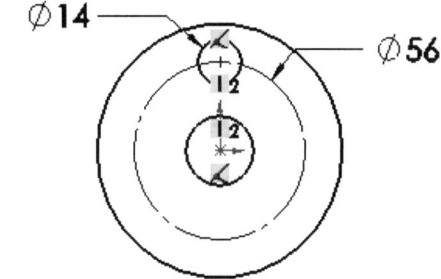

※ Apply construction reference geometry to assist in creating the sketch entities and geometry that are incorporated into the part. Construction reference geometry is ignored when the sketch is used to create a feature. Construction reference geometry uses the same line style as centerlines.

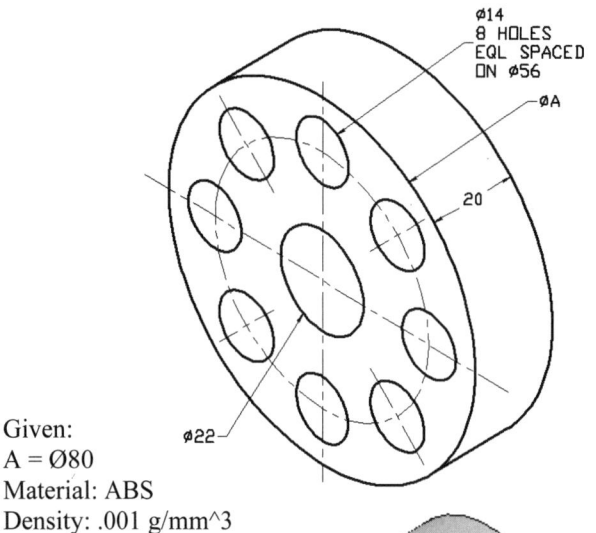

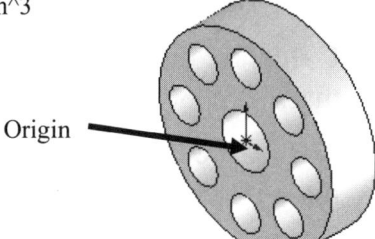

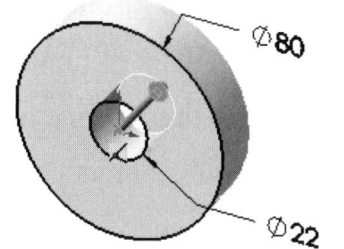

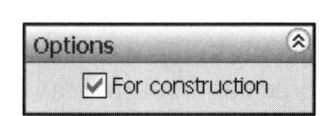

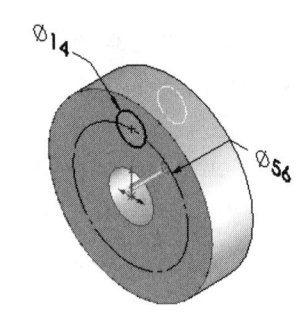

Part Modeling

7. Create the **Extruded Cut** feature. Extrude2 is the seed feature for the Circular Pattern. Select Through All for End Condition in Direction 1.

8. Create the **Circular Pattern** feature. Create a Circular Pattern of the Extrude2 feature. Use the View, Temporary Axes command to select the Pattern Axis for the CirPattern1 feature. Instances = 8. Equal spacing is selected by default.

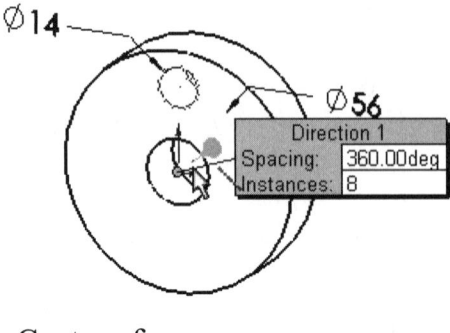

☼ Apply a circular pattern feature to create multiple instances of one or more features that you can space uniformly about an axis.

9. **Assign** the defined material to the part.

10. **Calculate** the overall mass. The overall mass = 69.66 grams.

11. **Locate** the Center of mass. The location of the Center of mass is derived from the part Origin.

- X = 0.00 millimeters
- Y = 0.00 millimeters
- Z = -10.00 millimeters

12. **Save** the part and name it Mass-Volume 3-16.

13. **Close** the model.

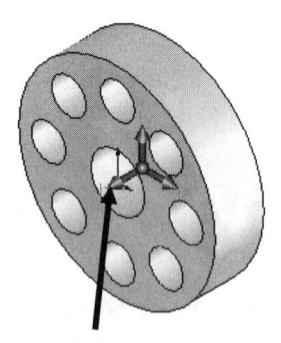

As an exercise, select the Top Plane for the Sketch plane to create Sketch1. Recalculate the location of the Center of mass with respect to the part Origin: X = 0.00 millimeters, Y = -10.00 millimeters, and Z = 0.00 millimeters. Save the part and name it Mass-Volume 3-16-TopPlane.

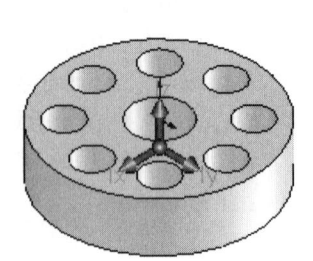

The complexity of the models along with the features increases throughout this chapter to simulate the final types of parts that would be provided on the CSWA exam. In the next section, the parts represent the feature types and complexity that you would see in the Part Modeling category of the CSWA exam.

Page 3 - 46

Tutorial: Basic-part 3-1

Build this model. Calculate the overall mass of the part and locate the Center of mass with the provided information.

1. **Create** a New part in SolidWorks.

2. **Build** the illustrated model. Think about the various features that create the model. Insert seven features to build this model: Extruded Base, Extruded Cut, Extruded Boss, Fillet, second Extruded Cut, Mirror, and a second Fillet. Apply symmetry. Create the left half of the model first, and then apply the Mirror feature.

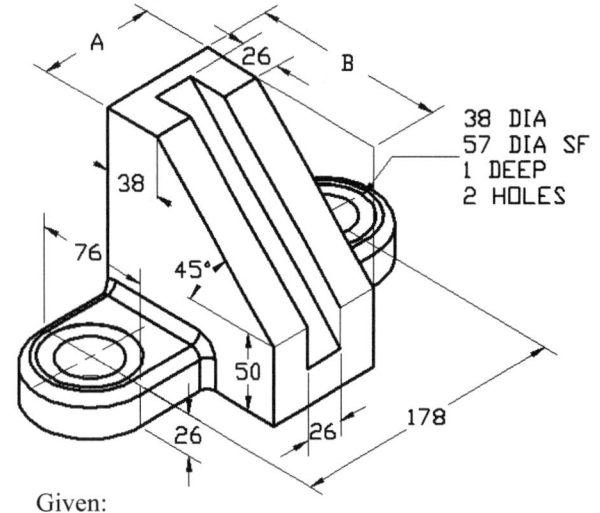

Given:
A = 76, B = 127
Material: 2014 Alloy
Density: .0028 g/mm^3
Units: MMGS
ALL ROUNDS EQUAL 6MM

There are numerous ways to build the models in this chapter. The goal is to display different design intents and techniques.

3. **Set** the document properties for the model.

4. Create **Sketch1**. Select the Front Plane as the Sketch plane. Create the main body of the part. The part Origin is located in the bottom left corner of the sketch. Insert the required geometric relations and dimensions.

5. Create the **Extruded Base** feature. Extrude1 is the Base feature. Select Mid Plane for End Condition in Direction 1. Depth = 76mm.

6. Create **Sketch2**. Select the top flat face of Extrude1 as the Sketch plane. Create the top cut on the Base feature. Apply construction geometry. Insert the required geometric relations and dimensions.

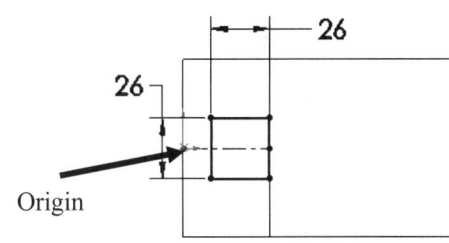

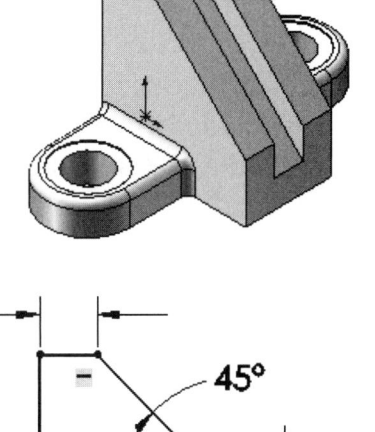

Page 3 - 47

Part Modeling

7. Create the first **Extruded Cut** feature. Select Through All for End Condition in Direction 1. Select the illustrated angled edge for the Direction of Extrusion.

8. Create **Sketch3**. Select the bottom face of Extrude1 as the Sketch plane. Sketch the first tab with a single hole as illustrated. Insert the required geometric relations and dimensions.

9. Create the **Extruded Boss** feature. Blind is the default End Condition in Direction 1. Depth = 26mm.

10. Create the first **Fillet** feature. Fillet the top edge of the left tab. Radius = 6mm. Constant radius is selected by default.

11. Create **Sketch4**. Select the top face of Extrude3 as the Sketch plane. Sketch a circle. Insert the required dimension.

12. Create the second **Extruded Cut** feature. Blind is the default End Condition in Direction 1. Depth = 1mm. The model displayed an Ø57mm Spot Face hole with a 1mm depth.

13. Create the **Mirror** feature. Mirror about the Front Plane. Mirror the Extrude3, Fillet1, and Extrude4 feature.

14. Create the second **Fillet** feature. Fillet the top inside edge of the left tab and the top inside edge of the right tab. Radius = 6mm.

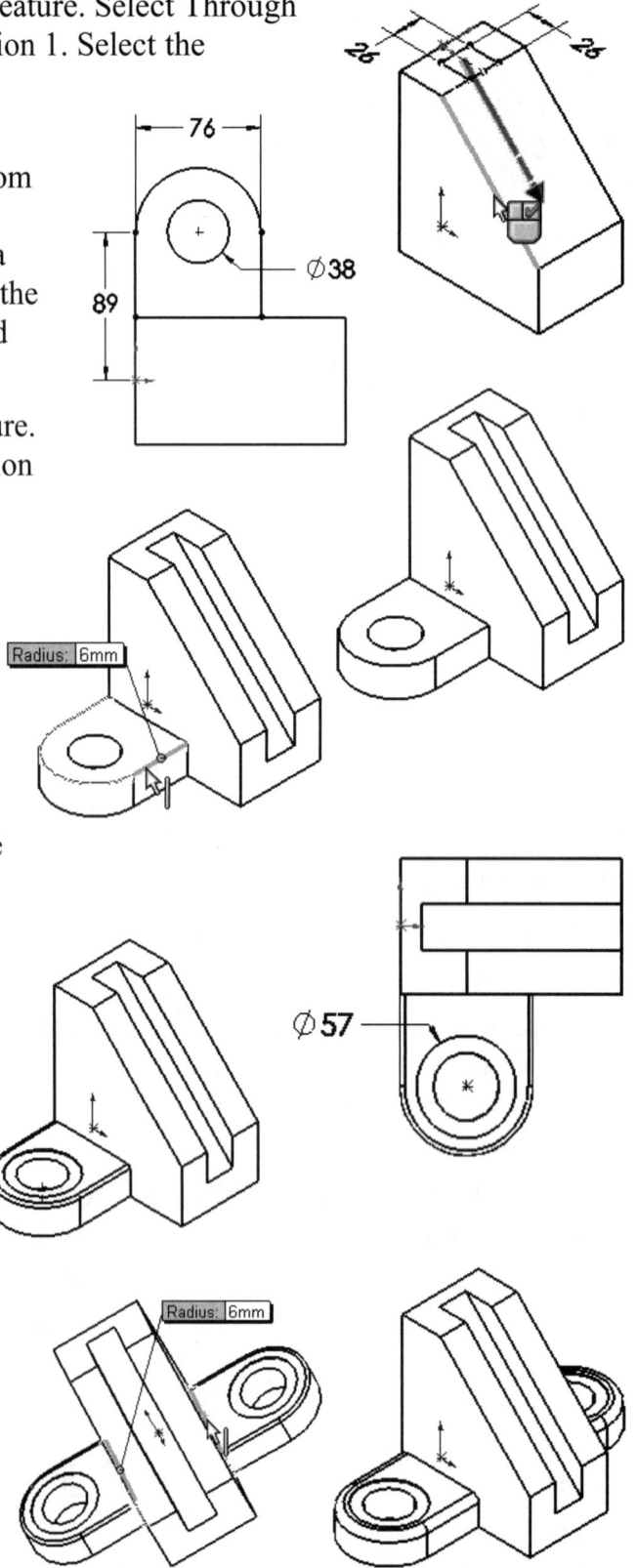

15. **Assign** the defined material to the part.
16. **Calculate** the overall mass of the part. The overall mass = 3437.29 grams.
17. **Locate** the Center of mass. The location of the Center of mass is derived from the part Origin.

   - X = 49.21 millimeters
   - Y = 46.88 millimeters
   - Z = 0.00 millimeters

18. **Save** the part and name it Part-Modeling 3-1.
19. **Close** the model.

In this category, Part Modeling an exam question could read: Build this model. What is the overall mass of the part?

   - A: 3944.44 grams
   - B: 4334.29 grams
   - C: 3437.29 grams
   - D: 2345.69 grams

The correct answer is C.

As an exercise, modify all ALL ROUNDS from 6MM to 8MM. Modify the material from 2014 Alloy to 6061 Alloy. Modify the Sketch1 angle from 45deg to 30deg. Modify the Extrude3 depth from 26mm to 36mm. Recalculate the location of the Center of mass with respect to the part Origin.

   - X = 49.76 millimeters
   - Y = 34.28 millimeters
   - Z = 0.00 millimeters

20. **Save** the part and name it Part-Modeling 3-1-Modify.

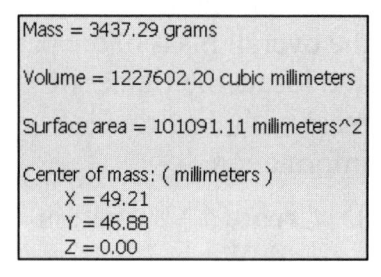

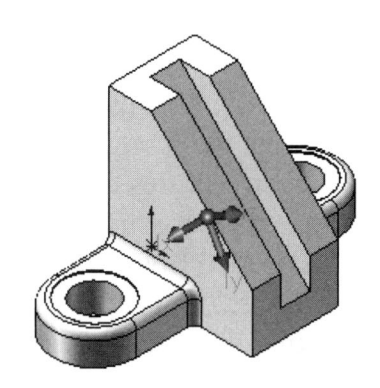

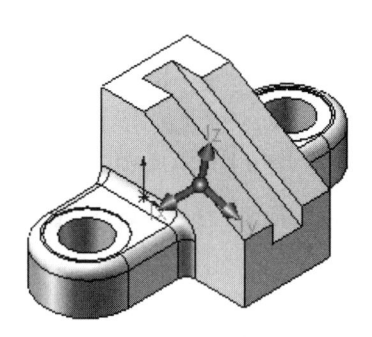

Part Modeling

Tutorial: Basic-part 3-2

Build this model. Calculate the overall mass of the part and locate the Center of mass with the provided information.

1. **Create** a New part in SolidWorks.

2. **Build** the illustrated model. Think about the various features that create the part. Insert seven features to build this part: Extruded-Thin, Extruded Boss, two Extruded Cuts, and three Fillets. Apply reference construction planes to build the circular features.

Given:
A = 52, B = 58
ALL-ROUNDS R 4MM
Material: 6061 Alloy
Density: .0027 g/mm^3
Units: MMGS

3. **Set** the document properties for the model.

4. Create **Sketch1**. Select the Front Plane as the Sketch plane. Apply construction geometry as the reference line for the 30deg angle. Insert the required geometric relations and dimensions. Note the location of the Origin.

5. Create the **Extrude-Thin1** feature. This is the Base feature. Apply symmetry. Select Mid Plane for End Condition in Direction 1 to maintain the location of the Origin. Depth = 52mm. Thickness = 12mm.

☼ Use the Thin Feature option to control the extrude thickness, not the Depth.

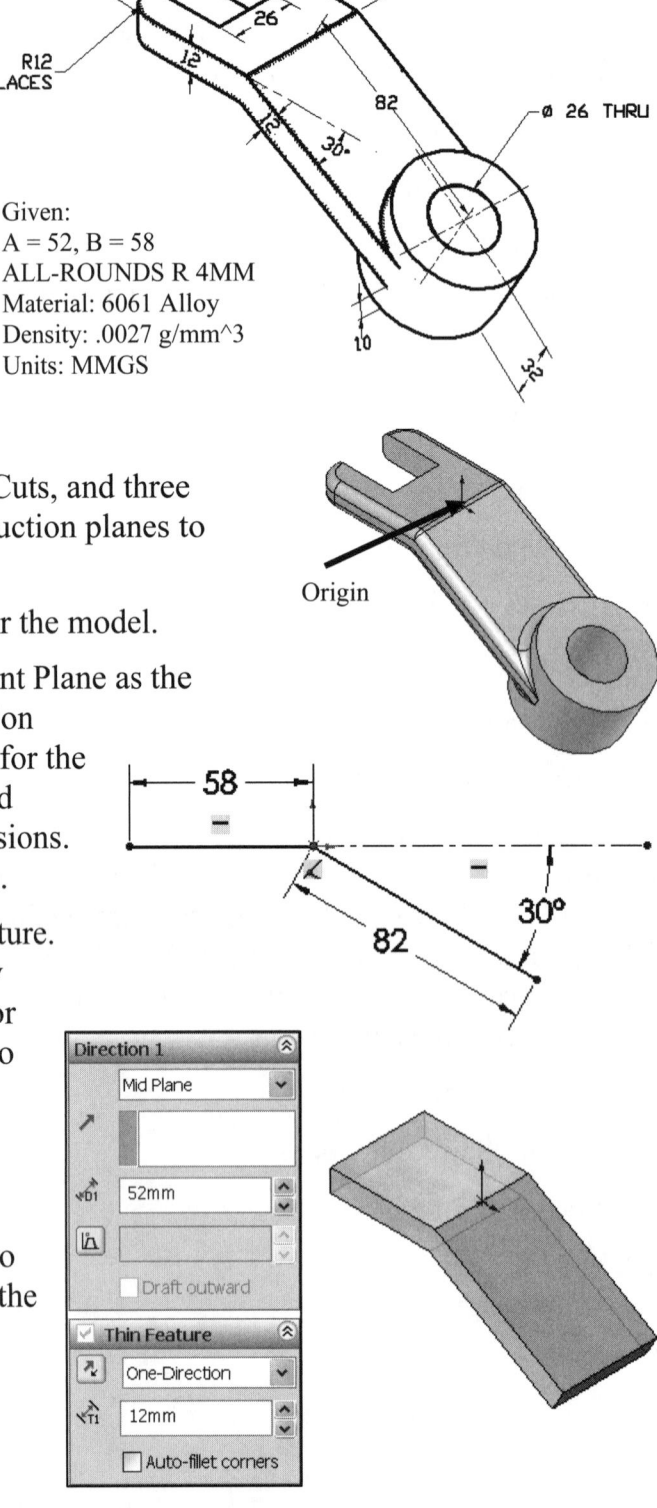

Page 3 - 50

6. Create **Plane1**. Plane1 is the Sketch plane for the Extruded Boss feature. Select the midpoint and the top face as illustrated. Plane1 is located in the middle of the top and bottom faces. Select Parallel Plane at Point for option.

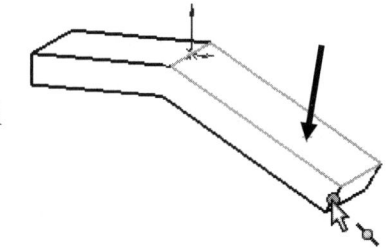

☼ Create Plane1 to use the Depth dimension of 32mm.

7. Create **Sketch2**. Select Plane1 as the Sketch plane. Use the Normal To view tool. Sketch a circle to create the Extruded Boss feature. Insert the required geometric relations.

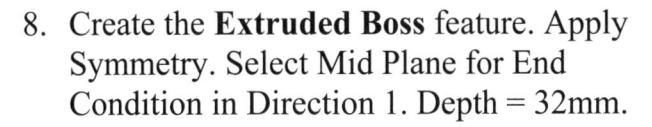

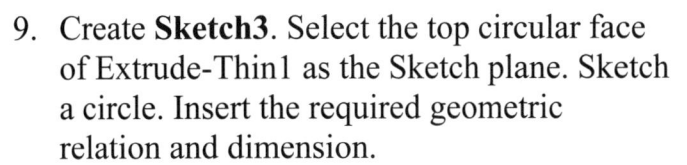

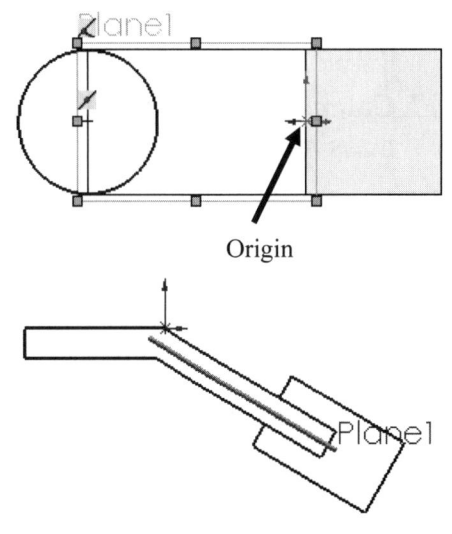

☼ The Normal To view tool rotates and zooms the model to the view orientation normal to the selected plane, planar face, or feature.

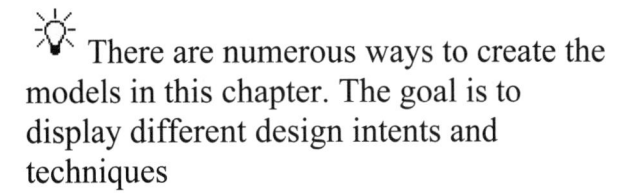

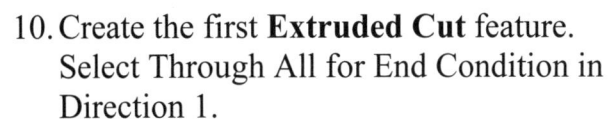

8. Create the **Extruded Boss** feature. Apply Symmetry. Select Mid Plane for End Condition in Direction 1. Depth = 32mm.

9. Create **Sketch3**. Select the top circular face of Extrude-Thin1 as the Sketch plane. Sketch a circle. Insert the required geometric relation and dimension.

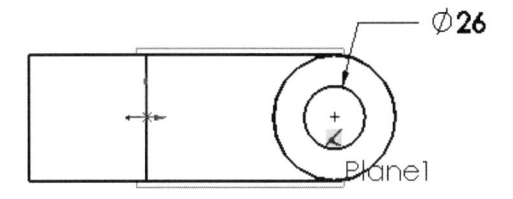

☼ There are numerous ways to create the models in this chapter. The goal is to display different design intents and techniques

10. Create the first **Extruded Cut** feature. Select Through All for End Condition in Direction 1.

11. Create **Sketch4**. Select the top face of Extrude-Thin1 as the Sketch plane. Apply construction geometry. Insert the required geometric relations and dimensions.

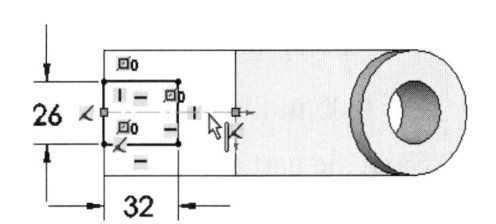

12. Create the second **Extruded Cut** feature. Select Through All for End Condition in Direction 1.

Page 3 - 51

Part Modeling

13. Create the **Fillet1** feature. Fillet the left and right edges of Extrude-Thin1 as illustrated. Radius = 12mm.

14. Create the **Fillet2** feature. Fillet the top and bottom edges of Extrude-Thin1 as illustrated. Radius = 4mm.

15. Create the **Fillet3** feature. Fillet the rest of the model; six edges as illustrated. Radius = 4mm.

16. **Assign** the defined material to the part.

17. **Calculate** the overall mass of the part. The overall mass = 300.65 grams.

18. **Locate** the Center of mass. The location of the Center of mass is derived from the part Origin.

- X: 34.26 millimeters
- Y: -29.38 millimeters
- Z: 0.00 millimeters

19. **Save** the part and name it Part-Modeling 3-2.

20. **Close** the model.

As an exercise, modify the Fillet2 and Fillet3 radius from 4mm to 2mm. Modify the Fillet1 radius from 12m to 10mm. Modify the material from 6061 Alloy to ABS. Modify the Sketch1 angle from 30deg to 45deg. Modify the Extrude1 depth from 32mm to 38mm. Recalculate the location of the Center of mass with respect to the part Origin.

- X = 27.62 millimeters
- Y = -40.44 millimeters
- Z = 0.00 millimeters

21. **Save** the part and name it Part-Modeling 3-2-Modify.

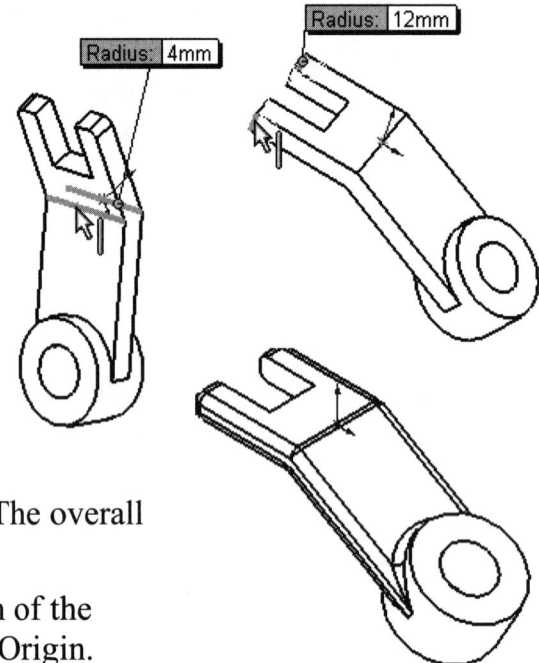

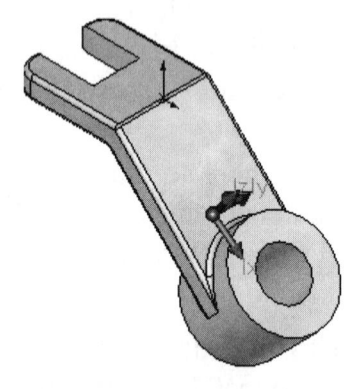

Part Modeling

Tutorial: Basic-part 3-3

Build this model. Calculate the volume of the part and locate the Center of mass with the provided information.

1. **Create** a New part in SolidWorks.

2. **Build** the illustrated model. Think about the various features that create this model. Insert five features to build this part: Extruded Base, two Extruded Bosses, Extruded Cut, and a Rib. Insert a reference plane to create the Extrude2 feature.

Given:
A = Ø3.00, B = 1.00
Material: 6061 Alloy
Density: .097 lb/in^3
Units: IPS
Decimal places = 2

3. **Set** the document properties for the model.

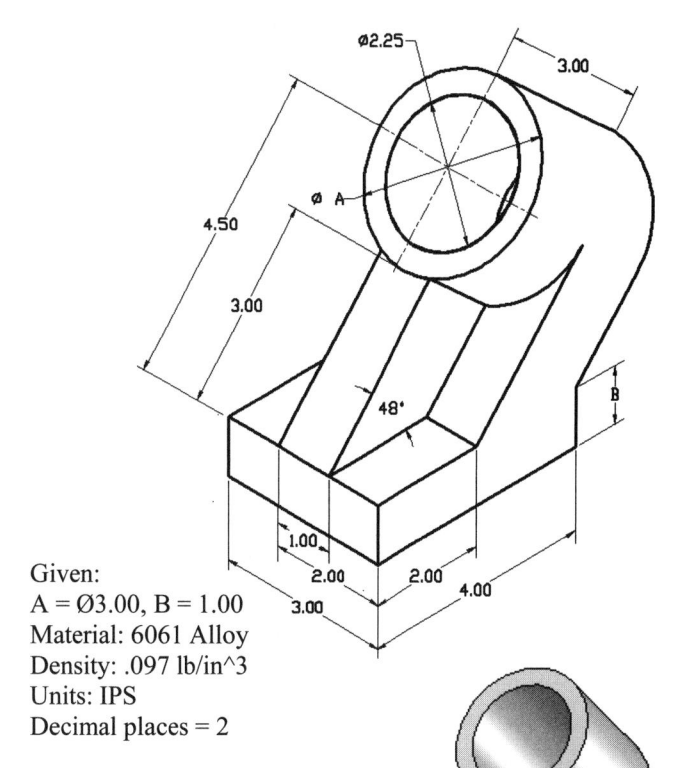

4. Create **Sketch1**. Select the Top Plane as the Sketch plane. Sketch a rectangle. Apply two construction lines for an Intersection relation. Use the horizontal construction line as the Plane1 reference. Insert the required relations and dimensions.

5. Create the **Extruded Base** feature. Blind is the default End Condition in Direction 1. Depth = 1.00in. Note the extrude direction is downward.

💡 You can create planes in part or assembly documents. You can use planes to sketch, to create a section view of a model, for a neutral plane in a draft feature, and so on.

💡 The created plane is displayed 5% larger than the geometry on which the plane is created, or 5% larger than the bounding box. This helps reduce selection problems when planes are created directly on faces or from orthogonal geometry.

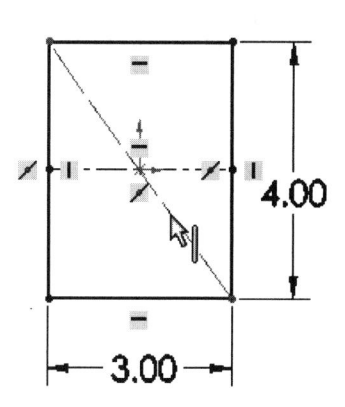

Page 3 - 53

Part Modeling

6. Create **Plane1**. Plane1 is the Sketch plane for the Extruded Boss feature. Show Sketch1. Select the horizontal construction line in Sketch1 and the top face of Extrude1. Angle = 48deg.

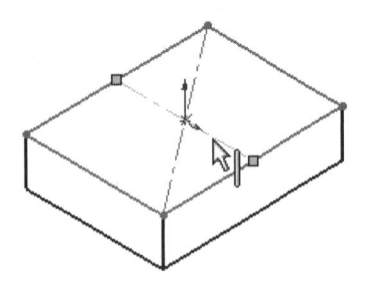

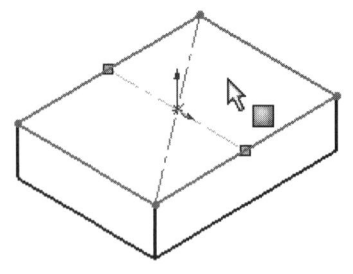

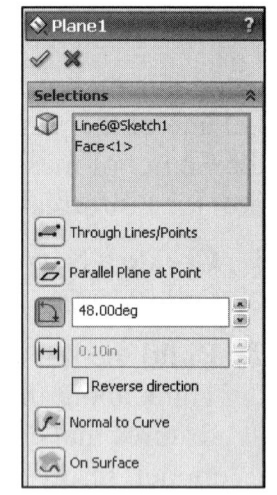

☼ Click **View**, **Sketches** from the Menu bar menu to displayed sketches in the Graphics window.

7. Create **Sketch2**. Select Plane1 as the Sketch plane. Create the Extruded Boss profile. Insert the required geometric relations and dimension. Note: Dimension to the front top edge of Extrude1 as illustrated.

8. Create the first **Extruded Boss** feature. Select the Up To Vertex End Condition in Direction 1. Select the back top right vertex point as illustrated.

9. Create **Sketch3**. Select the back angled face of Extrude2 as the Sketch plane. Sketch a circle. Insert the required geometric relations.

10. Create the second **Extruded Boss** feature. Blind is the default End Condition in Direction 1. Depth = 3.00in.

11. Create **Sketch4**. Select the front face of Extrude3 as the Sketch plane. Sketch a circle. Sketch4 is the profile for the Extruded Cut feature. Insert the required geometric relation and dimension.

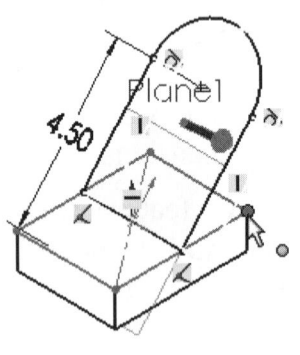

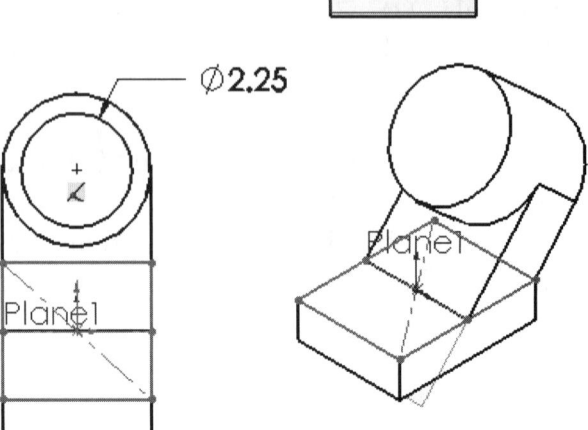

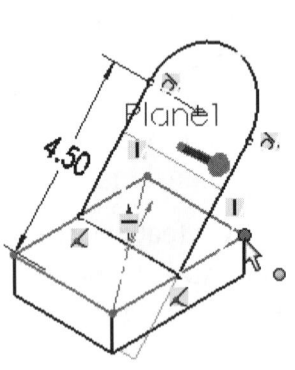

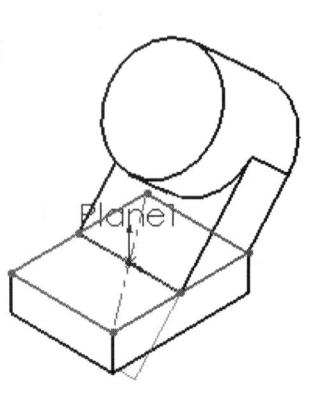

☼ The part Origin is displayed in blue.

12. Create the **Extruded Cut** feature. Select Through All for End Condition in Direction 1.

13. Create **Sketch5**. Select the Right Plane as the Sketch plane. Insert a Parallel relation to partially define Sketch5. Sketch5 is the profile for the Rib feature. Sketch5 does not need to be fully defined. Sketch5 locates the end conditions based on existing geometry.

14. Create the **Rib** feature. Thickness = 1.00in.

☼ The Rib feature is a special type of extruded feature created from open or closed sketched contours. The Rib feature adds material of a specified thickness in a specified direction between the contour and an existing part. You can create a rib feature using single or multiple sketches.

15. **Assign** 6061 Alloy material to the part.

16. **Calculate** the volume. The volume = 30.65 cubic inches.

17. **Locate** the Center of mass. The location of the Center of mass is derived from the part Origin.

- X: 0.00 inches
- Y: 0.73 inches
- Z: -0.86 inches

18. **Save** the part and name it Part-Modeling 3-3.

19. **Close** the model.

    As an exercise, modify the Rib1 feature from 1.00in to 1.25in. Modify the Extrude3 depth from 3.00in to 3.25in. Modify the material from 6061 Alloy to Copper. Modify the Plane1 angle from 48deg to 30deg. Recalculate the volume of the part. The new volume = 26.94 cubic inches.

20. **Save** the part and name it Part-Modeling 3-3-Modify.

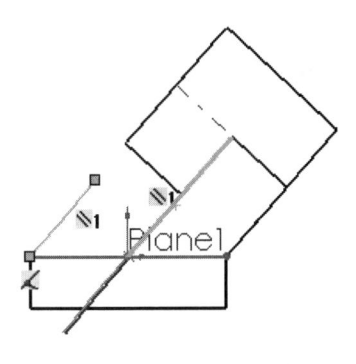

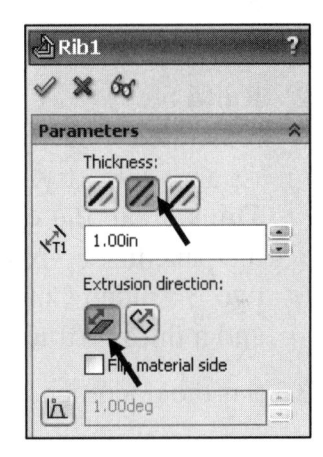

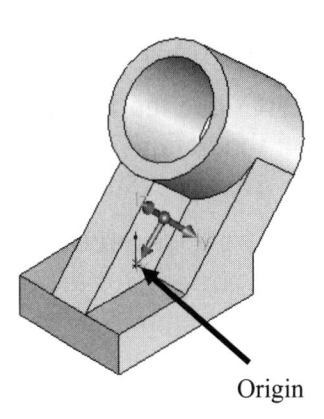

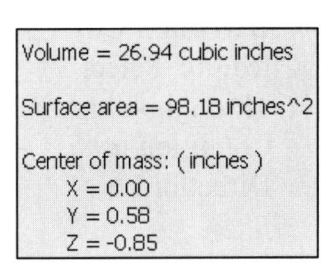

Origin

Volume = 26.94 cubic inches

Surface area = 98.18 inches^2

Center of mass: (inches)
 X = 0.00
 Y = 0.58
 Z = -0.85

Part Modeling

Tutorial: Basic-part 3-4

Build this model. Calculate the volume of the part and locate the Center of mass with the provided information.

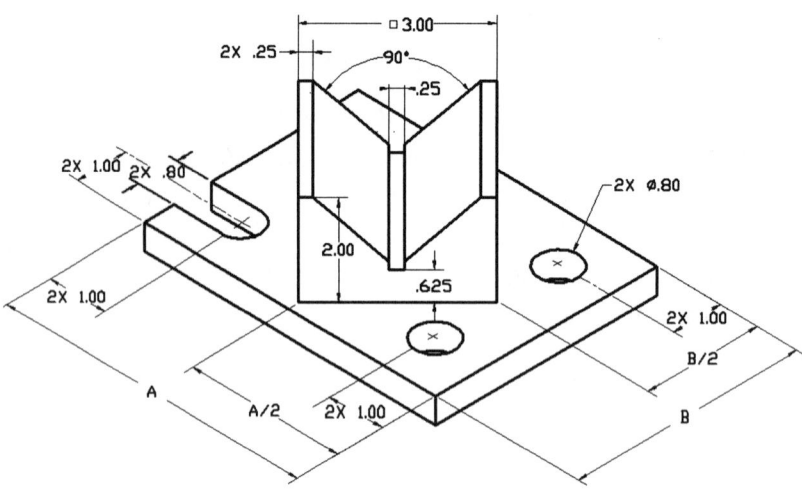

1. **Create** a New part in SolidWorks.

2. **Build** the illustrated model. Apply symmetry.
   Think about the various features that create the part. Insert six features: Extruded Base, two Extruded Cuts, Mirror, Extruded Boss, and a third Extruded Cut.

Given:
A = 6.00, B = 4.50
Material: 2014 Alloy
Plate thickness = .50
Units: IPS
Decimal places = 2

3. **Set** the document properties for the model.

4. Create **Sketch1**. Select the Top Plane as the Sketch plane. Apply symmetry. The part Origin is located in the center of the rectangle. Insert the required relations and dimensions.

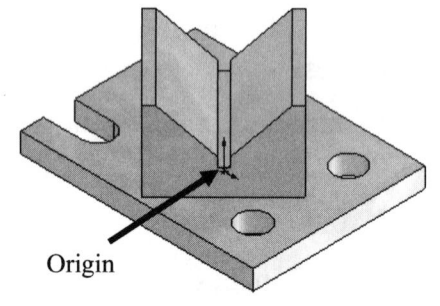

5. Create the **Extruded Base** feature. Blind is the default End Condition in Direction 1. Depth = .50in.

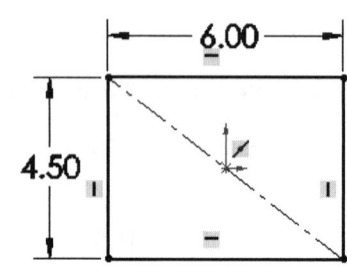

6. Create **Sketch2**. Select the top face of Extrude1 for the Sketch plane. Sketch a circle. Insert the required relations and dimensions.

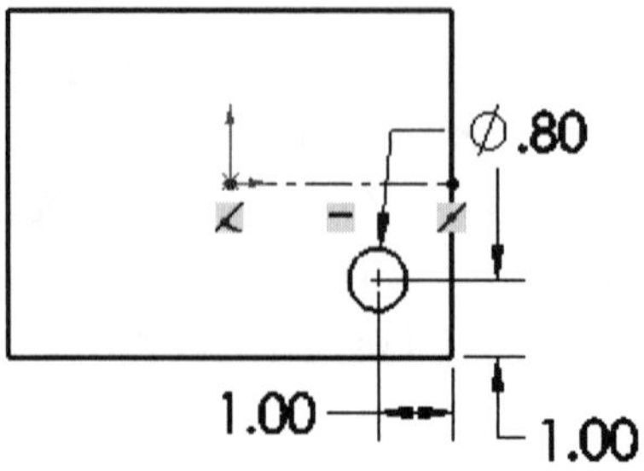

7. Create the first **Extruded Cut** feature. Select Through All as End Condition in Direction1.

Page 3 - 56

8. Create **Sketch3**. Select the top face of Extrude1 for the Sketch plane. Insert the required geometric relations and dimensions.

🔆 Click **View**, **Temporary** axes to view the part temporary axes in the Graphics window.

9. Create the second **Extruded Cut** feature. Select Through All as End Condition in Direction1.

10. Create the **Mirror** feature. Mirror the two Extruded Cut features about the Front Plane.

11. Create **Sketch4**. Select the top face of Extrude1 as the Sketch plane. Apply construction geometry to center the sketch. Insert the required relations and dimensions.

12. Create the **Extruded Boss** feature. Blind is the default End Condition in Direction 1. Depth = 2.00in.

13. Create **Sketch5**. Select the front face of Extrude4 as illustrated for the Sketch plane. Sketch5 is the profile for the third Extruded Cut feature. Apply construction geometry. Insert the required dimensions and relations.

14. Create the third **Extruded Cut** feature. Through All is selected for End Condition in Direction 1 and Direction 2.

15. **Assign** 2014 Alloy material to the part.

16. **Calculate** the volume of the part. The volume = 25.12 cubic inches.

17. **Locate** the Center of mass. The location of the Center of mass is derived from the part Origin.

- X: 0.06 inches
- Y: 0.80 inches
- Z: 0.00 inches

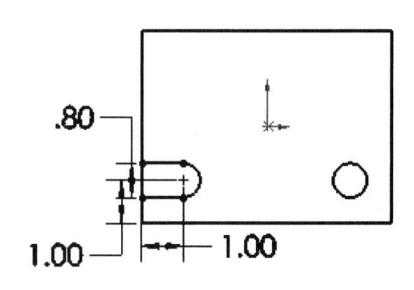

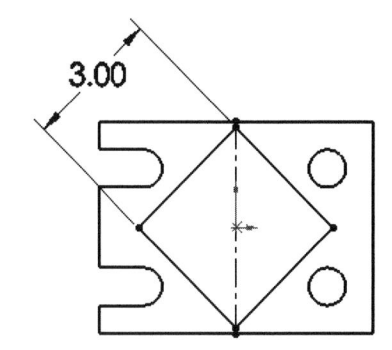

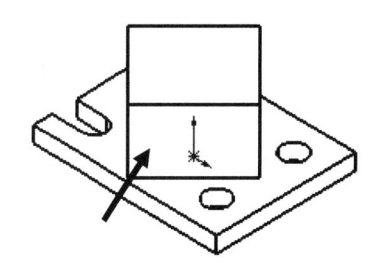

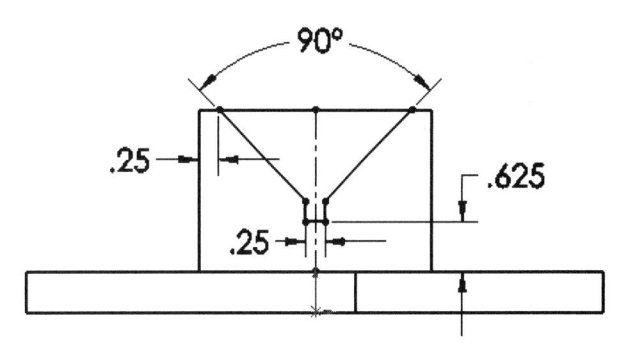

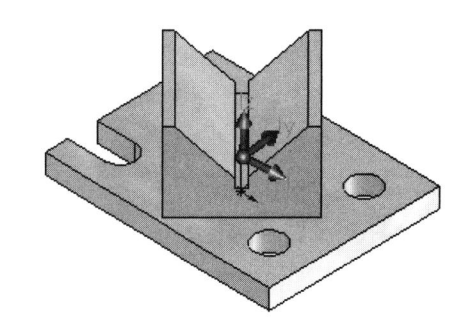

18. **Save** the part and name it Part-Modeling 3-4.
19. **Close** the model.

Summary

Part Modeling is one of the five categories on the CSWA exam. In this chapter you covered the knowledge to identify the part Origin, design intent, and key features to create a simple part from a detailed dimensioned illustration.

You are required to create a model, with six or more features and to answer a question either on the overall mass, volume, or the location of the Center of mass relative to the default part Origin. The complexity of the models along with the features increased throughout this chapter to simulate the final types of parts that would be provided on the CSWA exam. Spend no more than 40 minutes on the question in this category. This is a timed exam. Manage your time.

At this time, there are no modeling questions on the exam that requires you to use Sheet Metal, Loft, Swept, or Shell features.

Advanced Part Modeling is the next chapter in this book. There is one question on the CSWA exam in the Advanced Part Modeling category. The question is in a multiple choice single answer or fill in the blank format. The question is worth twenty points.

The main difference between the Part Modeling category and the Advanced Part modeling category is the complexity of the sketches and the number of dimensions and geometric relations along with an increase in the number of features.

Key terms

- *Annotation*. A text note or a symbol that adds specific design intent to a part, assembly, or drawing.
- *Base feature*. The first feature of a part is called the Base feature.
- *Base sketch*. The first sketch of a part is called the Base sketch. The Base sketch is the foundation for the 3D model. Create a 2D sketch on a default plane: Front, Top, and Right in the FeatureManager design tree, or on a created plane.
- *Constraints*. Geometric relations such as Perpendicular, Horizontal, Parallel, Vertical, Coincident, Concentric, etc. Insert constraints to your model to incorporate design intent.

- *Construction geometry.* The characteristic of a sketch entity that the entity is used in creating other geometry, but is not itself used in creating features. Construction geometry is also called reference geometry.

- *Coordinate system.* A system of planes used to assign Cartesian coordinates to features, parts, and assemblies. Part and assembly documents contain default coordinate systems; other coordinate systems can be defined with reference geometry. Coordinate systems can be used with measurement tools and for exporting documents to other file formats.

- *Features.* Individual shapes created by Sketch Entities tools: lines, circles, rectangles, etc. that when combined, creates the part. Features are geometry building blocks, they add or remove material, and are created from 2D or 3D sketched profiles or from edges and faces of existing geometry.

- *Fillet.* An internal rounding of a corner or edge in a sketch, or an edge on a surface or solid.

- *Fit tolerance.* The tolerance between a hole and a shaft.

- *Fly-out FeatureManager design tree.* Allows you to view and select items in the PropertyManager and the FeatureManager design tree at the same time.

- *Geometric tolerance.* A set of standard symbols that specify the geometric characteristics and dimensional requirements of a feature.

- *Mass Properties tool.* Displays the mass properties of a part or assembly model, or the section properties of faces or sketches.

- *Mirror Feature.* Creates a copy of a feature, (or multiple features), mirrored about a selected face or a plane. You can select the feature or you can select the faces that comprise the feature.

- *Pattern.* A pattern repeats selected sketch entities, features, or components in an array, which can be linear, circular, or sketch-driven. If the seed entity is changed, the other instances in the pattern update.

- *Reference Axis.* A reference axis is also called a construction axis. Reference geometry defines the shape or form of a surface or a solid. Reference geometry includes planes, axes, coordinate systems, and points.

- *Reference Plane.* Insert a plane as a reference to apply restraints. Reference geometry defines the shape or form of a surface or a solid. Reference geometry includes planes, axes, coordinate systems, and points.

- *Revolved Feature.* Revolves add or remove material by revolving one or more profiles about a centerline. You can create Revolved Boss/Base, Revolved Cuts, or Revolved Surfaces. The Revolved feature can be a solid, a thin feature, or a surface.

Part Modeling

- *Seed.* A sketch or an entity that is the basis for a pattern. If you edit the seed, the other entities in the pattern are updated.
- *Vertex.* A point at which two or more lines or edges intersect. Vertices can be selected for sketching, dimensioning, and many other operations.

Check your understanding

1: In Tutorial: Volume / Center of mass 3-2 you built the model using the FeatureManager that had three features vs. four features in the FeatureManager.

Calculate the overall mass of the part, volume, and locate the Center of mass with the provided information using the Option1 FeatureManager.

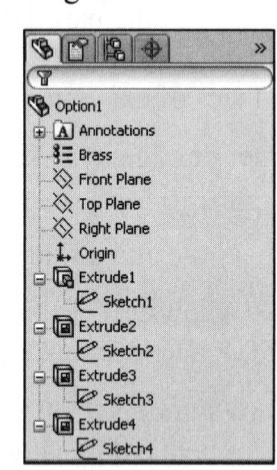

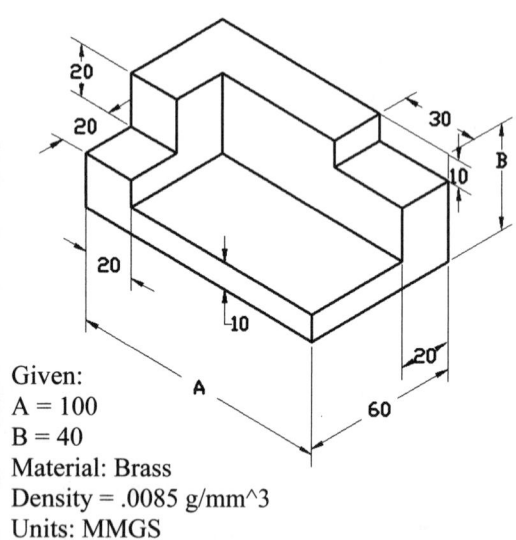

Given:
A = 100
B = 40
Material: Brass
Density = .0085 g/mm^3
Units: MMGS

2. In Tutorial: Mass / Volume 3-4 you built the model using the FeatureManager that had three features vs. four features in the FeatureManager.

Calculate the overall mass of the part, volume, and locate the Center of mass with the provided information using the Option3 FeatureManager.

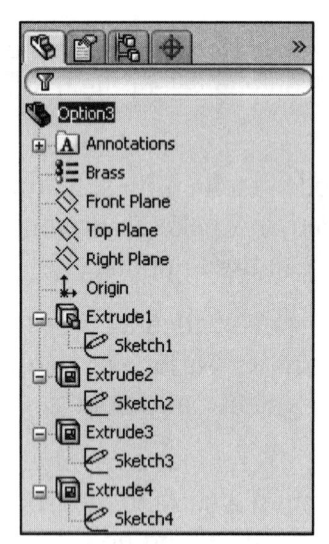

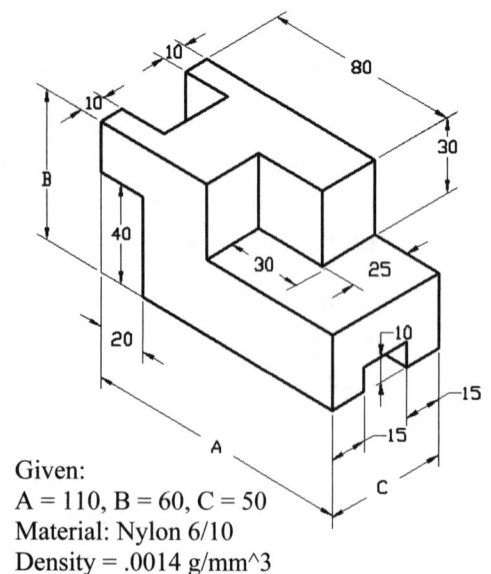

Given:
A = 110, B = 60, C = 50
Material: Nylon 6/10
Density = .0014 g/mm^3
Units: MMGS

Part Modeling

3. In Tutorial: Base Part 3-4 you built the illustrated model. Modify the plate thickness from .50in to .25in. Modify the Sketch5 angle from 90deg to 75deg. Re-assign the material from 2014 Alloy to 6061 Alloy.

Calculate the overall mass of the part, volume, and locate the Center of mass with the provided information.

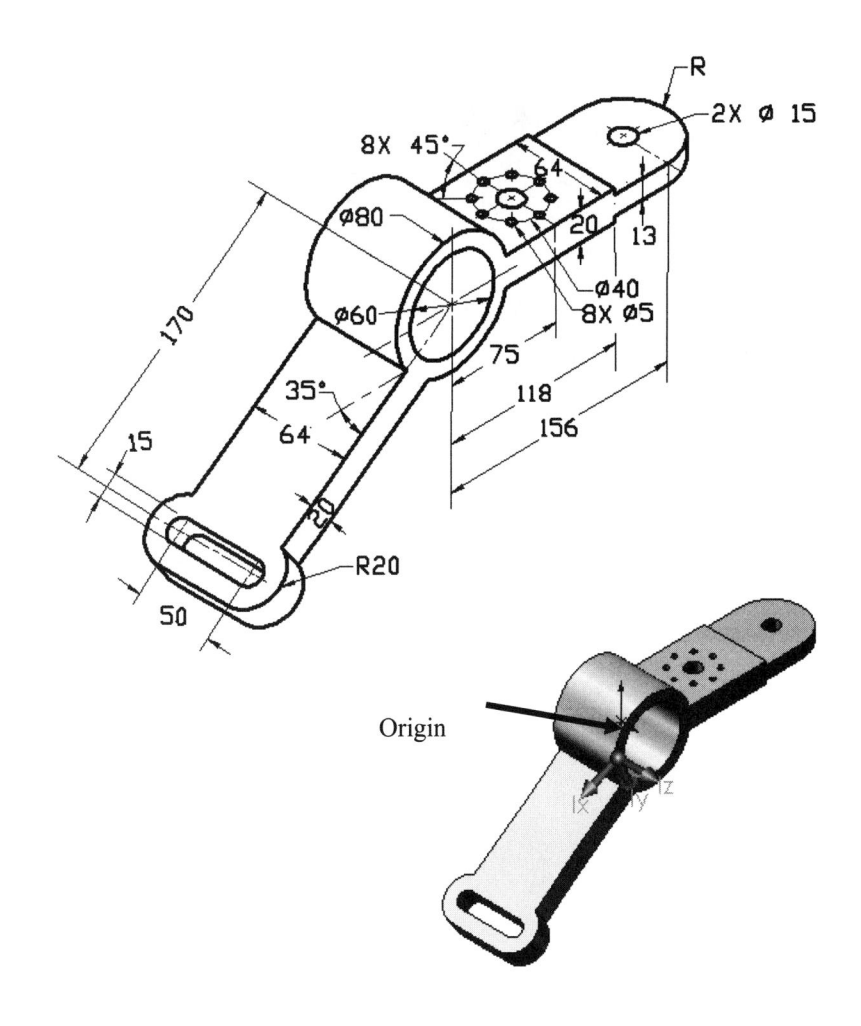

Given:
A = 6.00, B = 4.50
Material: 2014 Alloy
Plate thickness = .50
Units: IPS
Decimal places = 2

4. Build this model: Set document properties, identify the correct Sketch planes, apply the correct Sketch and Feature tools, and apply material.

Calculate the overall mass of the part, volume, and locate the Center of mass with the provided illustrated information.

- Material: 6061 Alloy
- Units: MMGS

Origin

Page 3 - 61

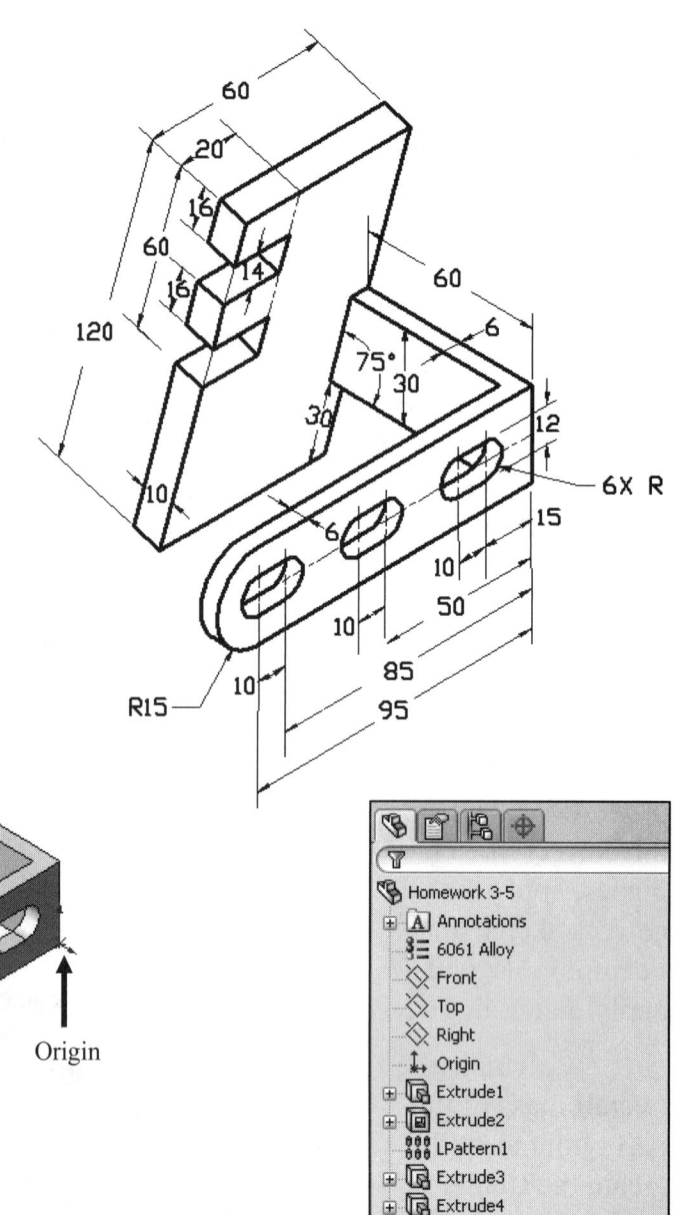

Part Modeling

5. Build this model. Set document properties and identify the correct Sketch planes. Apply the correct Sketch and Feature tools, and apply material.

   Calculate the overall mass of the part, volume, and locate the Center of mass with the provided information.

   - Material: 6061 Alloy
   - Units: MMGS

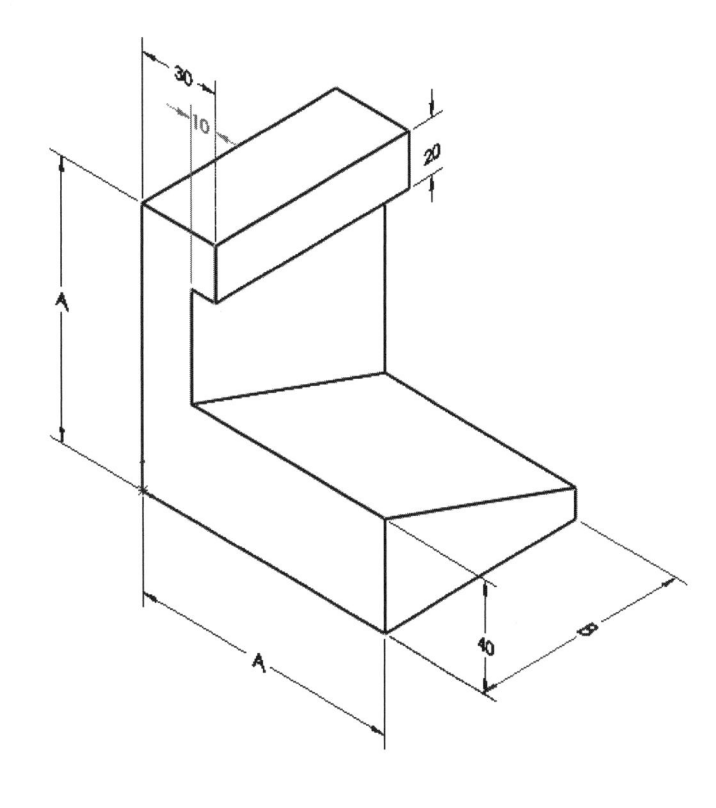

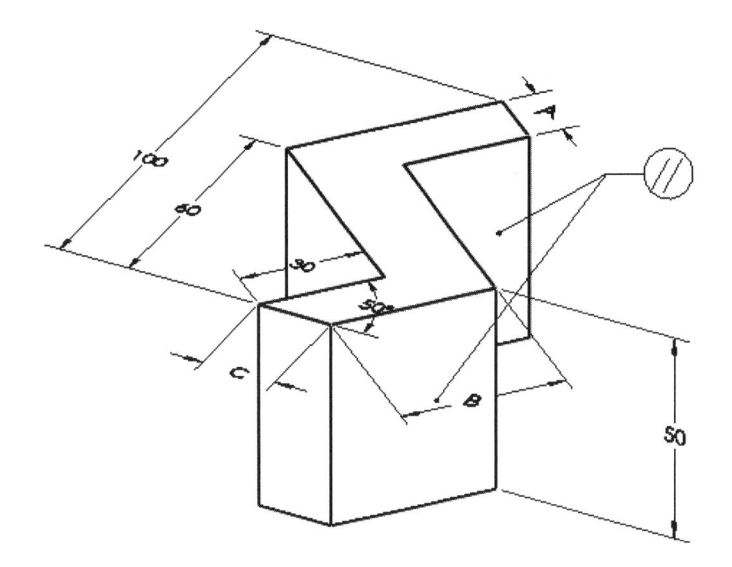

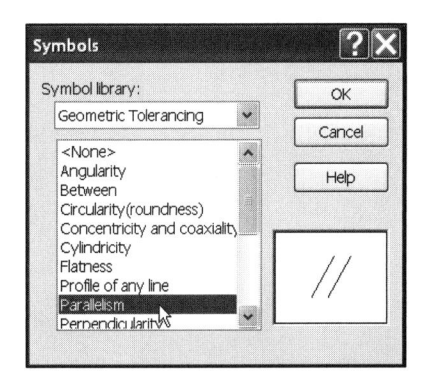

Part Modeling

6. Build this model. Set document properties and identify the correct Sketch planes. Apply the correct Sketch and Feature tools, and apply material.

Calculate the overall mass of the part with the provided information. Note: The Origin is arbitrary.

- Material: Copper
- Units: MMGS
- A = 100
- B = 80

7. Build this model. Set document properties and identify the correct Sketch planes. Apply the correct Sketch and Feature tools, and apply material.

Calculate the overall mass of the part with the provided information. The location of the Origin is arbitrary.

- Material: 6061
- Units: MMGS
- A = 16
- B = 40
- Side A is perpendicular to side B
- C = 16

Notes:

Chapter 4: Advanced Part Modeling

Chapter Objective

Advanced Part Modeling is one of the five categories on the CSWA exam. The main difference between the Advanced Part modeling and the Part Modeling category is the complexity of the sketches and the number of dimensions and geometric relations along with an increase number of features.

There is one question on the CSWA exam in this category. The question is worth twenty points and is in a multiple choice single answer or fill in the blank format. The question is either on the location of the Center of mass relative to the default part Origin or to a new created coordinate system and all of the mass properties located in the Mass Properties dialog box: total overall mass, volume, etc.

There are numerous ways to create the models in this chapter. The goal is to display different design intents and techniques

On the completion of the chapter, you will be able to:

- Specify Document Properties
- Interpret engineering terminology:
  - Create and manipulate a model coordinate system
- Build an advanced part from a detailed dimensioned illustration using the following tools and features:
  - 2D & 3D Sketch tools
  - Extruded Boss/Base
  - Extruded Cut
  - Fillet
  - Mirror
  - Revolved Boss/Base
  - Linear & Circular Pattern
  - Chamfer
  - Revolved Cut

Advanced Part Modeling

- Locate the Center of mass relative to the default part Origin
- Create a new coordinate system location for a part
- Locate the Center of mass relative to the default or created coordinate system

Build an Advanced part from a detailed dimensioned illustration

Tutorial: Advanced Part 4-1

An exam question in this category could read: Build this part. Calculate the overall mass and locate the Center of mass of the illustrated model.

1. **Create** a New part in SolidWorks.
2. **Build** the illustrated model. Insert seven features: Extruded Base, two Extruded Bosses, two Extruded Cuts, Chamfer and a Fillet.

Think about the steps that you would take to build the illustrated part. Identify the location of the part Origin. Start with the back base flange. Review the provided dimensions and annotations in the part illustration.

Given:
A = 2.00, B = Ø.35
Material: 1060 Alloy
Density: 0.097 lb/in^3
Units: IPS
Decimal places = 2

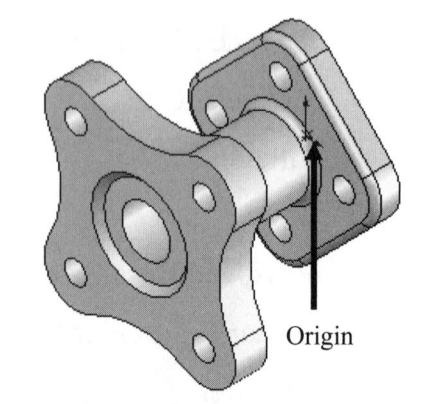

☼ The key difference between the Advanced Part Modeling and the Part Modeling category is the complexity of the sketches and the number of features, dimensions, and geometric relations.

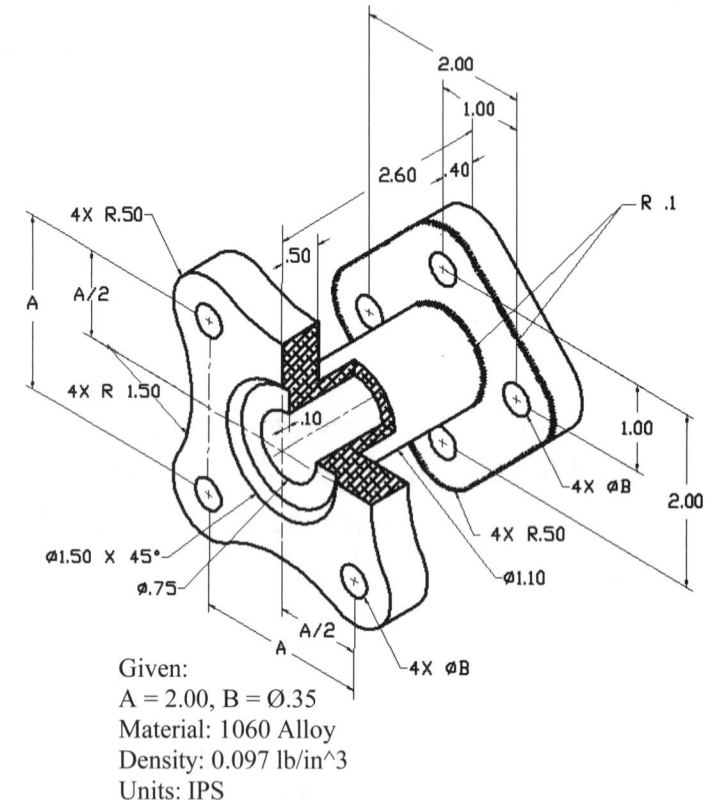

3. **Set** the document properties for the model.

4. Create **Sketch1**. Sketch1 is the Base sketch. Select the Front Plane as the Sketch plane. Apply construction geometry. Sketch a horizontal and vertical centerline. Sketch four circles. Insert an Equal relation. Insert a Symmetric relation about the vertical and horizontal centerlines. Sketch two top angled lines and a tangent arc. Apply the Mirror Sketch tool. Complete the sketch. Insert the required geometric relations and dimensions.

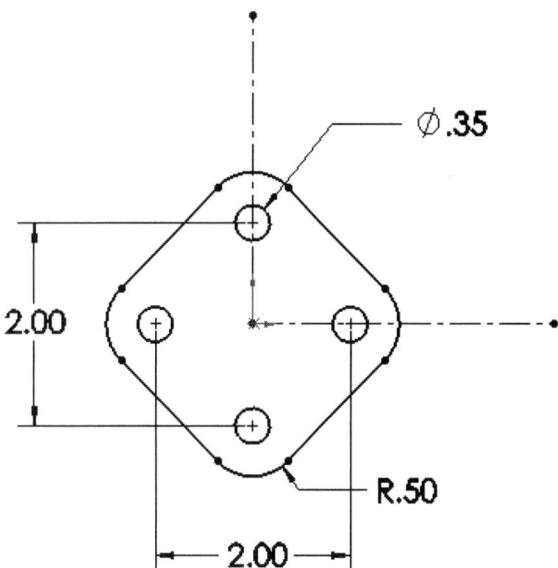

☼ In a Symmetric relation, the selected items remain equidistant from the centerline, on a line perpendicular to the centerline. Sketch entities to select: a centerline and two points, lines, arcs, or ellipses.

☼ The Sketch Fillet tool rounds the selected corner at the intersection of two sketch entities, creating a tangent arc.

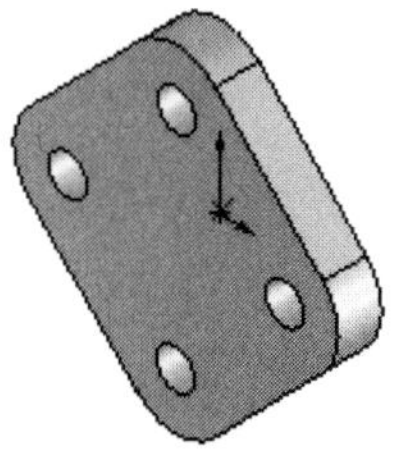

5. Create the **Extruded Base** feature. Extrude1 is the Base feature. Blind is the default End Condition in Direction 1. Depth = .40in.

6. Create **Sketch2**. Select the front face of Extrude1 as the Sketch plane. Sketch a circle. Insert the required geometric relation and dimension.

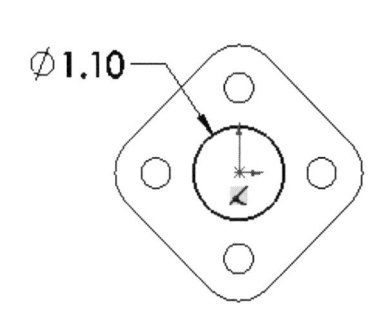

7. Create the first **Extruded Boss** feature. Blind is the default End Condition in Direction 1. The Extrude2 feature is the tube between the two flanges. Depth = 1.70in. Note: 1.70in = 2.60in - (.50in + .40in).

☼ The complexity of the models along with the features progressively increases throughout this chapter to simulate the final types of parts that could be provided on the CSWA exam.

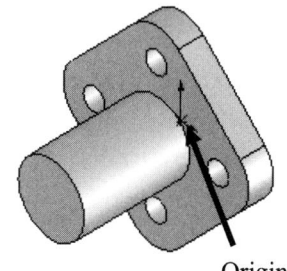

Origin

Advanced Part Modeling

8. Create **Sketch3**. Select the front circular face of Extrude2 as the Sketch plane. Sketch a horizontal and vertical centerline. Sketch the top two circles. Insert an Equal and Symmetric relation between the two circles. Mirror the top two circles about the horizontal centerline. Insert dimensions to locate the circles from the Origin. Apply either the 3 Point Arc or the Centerpoint Arc Sketch tool. The centerpoint of the Tangent Arc is aligned with a Vertical relation to the Origin. Complete the sketch.

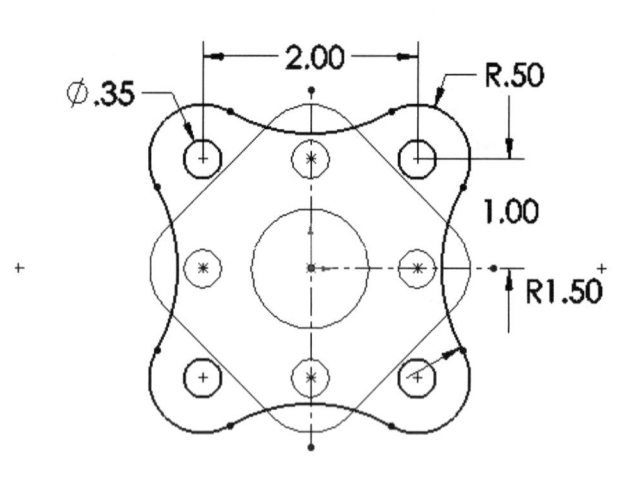

🔅 There are numerous ways to create the models in this chapter. The goal is to display different design intents and techniques

🔅 Use the Centerpoint Arc Sketch tool to create an arc from a: centerpoint, a start point, and an end point.

🔅 Apply the Tangent Arc Sketch tool to create an arc, tangent to a sketch entity.

🔅 The Arc PropertyManager controls the properties of a sketched Centerpoint Arc, Tangent Arc, and 3 Point Arc.

9. Create the second **Extruded Boss** feature. Blind is the default End Condition in Direction 1. Depth = .50in.

10. Create **Sketch4**. Select the front face of Extrude3 as the Sketch plane. Sketch a circle. Insert the required geometric relation and dimension.

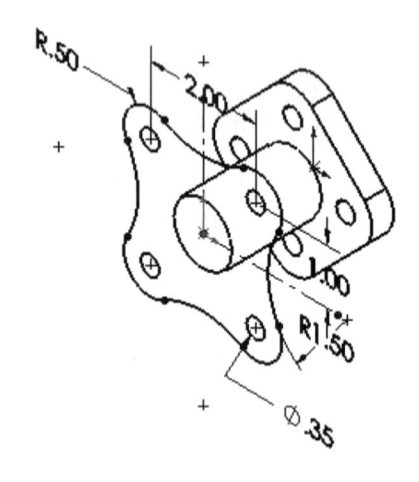

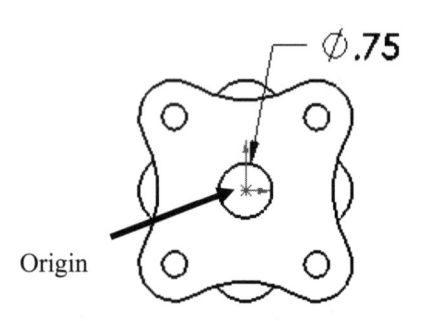

Advanced Part Modeling

11. Create the first **Extruded Cut** feature. Select the Through All End Condition for Direction 1.

12. Create **Sketch5**. Select the front face of Extrude3 as the Sketch plane. Sketch a circle. Insert the required geometric relation and dimension.

13. Create the second **Extruded Cut** feature. Blind is the default End Condition for Direction1. Depth = .10in.

14. Create the **Chamfer** feature. In order to have the outside circle 1.50in, select the inside edge of the sketched circle. Create an Angle distance chamfer. Distance = .10in. Angle = 45deg.

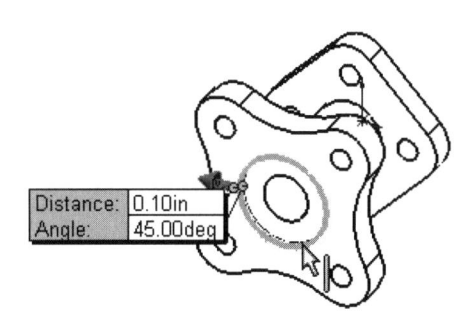

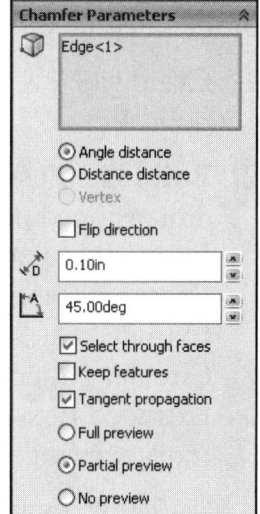

☼ The Chamfer feature creates a beveled feature on selected edges, faces, or a vertex.

15. Create the **Fillet** feature. Fillet the two edges as illustrated. Radius = .10in.

16. **Assign** 1060 Alloy material to the part. Material is required to calculate the overall mass of the part.

17. **Calculate** the overall mass. The overall mass = 0.59 pounds.

18. **Locate** the Center of mass. The location of the Center of mass is relative to the part Origin.

- X: 0.00
- Y: 0.00
- Z: 1.51

19. **Save** the part and name it Advanced Part 4-1.

20. **Close** the model.

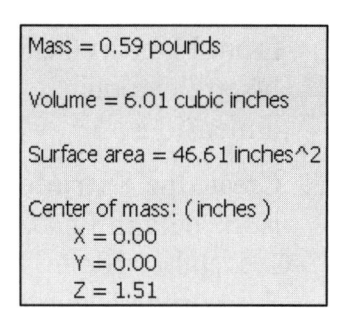

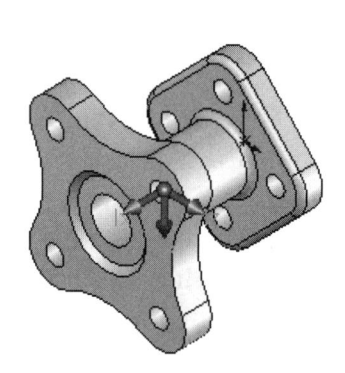

Advanced Part Modeling

Tutorial: Advanced Part 4-2

An exam question in this category could read: Build this part. Calculate the overall mass and locate the Center of mass of the illustrated model.

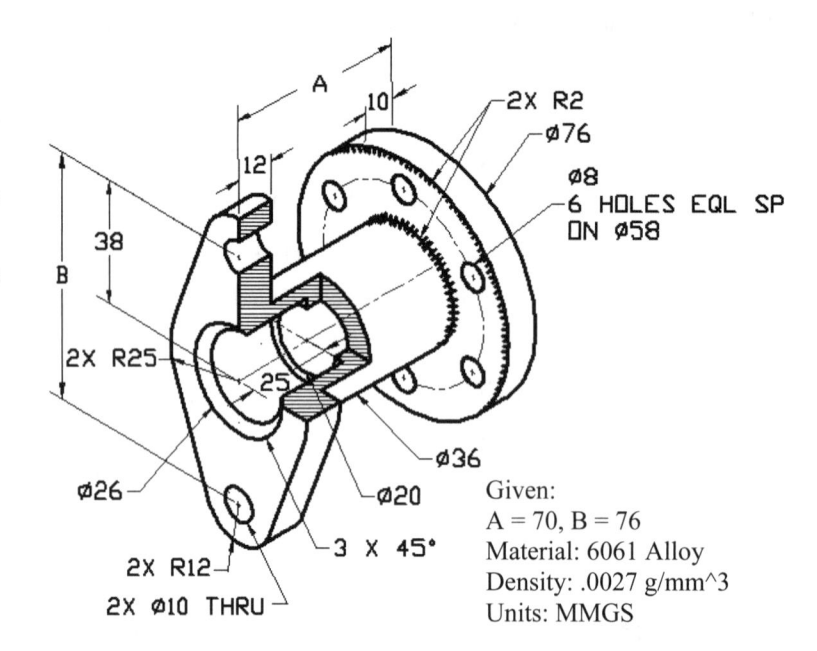

1. **Create** a New part in SolidWorks.

2. **Build** the illustrated dimensioned model. Insert eight features:
   Extruded Base,
   Extruded Cut,
   Circular Pattern, two
   Extruded Bosses,
   Extruded Cut, Chamfer, and Fillet.

Given:
A = 70, B = 76
Material: 6061 Alloy
Density: .0027 g/mm^3
Units: MMGS

   Think about the steps that you would take to build the illustrated part. Review the provided information. Start with the six hole flange.

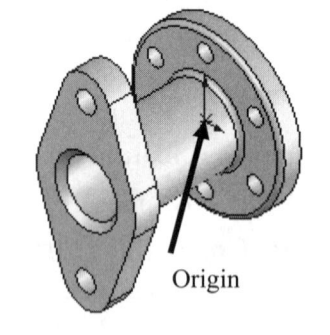

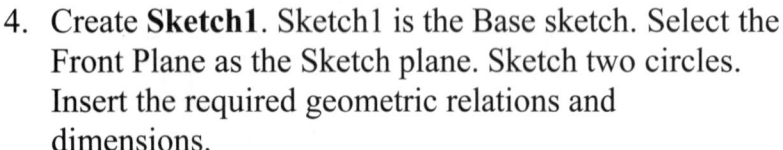

3. **Set** the document properties for the model.

4. Create **Sketch1**. Sketch1 is the Base sketch. Select the Front Plane as the Sketch plane. Sketch two circles. Insert the required geometric relations and dimensions.

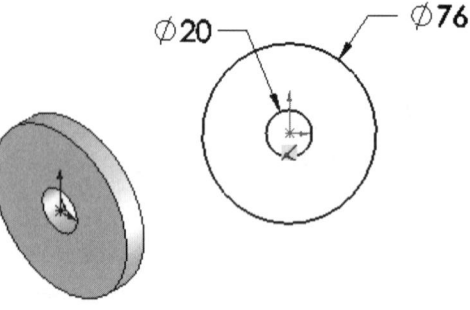

5. Create the **Extruded Base** feature. Blind is the default End Condition in Direction 1. Depth = 10mm. Note the direction of the extrude feature to maintain the Origin location.

6. Create **Sketch2**. Select the front face of Extrude1 as the Sketch plane. Sketch2 is the profile for first Extruded Cut feature. The Extruded Cut feature is the seed feature for the Circular Pattern. Apply construction reference geometry. Insert the required geometric relations and dimensions.

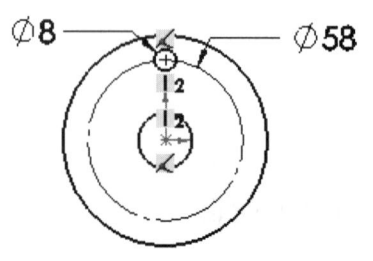

Advanced Part Modeling

7. Create the **Extruded Cut** feature. Extrude2 is the first bolt hole. Select Through All for End Condition in Direction 1.

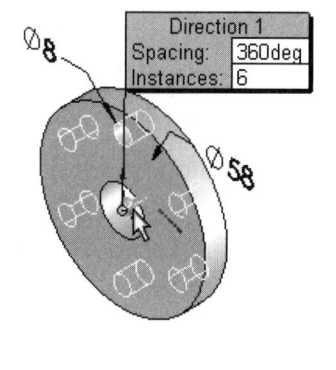

8. Create the **Circular Pattern** feature. Default Angle = 360deg. Number of instances = 6. Select the center axis for the Pattern Axis box.

☼ The Circular Pattern PropertyManager is displayed when you pattern one or more features about an axis.

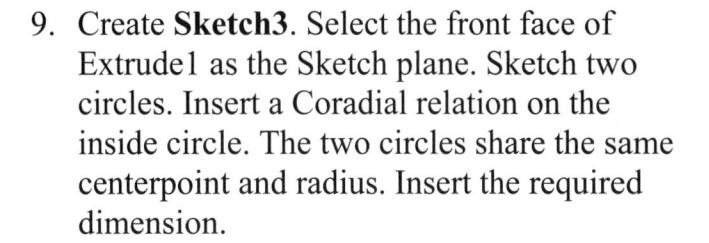

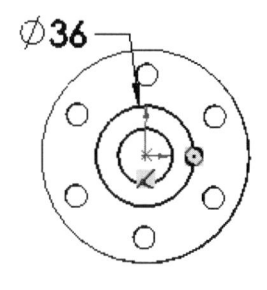

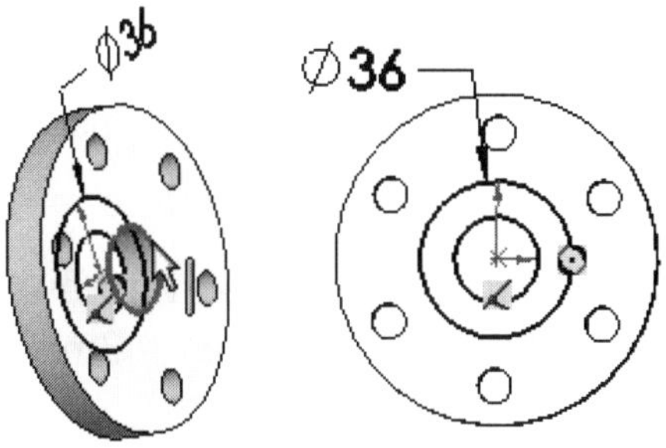

9. Create **Sketch3**. Select the front face of Extrude1 as the Sketch plane. Sketch two circles. Insert a Coradial relation on the inside circle. The two circles share the same centerpoint and radius. Insert the required dimension.

10. Create the first **Extruded Boss** feature. The Extrude3 feature is the connecting tube between the two flanges. Blind is the default End Condition in Direction 1. Depth = 48mm.

11. Create **Sketch4**. Select the front circular face of Extrude3 as the Sketch plane. Sketch a horizontal and vertical centerline from the Origin. Sketch the top and bottom circles symmetric about the horizontal centerline. Dimension the distance between the two circles and their diameter. Create the top centerpoint arc with the centerpoint Coincident to the top circle. The start point and the end point of the arc are horizontal. Sketch the two top angled lines symmetric about the vertical centerline. Apply symmetry. Mirror the two lines and the centerpoint arc about the horizontal centerline. Insert the left and right tangent arcs with a centerpoint Coincident with the Origin. Complete the sketch.

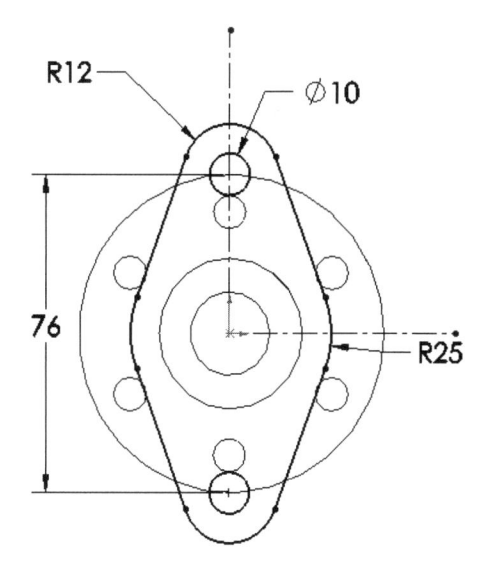

12. Create the second **Extruded Boss** feature. Blind is the default End Condition in Direction 1. Depth = 12mm.

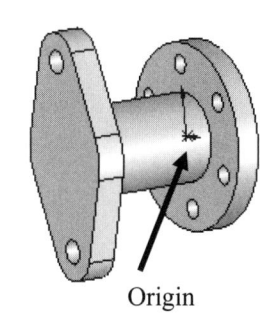

Origin

Page 4 - 7

Advanced Part Modeling

13. Create **Sketch5**. Select the front face of Extrude4 as the Sketch plane. Sketch a circle. The part Origin is located in the center of the model. Insert the required dimension.

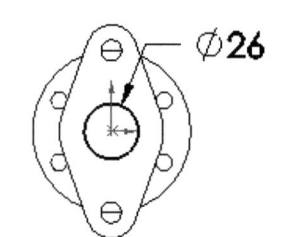

14. Create the second **Extruded Cut** feature. Blind is the default End Condition in Direction 1. Depth = 25mm.

15. Create the **Chamfer** feature. Chamfer1 is an Angle distance chamfer. Chamfer the inside edge of Extrude4 as illustrated. Distance = 3mm. Angle = 45deg.

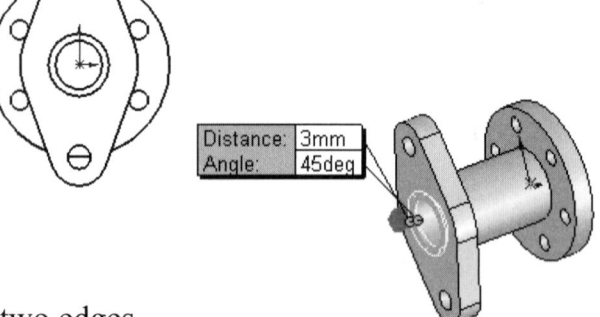

16. Create the **Fillet** feature. Fillet the two edges of Extrude1. Radius = 2mm.

17. **Assign** 6061 Alloy material to the part.

18. **Calculate** the overall mass of the part. The overall mass = 276.97 grams.

19. **Locate** the Center of mass. The location of the Center of mass is relative to the part Origin.

- X: 0.00 millimeters
- Y: 0.00 millimeters
- Z: 21.95 millimeters

20. **Save** the part and name it Advanced Part 4-2.

21. **Close** the model.

In the Advanced Part Modeling category, an exam question could read: Build this model. Locate the Center of mass with respect to the part Origin.

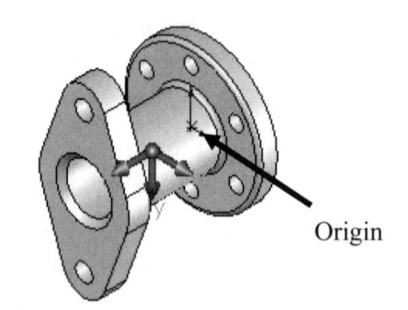

- A: X = 0.00 millimeters, Y = 0.00 millimeters, Z = 21.95 millimeters
- B: X = 21.95 millimeters, Y = 10.00 millimeters, Z = 0.00 millimeters
- C: X = 0.00 millimeters, Y = 0.00 millimeters, Z = -27.02 millimeters
- D: X = 1.00 millimeters, Y = -1.01 millimeters, Z = -0.04 millimeters

The correct answer is A.

Tutorial: Advanced Part 4-3

An exam question in this category could read: Build this part. Calculate the volume and locate the Center of mass of the illustrated model.

1. **Create** a New part in SolidWorks.

2. **Build** the illustrated dimensioned model. Insert five sketches, five features and a Reference plane: Extruded Base, Plane1, Extruded Boss, Extruded Cut, Fillet, and Extruded Cut.

   Think about the steps that you would take to build the illustrated part. Insert a Reference plane to create the Extruded Boss feature. Create Sketch2 for Plane1. Plane1 is the Sketch plane for Sketch3. Sketch3 is the profile for Extrude2.

Given:
A = .700, B = 4.000,
C = 2.700, D = .900
Material: 6061 Alloy
Density: 0.097 lb/in^3
Units: IPS
Decimal places = 3

3. **Set** the document properties for the model.

4. Create **Sketch1**. Sketch1 is the Base sketch. Select the Top Plane as the Sketch plane. Sketch a rectangle. Insert the required geometric relations and dimensions.

5. Create the **Extruded Base** feature. Blind is the default End Condition in Direction 1. Depth = .700in.

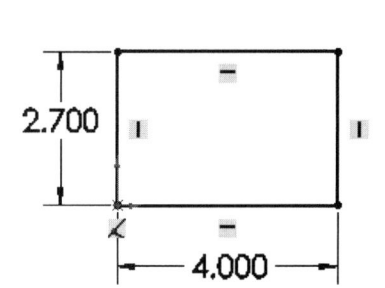

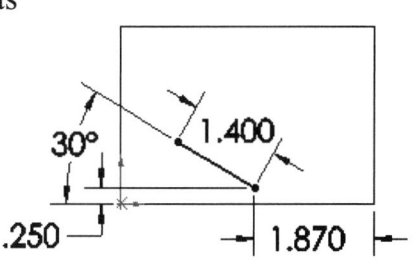

6. Create **Sketch2**. Select the top face of Extrude1 as the Sketch plane. Sketch a diagonal line as illustrated. Plane1 is the Sketch plane for Sketch3. Sketch3 is the sketch profile for Extrude2. The Origin is located in the bottom left corner of the sketch. Complete the sketch.

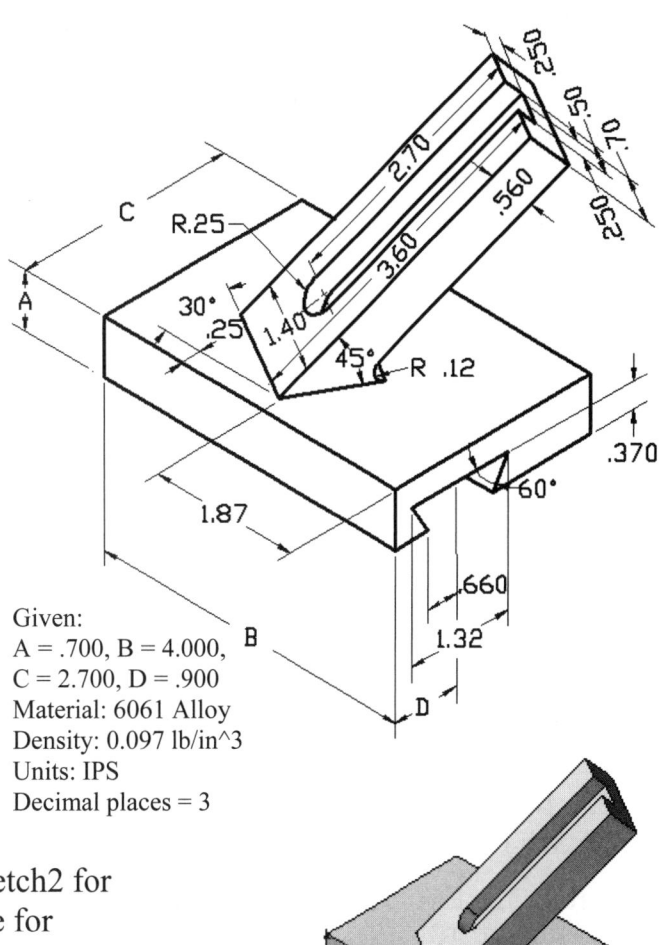

Page 4 - 9

Advanced Part Modeling

7. Create **Plane1**. Show Sketch2. Select the top face of Extrude1 and Sketch2. Sketch2 and face<1> are the Reference Entities. Angle = 45 deg.

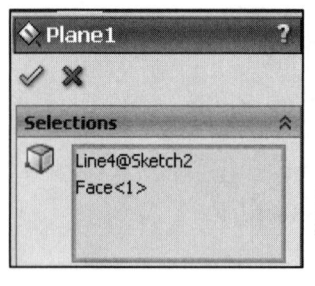

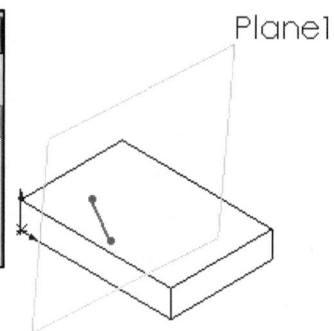

💡 Activate the Plane PropertyManager. Click **Plane** from the Reference Geometry Consolidated toolbar, or click **Insert**, **Reference Geometry**, **Plane** from the Menu bar.

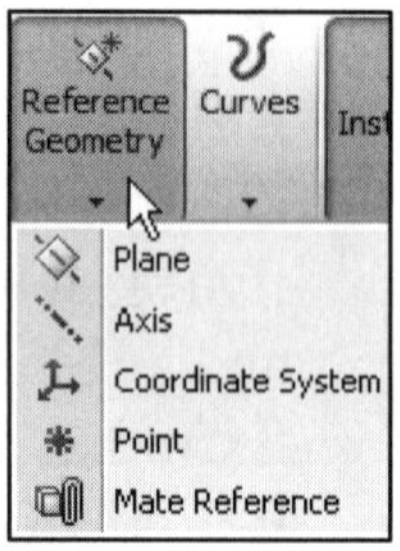

💡 View Sketch2. Click **View**, **Sketches** from the Menu bar.

💡 View Plane1. Click **View**, **Planes** from the Menu bar.

8. Create **Sketch3**. Select Plane1 as the Sketch plane. Select the Line Sketch tool. Use Sketch2 as a reference for the width dimension of the rectangle. Insert the required geometric relations and dimension. Sketch3 is the sketch profile for Extrude2.

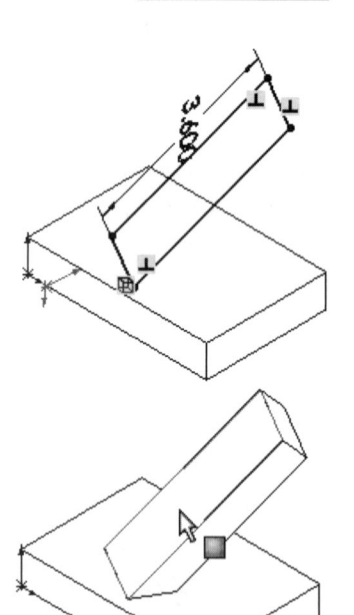

9. Create the **Extruded Boss** feature. Extrude2 is located on Plane1. Blind is the default End Condition in Direction 1. Depth = .560in.

10. Create **Sketch4**. Select the top angle face of Extrude2 as the Sketch plane. Sketch4 is the profile for the first Extruded Cut feature. Apply a Mid point relation with the Centerline Sketch tool. Insert a Parallel, Symmetric, Perpendicular, and Tangent relation. Insert the required dimensions.

11. Create the first **Extruded Cut** feature. Blind is the default End Condition.
Depth = .250in.

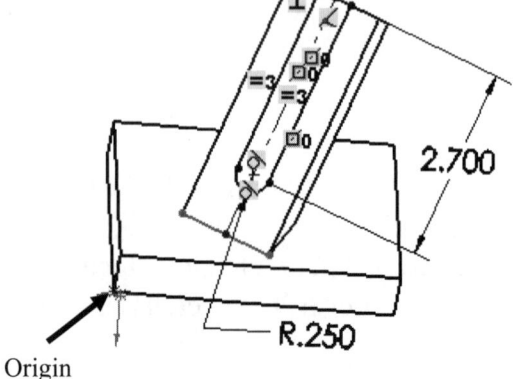

Origin

Page 4 - 10

12. Create the **Fillet** feature. Fillet the illustrated edge. Edge<1> is displayed in the Items To Fillet box. Radius = .12in.

13. Create **Sketch5**. Select the right face of Extrude1 as the Sketch plane. Insert the required relations and dimensions.

14. Create the second **Extruded Cut** feature. Select Through All as the End Condition in Direction 1.

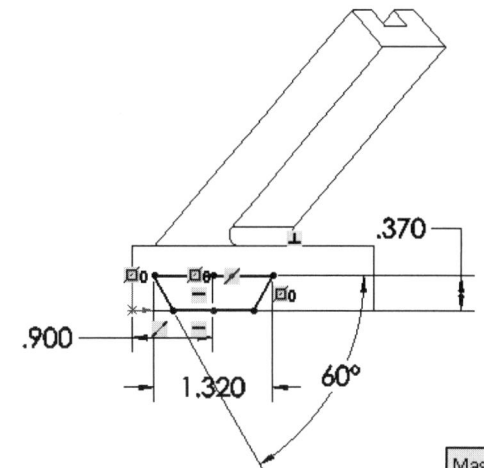

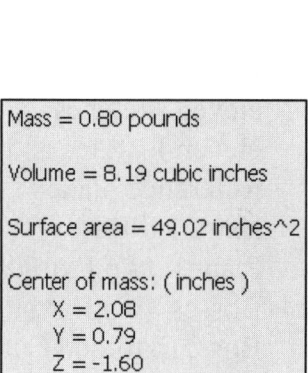

15. **Assign** 6061 Alloy material to the part.

16. **Calculate** the volume of the part. The volume = 8.19 cubic inches.

17. **Locate** the Center of mass. The location of the Center of mass is relative to the part Origin.

- X: 2.08 inches
- Y: 0.79 inches
- Z: -1.60 inches

18. **Save** the part and name it Advanced Part 4-3.

19. **Close** the model.

As an exercise, apply the MMGS unit system to the part. Modify the material from 6061 Alloy to ABS. Modify the Plane1 angle from 45deg to 30deg. Calculate the total mass of the part and the location of the Center of mass relative to the part Origin. Save the part and name it Advanced Part 4-3 MMGS System.

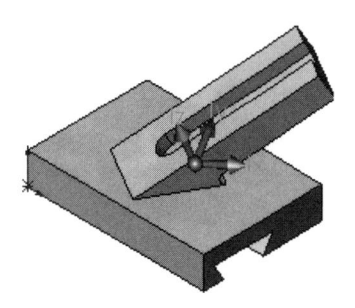

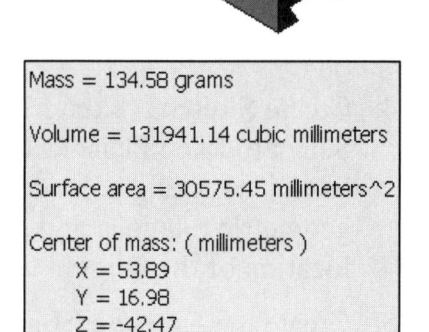

Advanced Part Modeling

Tutorial: Advanced Part 4-4

An exam question in this category could read: Build this part. Calculate the volume and locate the Center of mass of the illustrated model.

1. **Create** a New part in SolidWorks.

2. **Build** the illustrated dimensioned model. Create the part with eleven sketches, eleven features and a Reference plane: Extruded Base, Plane1, two Extruded Bosses, two Extruded Cuts, Extruded Boss, Extruded Cut, Extruded-Thin, Mirror, Extruded Cut, and Extruded Boss.

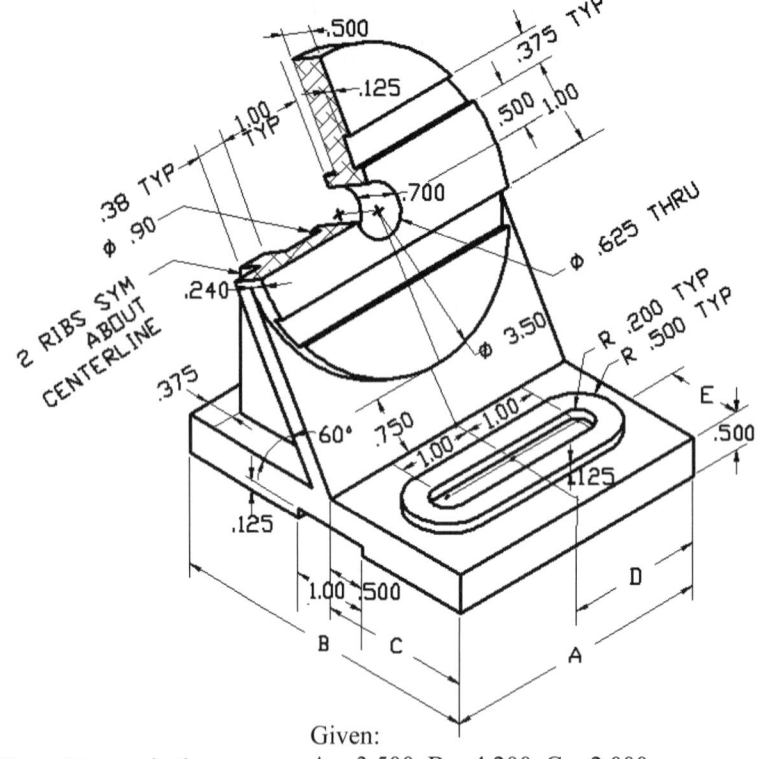

   Think about the steps that you would take to build the illustrated part. Create the rectangular Base feature. Create Sketch2 for Plane1. Insert Plane1 to create the Extruded Boss feature: Extrude2. Plane1 is the Sketch plane for Sketch3. Sketch3 is the sketch profile for Extrude2.

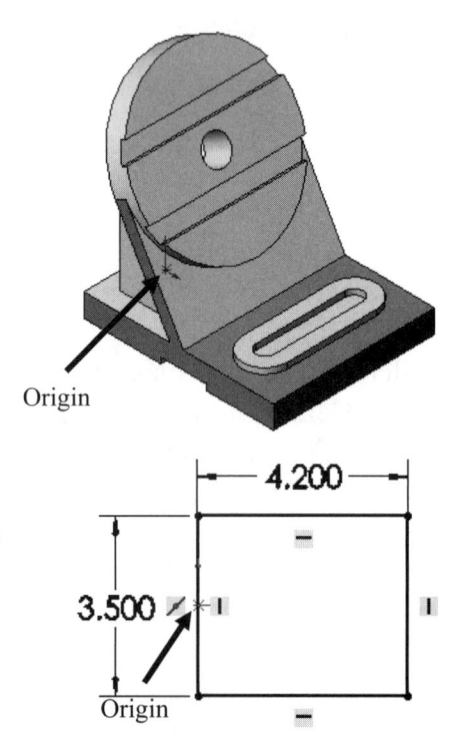

Given:
A = 3.500, B = 4.200, C = 2.000,
D = 1.750, E = 1.000
Material: 6061 Alloy
Density: 0.097 lb/in^3
Units: IPS
Decimal places = 3

3. **Set** the document properties for the model.

4. Create **Sketch1**. Sketch1 is the Base sketch. Select the Top Plane as the Sketch plane. Sketch a rectangle. Insert the required geometric relations and dimensions. Note the location of the Origin.

5. Create the **Extruded Base** feature. Blind is the default End Condition in Direction 1. Depth = .500in.

Advanced Part Modeling

6. Create **Sketch2**. Sketch2 is the sketch profile for Plane1. Select the top face of Extrude1 as the Sketch plane. Sketch a centerline. Show Sketch2.

7. Create **Plane1**. Select the top face of Extrude1 and Sketch2. Face<1> and Line1@Sketch2 are displayed in the Selections box. Angle = 60deg.

8. Create **Sketch3**. Select Plane1 as the Sketch plane. Sketch3 is the sketch profile for the Extruded Boss feature. Utilize the Convert Entities Sketch tool to convert the Sketch2 line to Plane1. Sketch two equal vertical lines Collinear with the left and right edges. Sketch a construction circle with a diameter Coincident to the left and right vertical lines. Create an 180deg tangent arc between the two vertical lines. Insert the required geometric relations and dimensions. Complete the sketch. Utilize the First arc condition from the Leaders tab in the Dimension PropertyManager to minimum the dimension to the bottom of the circle, Sketch2.

Insert a construction circle when dimensions are reference to a minimum or maximum arc condition.

9. Create the first **Extruded Boss** feature. Blind is the default End Condition. Depth = .260in. Note: .260in = (.500in - .240in). The extrude direction is towards the back.

10. Create **Sketch4**. Select the right angled face of Extrude2 as the Sketch plane. Wakeup the centerpoint of the tangent Arc. Sketch a circle. The circle is Coincident and Coradial to the Extrude2 feature.

11. Create the second **Extruded Boss** feature. Blind is the default End Condition in Direction 1. Depth = .240in.

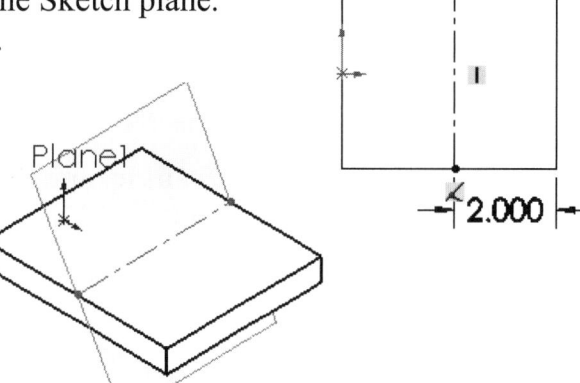

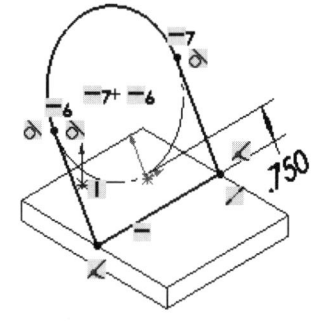

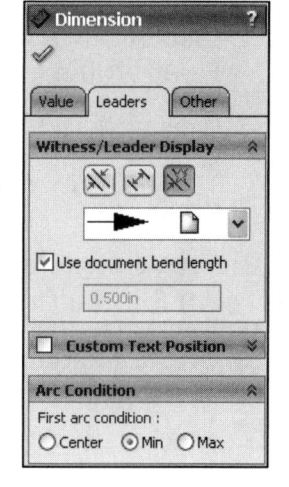

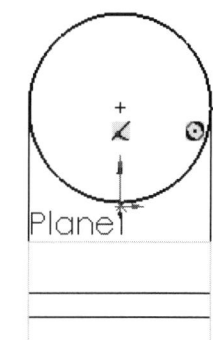

Advanced Part Modeling

12. Create **Sketch5**. Sketch5 is the profile for the Extruded Cut feature. Select the right angled face of Extrude3 as the Sketch plane. Apply the Convert Entities and Trim Sketch tools. Insert the required geometric relations and dimensions.

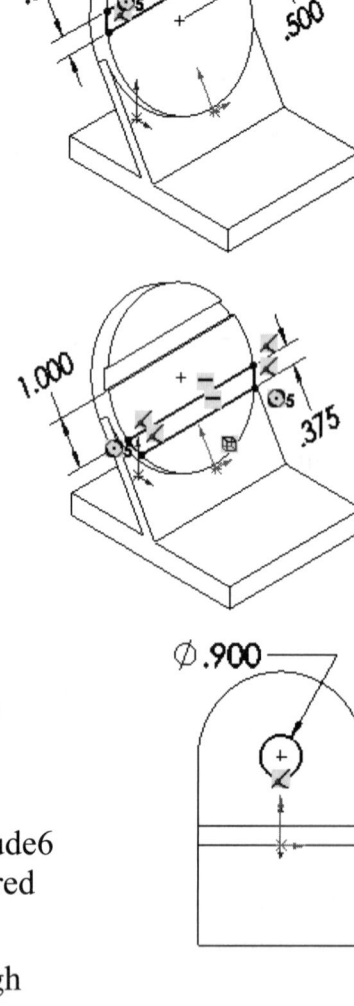

13. Create the first **Extruded Cut** feature. Blind is the default End Condition. Depth = .125in.

14. Create **Sketch6**. Select the right angled face of Extrude3 as the Sketch plane. Apply the Convert Entities and Trim Sketch tools. Insert the required geometric relations and dimensions.

15. Create the second **Extruded Cut** feature. Blind is the default End Condition. Depth = .125in.

16. Create **Sketch7**. Select the left angled face of Extrude2 as the Sketch plane. Sketch a circle. Insert the required geometric relation and dimension.

17. Create the third **Extruded Boss** feature. Blind is the default End Condition. Depth = .200in.
Note: .200in = (.700in - .500in).

18. Create **Sketch8**. Select the flat circular face of Extrude6 as the Sketch plane. Sketch a circle. Insert the required dimension.

19. Create the third **Extrude Cut** feature. Select Through All for End Condition in Direction 1.

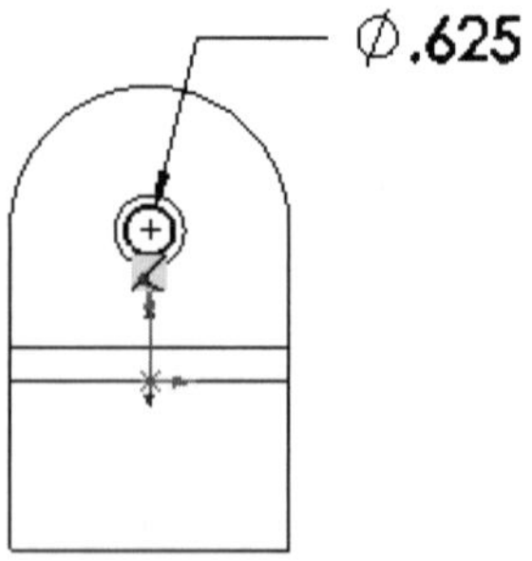

20. Create **Sketch9**. Select the left flat top face of Extrude1 as the Sketch Plane. Sketch a line parallel to the front edge as illustrated. Insert the required geometric relations and dimensions.

Page 4 - 14

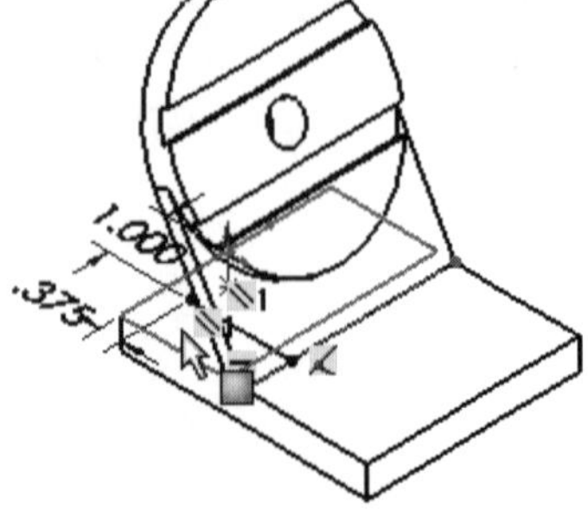

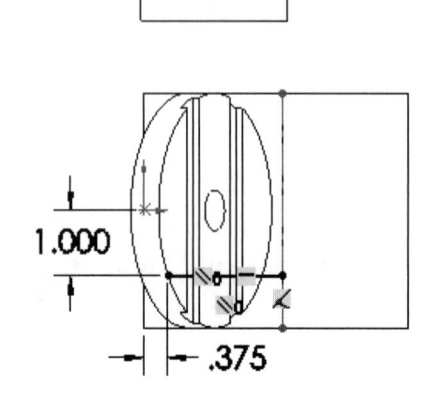

Advanced Part Modeling

21. Create the **Extrude-Thin1** feature. Extrude-Thin1 is the left support feature. Select Up To Surface for End Condition in Direction 1. Select face<1> for direction as illustrated. Thickness = .38in. Select One-Direction.

22. Create the **Mirror** feature. Mirror the Extrude-Thin1 feature about the Front Plane.

23. Create **Sketch10**. Select the bottom front flat face of Extrude1 as the Sketch plane. Sketch10 is the profile for the forth Extruded Cut feature. Insert the required geometric relations and dimensions.

24. Create the forth **Extruded Cut** feature. Select Through All for End Condition in Direction 1.

25. Create **Sketch11**. Select the top face of Extrude1 as the Sketch plane. Apply construction geometry. Sketch11 is the profile for Extrude9. Insert the required geometric relations and dimensions.

26. Create the **Extruded Boss** feature. Blind is the default End Condition in Direction 1. Depth = .125in.

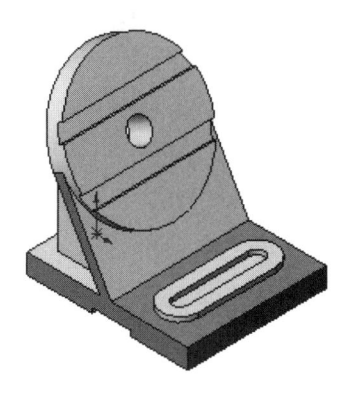

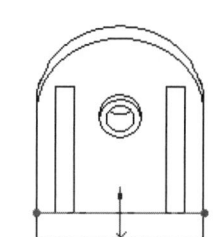

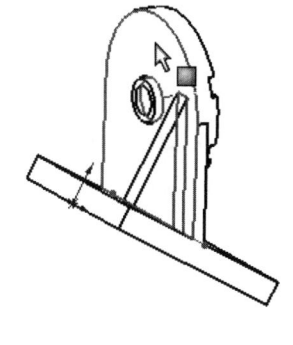

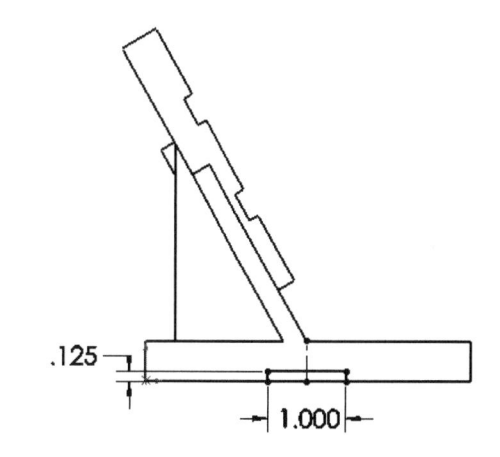

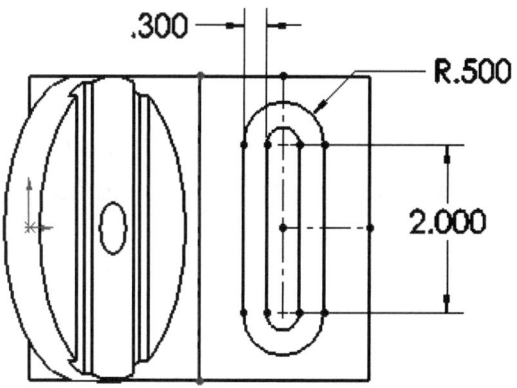

Page 4 - 15

Advanced Part Modeling

27. **Assign** 6061 Alloy material to the part.

28. **Calculate** the volume of the part. The volume = 14.05 cubic inches.

29. **Locate** the Center of mass. The location of the Center of mass is relative to the part Origin.

- X: 1.59 inches
- Y: 1.19 inches
- Z: 0.00 inches

30. **Save** the part and name it Advanced Part 4-4.

31. **Close** the model.

   In the Advanced Part Modeling category, an exam question could read: Build this model. Calculate the volume of the part.

- A: 14.05 cubic inches
- B: 15.66 cubic inches
- C: 13.44 cubic inches
- D: 12.71 cubic inches

The correct answer is A.

   As an exercise, modify A from 3.500in to 3.600in. Modify B from 4.200in to 4.100in. Modify the Plane1 angle from 60deg to 45deg. Modify the system units from IPS to MMGS. Calculate the mass and locate the Center of mass. The mass = 597.09 grams.

- X: 34.27 millimeters
- Y: 26.70 millimeters
- Z: 0.00 millimeters

32. **Save** the part and name it Advanced Part 4-4 Modified.

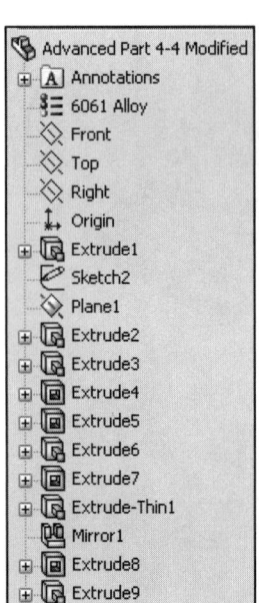

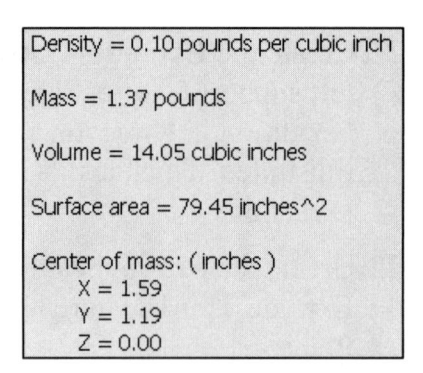

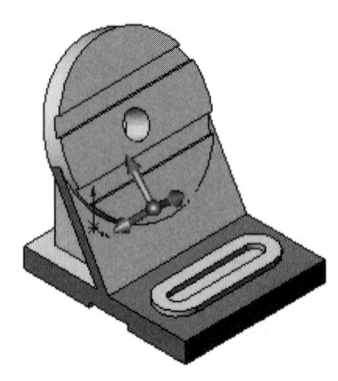

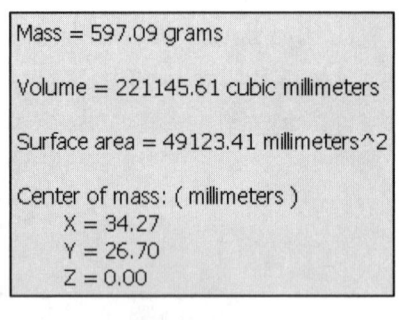

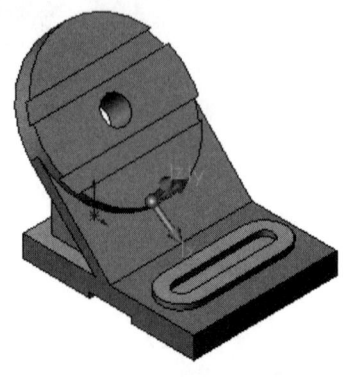

```
Density = 0.10 pounds per cubic inch
Mass = 1.37 pounds
Volume = 14.05 cubic inches
Surface area = 79.45 inches^2
Center of mass: ( inches )
   X = 1.59
   Y = 1.19
   Z = 0.00
```

```
Mass = 597.09 grams
Volume = 221145.61 cubic millimeters
Surface area = 49123.41 millimeters^2
Center of mass: ( millimeters )
   X = 34.27
   Y = 26.70
   Z = 0.00
```

Calculate the Center of mass relative to a created coordinate system location

In Chapter 3, you located the Center of mass relative to the default part Origin. In the Advanced Part Modeling category you may need to locate the Center of mass relative to a created coordinate system location. The exam model may display a created coordinate system location. Example:

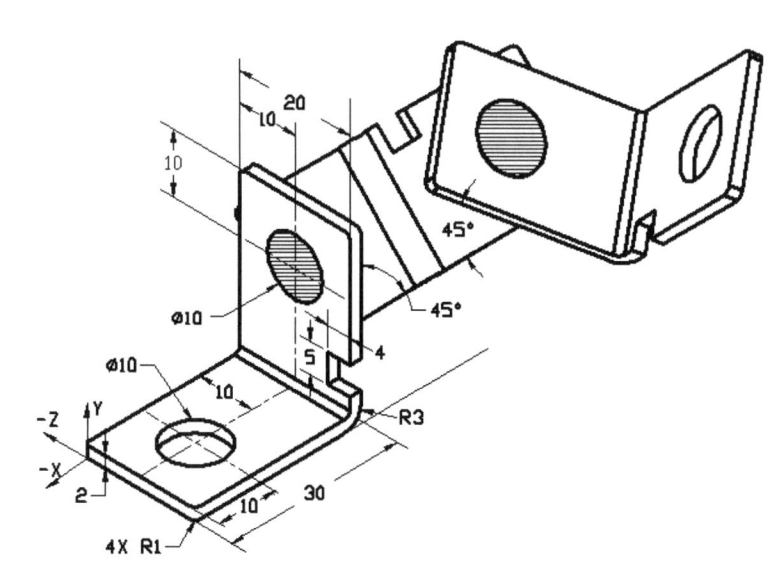

☼ The SolidWorks software displays positive values for (X, Y, Z) coordinates for a reference coordinate system. The CSWA exam displays either a positive or negative sign in front of the (X, Y, Z) coordinates to indicate direction as illustrated, (-X, +Y, -Z).

The following section reviews creating a coordinate system location for a part.

Tutorial: Coordinate location 4-1

Use the Mass Properties tool to calculate the Center of mass for a part located at a new coordinate location through a point.

1. **Open** the Plate-3-Point part from the SolidWorks CSWA Folder\Chapter4 location. View the location of the part Origin.

2. **Locate** the Center of mass. The location of the Center of mass is relative to the part Origin.

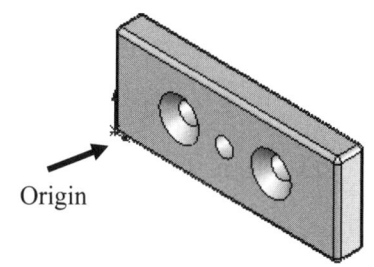

Advanced Part Modeling

- X = 28 millimeters
- Y = 11 millimeters
- Z = -3 millimeters

Create a new coordinate system location. Locate the new coordinate system location at the center of the center hole as illustrated.

3. Right-click the **front face** of Base-Extrude.
4. Click **Sketch** from the Context toolbar.
5. Click the **edge** of the center hole as illustrated.
6. Click **Convert Entities** from the Sketch toolbar. The center point for the new coordinate location is displayed.
7. **Exit** the sketch. Sketch4 is displayed.
8. Click the **Coordinate System** tool from the Consolidated Reference Geometry toolbar. The Coordinate System PropertyManager is displayed.
9. Click the **center point** of the center hole in the Graphics window. Point2@Sketch4 is displayed in the Selections box as the Origin.
10. Click **OK** from the Coordinate System PropertyManager. Coordinate System1 is displayed.
11. **View** the new coordinate location at the center of the center hole.

View the Mass Properties of the part with the new coordinate location.

12. Click the **Mass Properties** tool from the Evaluate tab.
13. Select **Coordinate System1** from the Output box. The Center of mass relative to the new location is located at the following coordinates: X = 0 millimeters, Y = 0 millimeters, Z = -3 millimeters

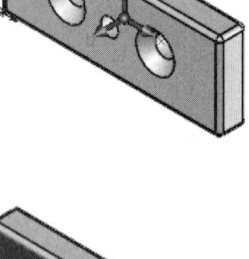

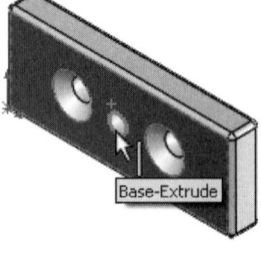

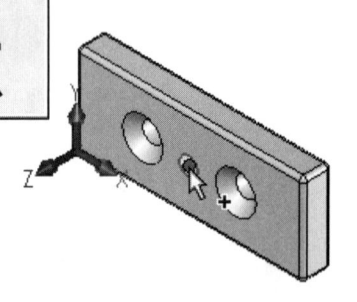

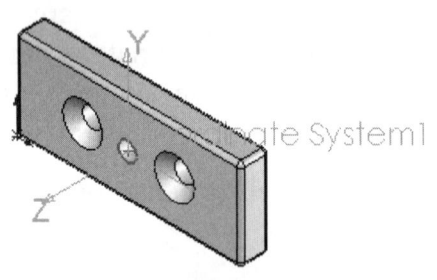

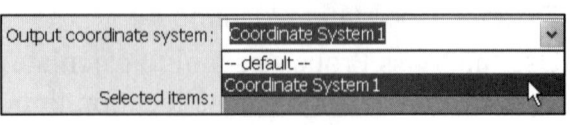

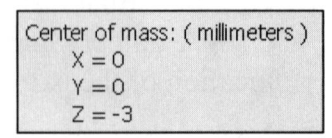

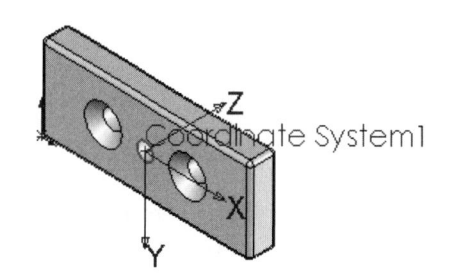

Advanced Part Modeling

14. **Reverse** the direction of the axes as illustrated. On the CSWA exam, the coordinate system axes could be represented by: (+X, -Y, -Z).

15. **Close** the model.

☼ To reverse the direction of an axis, click its **Reverse Axis Direction** button in the Coordinate System PropertyManager.

Tutorial: Coordinate location 4-2

Create a new coordinate system location. Locate the new coordinate system at the top back point as illustrated.

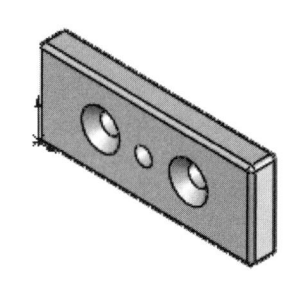

1. **Open** the Plate-X-Y-Z part from the SolidWorks CSWA Folder\Chapter4 location.

2. **View** the location of the part Origin.

3. Drag the **Rollback bar** under the Base-Extrude feature in the FeatureManager.

4. Click the **Coordinate System** tool from the Consolidated Reference Geometry toolbar. The Coordinate System PropertyManager is displayed.

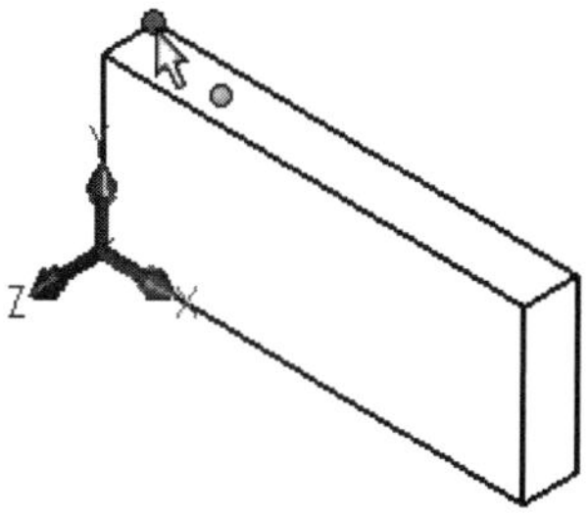

5. Click the **back left vertex** as illustrated.

6. Click the **top back horizontal** edge as illustrated. Do not select the midpoint.

7. Click the **back left vertical** edge as illustrated.

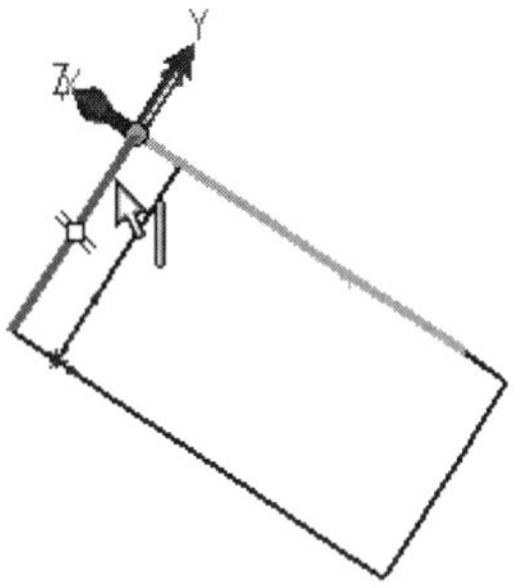

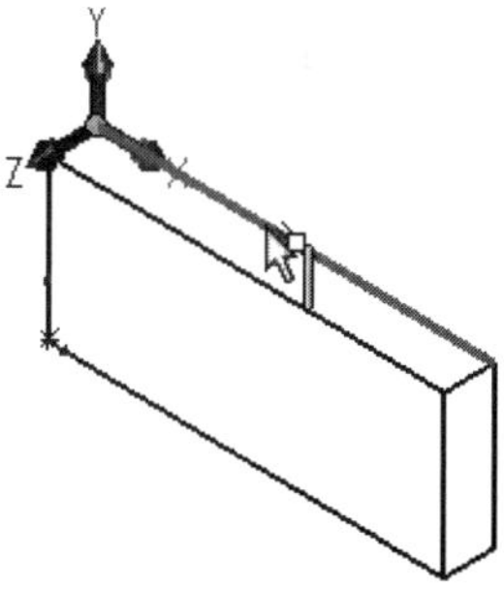

8. Click **OK** from the Coordinate System PropertyManager. Coordinate System1 is displayed in the FeatureManager and in the Graphics window.

9. Drag the **Rollback bar** to the bottom of the FeatureManager.

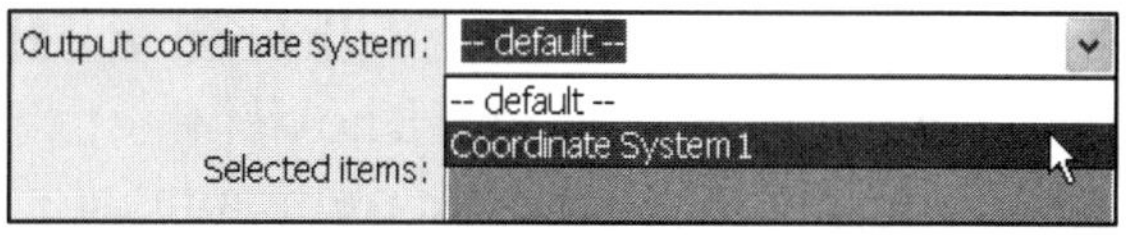

10. **Calculate** the Center of mass relative to the new coordinate system.

11. Select **Coordinate System1**. The Center of mass relative to the new location is located at the following coordinates:

Page 4 - 19

Advanced Part Modeling

- X = -28 millimeters
- Y = -11 millimeters
- Z = -4 millimeters

12. **Reverse** the direction of the axes as illustrated.

13. **Close** the model.

☼ You can define a coordinate system for a part or assembly. Use the coordinate system with the Measure and Mass Properties tools.

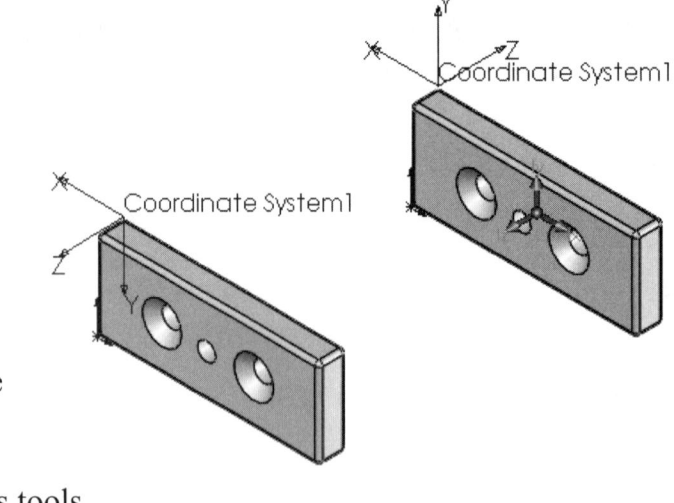

Tutorial: Advanced part 4-5

An exam question in this category could read: Build this part. Calculate the overall mass and locate the Center of mass of the illustrated model.

1. **Create** a New part in SolidWorks.

2. **Build** the illustrated dimensioned model. Insert thirteen features: Extrude-Thin1, Fillet, two Extruded Cuts, Circular Pattern, two Extruded Cuts, Mirror, Chamfer, Extruded Cut, Mirror, Extruded Cut, and Mirror.

Given:
A = 110, B = 65,
C = 5 X 45Ø CHAMFER
Material: 5MM, 6061 Alloy
Density: .0027 g/mm^3
Units: MMGS
ALL HOLES 6MM

    Think about the steps that you would take to build the illustrated part. Review the provided information. The depth of the left side is 50mm. The depth of the right side is 60mm.

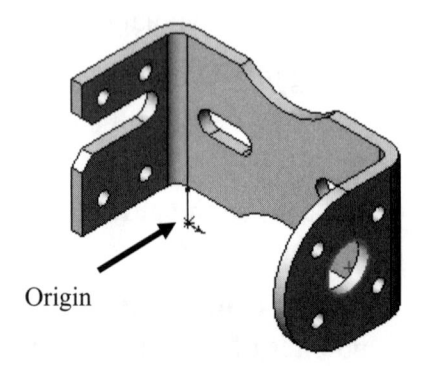

Page 4 - 20

Advanced Part Modeling

☼ If the inside radius = 5mm and the material thickness = 5mm, then the outside radius = 10mm.

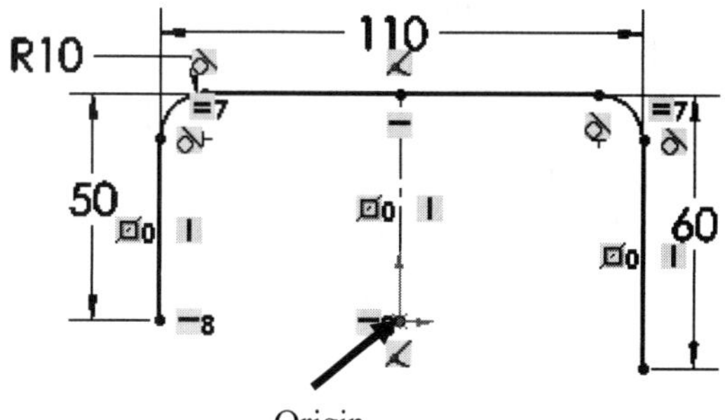

3. **Set** the document properties for the model.

4. Create **Sketch1**. Sketch1 is the Base sketch. Select the Top Plane as the Sketch plane. Apply the Line and Sketch Fillet Sketch tools. Apply construction geometry. Insert the required geometric relations and dimensions.

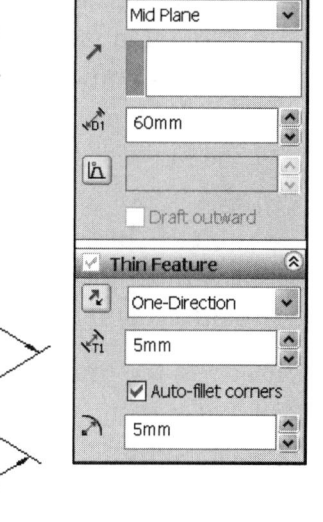

5. Create the **Extrude-Thin1** feature. Extrude1 is the Base feature. Apply symmetry in Direction 1. Depth = 60mm. Thickness = 5mm. Check the Auto-fillet corners box. Radius = 5mm.

☼ The Auto-fillet corners option creates a round at each edge where lines meet at an angle.

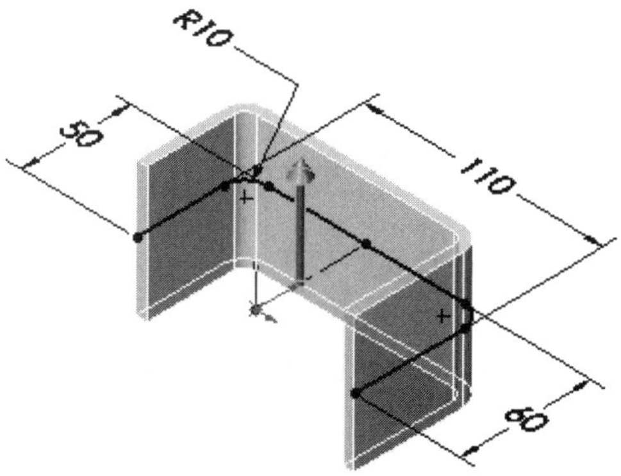

6. Create the **Fillet** feature. Fillet1 is a full round fillet. Fillet the three illustrated faces: top, front, and bottom.

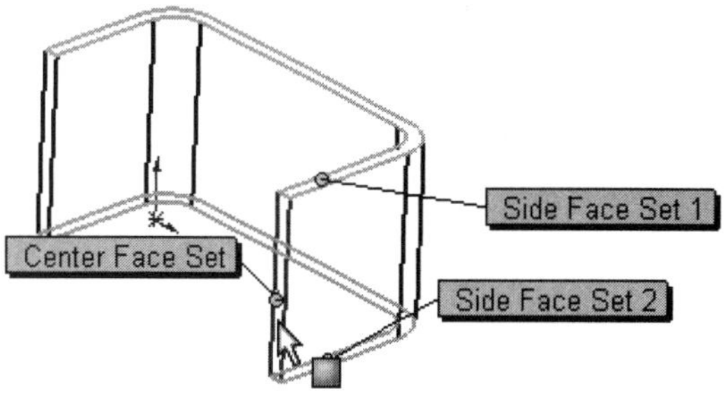

7. Create **Sketch2**. Select the right face as the Sketch plane. Wake-up the centerpoint. Sketch a circle. Insert the required relation and dimension.

8. Create the first **Extruded Cut** feature. Select Up To Next for the End Condition in Direction 1.

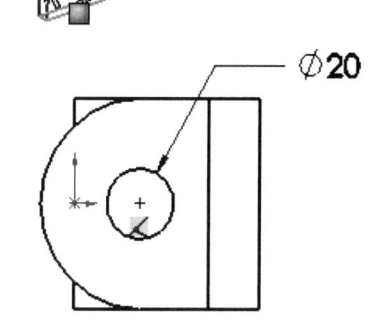

☼ The Up To Next End Condition extends the feature from the sketch plane to the next surface that intercepts the entire profile. The intercepting surface must be on the same part.

Advanced Part Modeling

9. Create **Sketch3**. Select the right face as the Sketch plane. Create the profile for the second Extruded Cut feature. Extrude2 is the seed feature for CirPattern1. Apply construction geometry to locate the center point of Sketch3. Insert the required relations and dimensions.

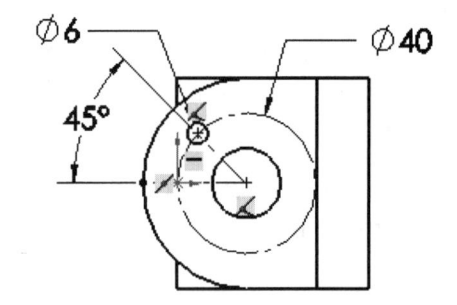

10. Create the second **Extruded Cut** feature. Select Up To Next for the End Condition in Direction 1.

11. Create the **Circular Pattern** feature. Number of Instances = 4. Default angle = 360deg.

12. Create **Sketch4**. Select the left outside face of Extrude-Thin1 as the Sketch plane. Apply the Line and Tangent Arc Sketch tool to create Sketch4. Insert the required geometric relations and dimensions.

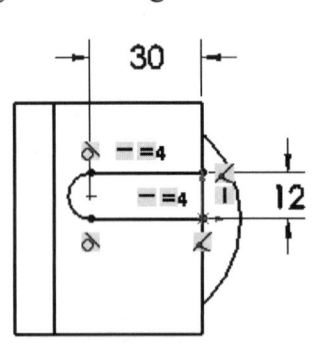

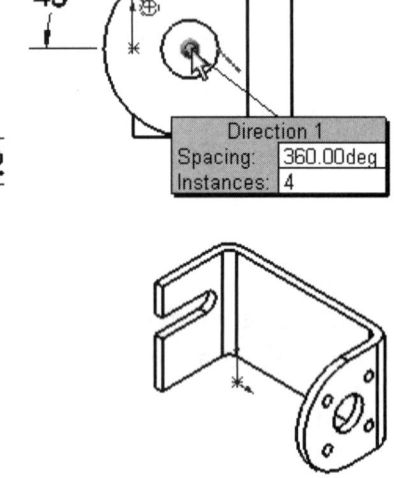

13. Create the third **Extruded Cut** feature. Select Up To Next for End Condition in Direction 1. The Slot on the left side of Extrude-Thin1 is created.

14. Create **Sketch5**. Select the left outside face of Extrude-Thin1 as the Sketch plane. Sketch two circles. Insert the required geometric relations and dimensions.

15. Create the forth **Extruded Cut** feature. Select Up To Next for End Condition in Direction 1.

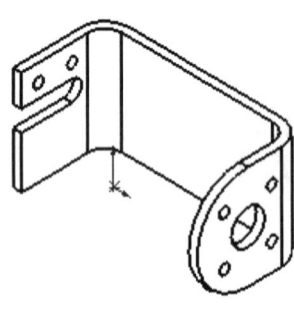

☼ There are numerous ways to create the models in this chapter. The goal is to display different design intents and techniques

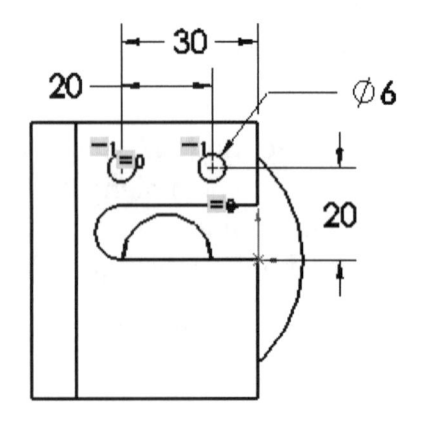

Page 4 - 22

Advanced Part Modeling

16. Create the first **Mirror** feature. Mirror the top two holes about the Top Plane.

17. Create the **Chamfer** feature. Create an Angle distance chamfer. Chamfer the selected edges as illustrated. Distance = 5mm. Angle = 45deg.

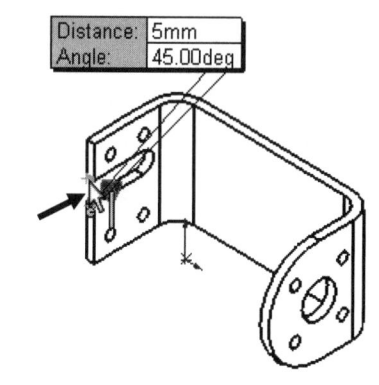

18. Create **Sketch6**. Select the front face of Extrude-Thin1 as the Sketch plane. Insert the required geometric relations and dimensions.

19. Create the fifth **Extruded Cut** feature. Select Thought All for End Condition in Direction 1.

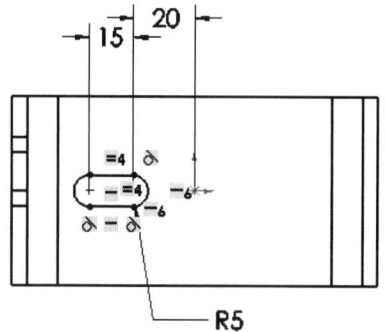

20. Create the second **Mirror** feature. Mirror Extrude5 about the Right Plane.

21. Create **Sketch7**. Select the front face of Extrude-Thin1 as the Sketch plane. Apply the 3 Point Arc Sketch tool. Apply the min First Arc Condition option. Insert the required geometric relations and dimensions.

22. Create the last **Extruded Cut** feature. Through All is the End Condition in Direction 1 and Direction 2.

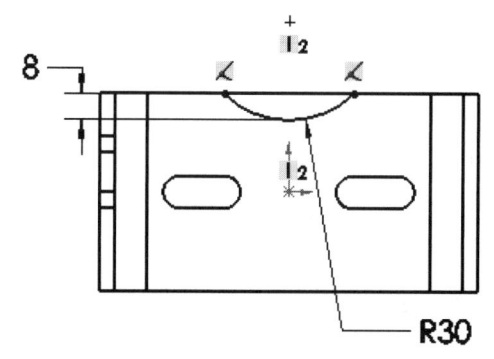

23. Create the third **Mirror** feature. Mirror Extrude6 about the Top Plane.

24. **Assign** the material to the part.

25. **Calculate** the overall mass of the part. The overall mass = 134.19 grams.

26. **Locate** the Center of mass relative to the part Origin:

- X: 1.80 millimeters
- Y: -0.27 millimeters
- Z: -35.54 millimeters

27. **Save** the part and name it Advanced Part 4-5.

28. **Close** the model.

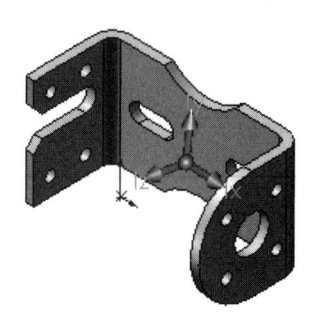

Page 4 - 23

Advanced Part Modeling

All questions on the exam are in a multiple choice single answer or fill in the blank format. In the Advanced Part Modeling category, an exam question could read: Build this model. Calculate the overall mass of the part with the provided information.

- A: 139.34 grams
- B: 155.19 grams
- C: 134.19 grams
- D: 143.91 grams

The correct answer is C.

☼ Use the Options button in the Mass Properties dialog box to apply custom settings to units.

Tutorial: Advanced part 4-5A

An exam question in this category could read: Build this part. Locate the Center of mass. Note the coordinate system location of the model as illustrated.

Where do you start? Build the model as you did in the Tutorial: Advanced Part 4-5. Create Coordinate System1 to locate the Center of mass.

1. **Open** Advanced Part 4-5 from your SolidWorks folder.

Create the illustrated coordinate system location.

2. Show **Sketch2** from the FeatureManager design tree.

3. Click the **center point** of Sketch2 in the Graphics window as illustrated.

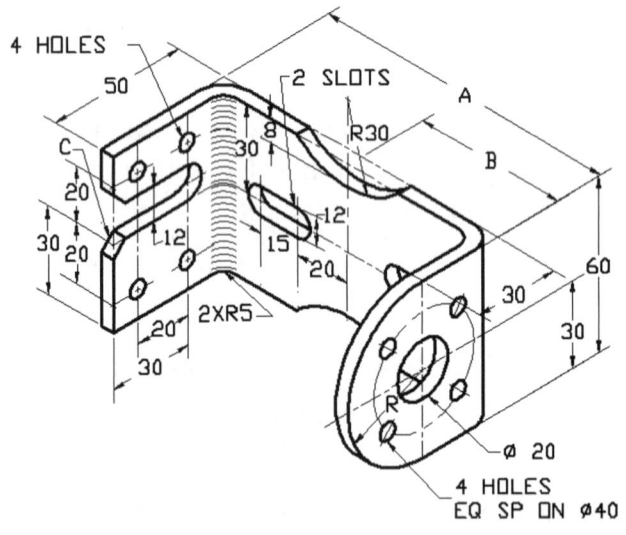

A = 110, B = 65, C = 5 X 45Ø CHAMFER
Material: 5MM, 6061 Alloy
Density: .0027 g/mm^3
Units: MMGS
ALL HOLES 6MM

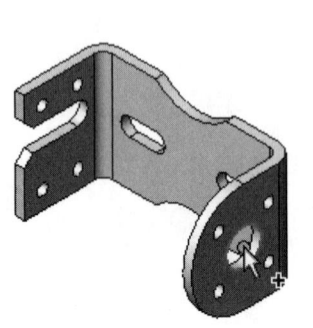

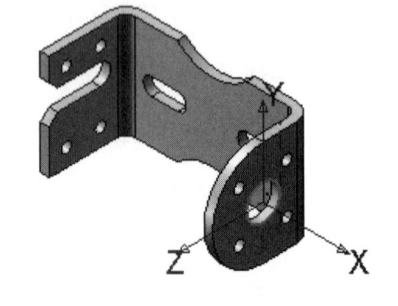

Coordinate system: +X, +Y. +Z

Page 4 - 24

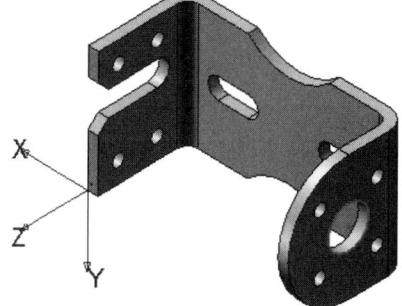

4. Click the **Coordinate System** tool from the Consolidated Reference Geometry toolbar. The Coordinate System PropertyManager is displayed. Point2@Sketch2 is displayed in the Origin box.

5. Click **OK** from the Coordinate System PropertyManager. Coordinate System1 is displayed

6. **Locate** the Center of mass based on the location of the illustrated coordinate system. Select Coordinate System1.

   - X: -53.20 millimeters
   - Y: -0.27 millimeters
   - Z: -15.54 millimeters

7. **Save** the part and name it Advanced Part 4-5A.

8. **Close** the model.

Tutorial: Advanced part 4-5B

Build this part. Locate the Center of mass. View the location of the coordinate system. The coordinate system is located at the left front point of the model.

Build the illustrated model as you did in the Tutorial: Advanced Part 4-5. Create Coordinate System1 to locate the Center of mass for the model.

1. **Open** Advance Part 4-5 from your SolidWorks folder.

Create the illustrated coordinate system.

2. Click the **vertex** as illustrated for the Origin location.

Given:
A = 110, B = 65,
C = 5 X 45Ø CHAMFER
Material: 5MM, 6061 Alloy
Density: .0027 g/mm^3
Units: MMGS
ALL HOLES 6MM

☼ To reverse the direction of an axis, click its **Reverse Axis Direction** button in the Coordinate System PropertyManager.

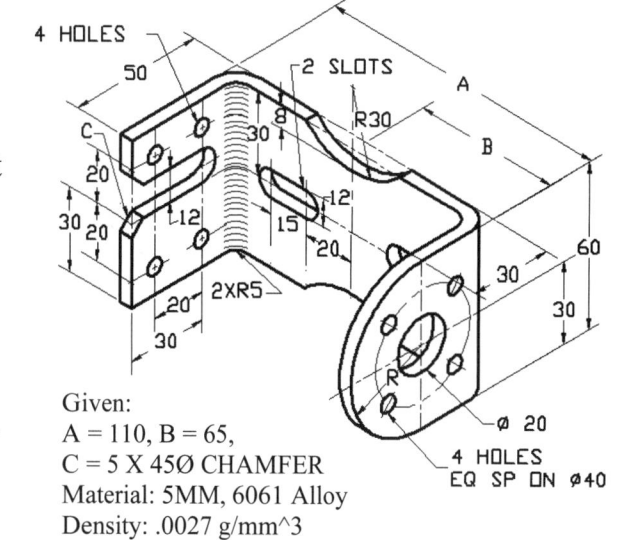

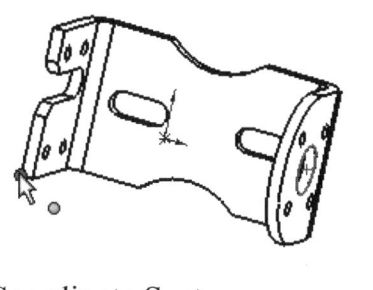

Advanced Part Modeling

3. Click the **Coordinate System** tool from the Consolidated Reference Geometry toolbar. The Coordinate System PropertyManager is displayed. Vertex<1> is displayed in the Origin box.

4. Click the **bottom horizontal edge** as illustrated. Edge<1> is displayed in the X Axis Direction box.

5. Click the **left back vertical edge** as illustrated. Edge<2> is displayed in the Y Axis Direction box.

6. Click **OK** from the Coordinate System PropertyManager. Coordinate System1 is displayed.

9. **Locate** the Center of mass based on the location of the illustrated coordinate system. Select Coordinate System1.

- X: -56.80 millimeters
- Y: -29.73 millimeters
- Z: -35.54 millimeters

10. **Save** the part and name it Advanced Part 4-5B.
11. **Close** the model.

In the Advanced Part Modeling category, an exam question could read: Build this model. Locate the Center of mass.

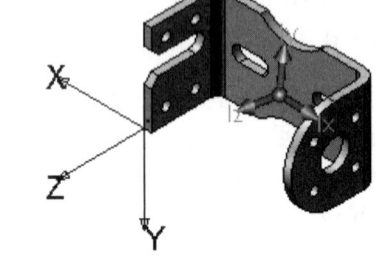

- A: X = -56.80 millimeters, Y = -29.73 millimeters, Z = -35.54 millimeters
- B: X = 1.80 millimeters, Y = -0.27 millimeters, Z = -35.54 millimeters
- C: X = -59.20 millimeters, Y = -0.27 millimeters, Z = -15.54 millimeters
- D: X= -1.80 millimeters, Y = 1.05 millimeters, Z = -0.14 millimeters

The correct answer is A.

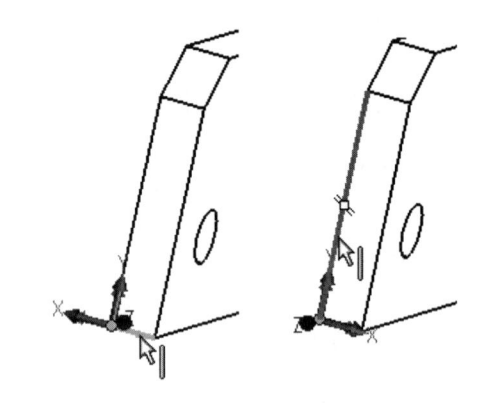

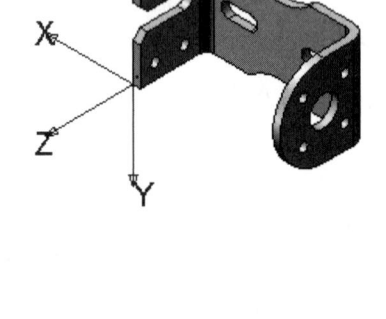

Tutorial: Advanced part 4-6

An exam question in this category could read: Build this part. Calculate the overall mass and locate the Center of mass of the illustrated model.

1. **Create** a New part in SolidWorks.

2. **Build** the illustrated dimensioned model. Insert twelve features and a Reference plane: Extrude-Thin1, two Extruded Bosses, Extruded Cut, Extruded Boss, Extruded Cut, Plane1, Mirror, and five Extruded Cuts.

   Think about the steps that you would take to build the illustrated part. Create an Extrude-Thin1 feature as the Base feature.

3. **Set** the document properties for the model. Review the given information.

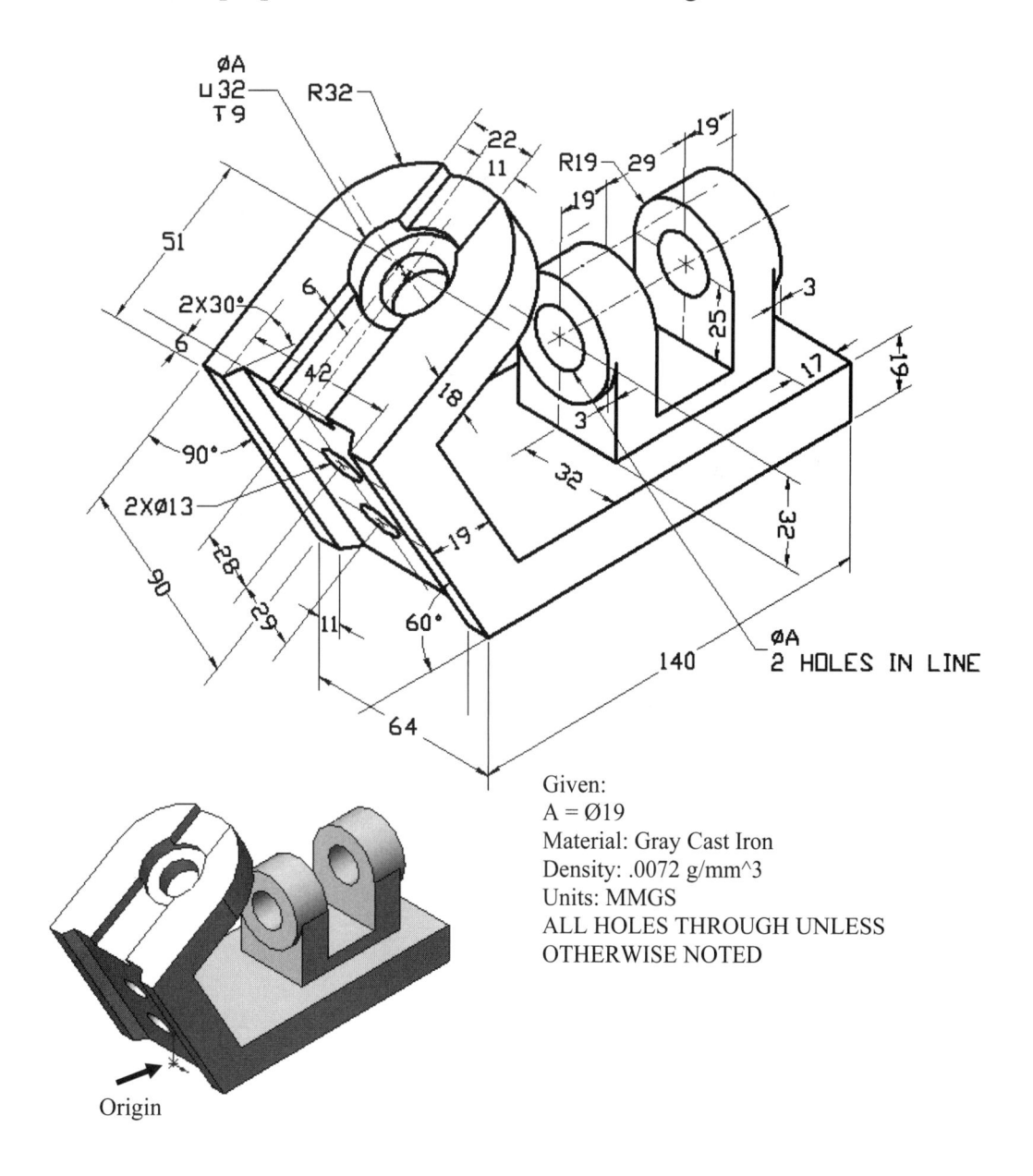

Given:
A = Ø19
Material: Gray Cast Iron
Density: .0072 g/mm^3
Units: MMGS
ALL HOLES THROUGH UNLESS OTHERWISE NOTED

Advanced Part Modeling

4. Create **Sketch1**. Sketch1 is the Base sketch. Select the Right Plane as the Sketch plane. Apply construction geometry. Insert the required geometric relations and dimensions. Sketch1 is the profile for Extrude-Thin1. Note the location of the Origin.

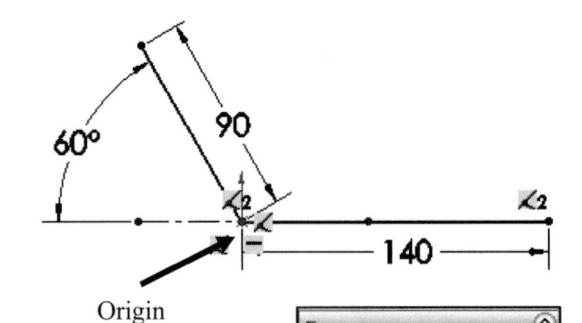

5. Create the **Extrude-Thin1** feature. Apply symmetry. Select Mid Plane as the End Condition in Direction 1. Depth = 64mm. Thickness = 19mm.

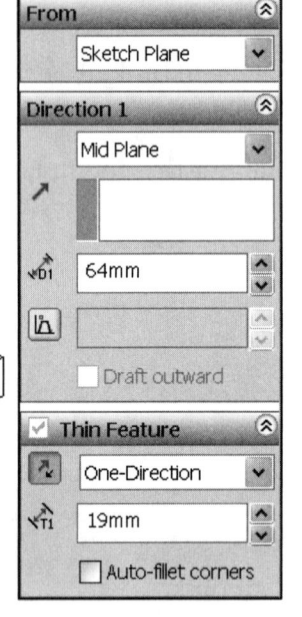

6. Create **Sketch2**. Select the top narrow face of Extrude-Thin1 as the Sketch plane. Sketch three lines: two vertical and one horizontal and a tangent arc. Insert the required geometric relations and dimensions.

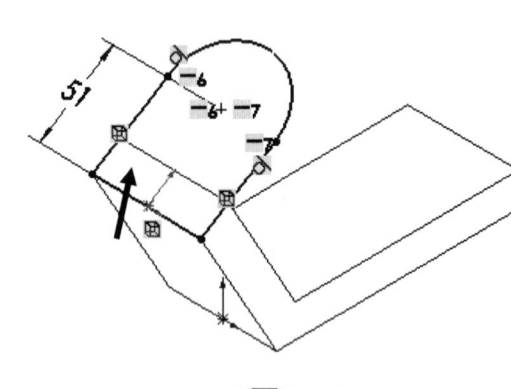

7. Create the **Extrude1** feature. Blind is the default End Condition in Direction 1. Depth = 18mm.

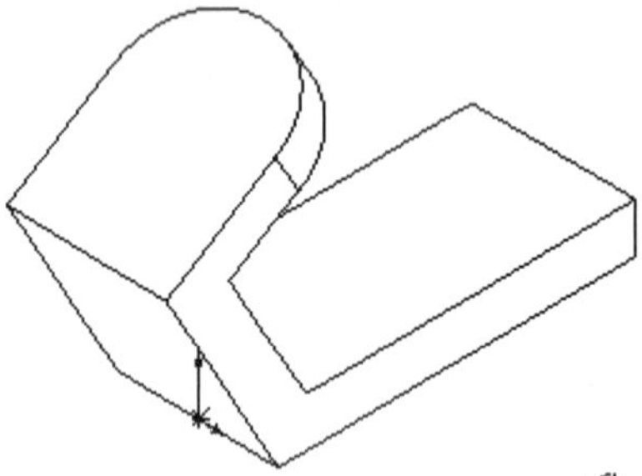

8. Create **Sketch3**. Select the Right Plane as the Sketch plane. Sketch a rectangle. Insert the required geometric relations and dimensions.
Note: 61mm = (19mm - 3mm) x 2 + 29mm.

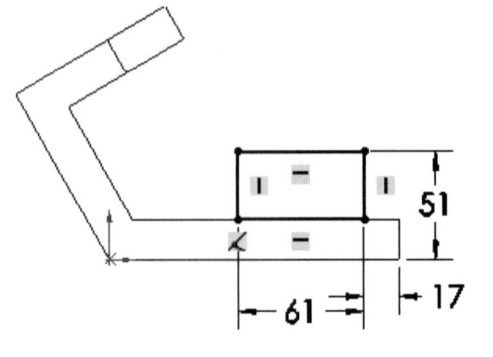

Advanced Part Modeling

9. Create the **Extrude2** feature. Select Mid Plane for End Condition in Direction 1. Depth = 38mm. Note: 2 x R19.

10. Create **Sketch4**. Select the Right Plane as the Sketch plane. Sketch a vertical centerline from the top midpoint of the sketch. The centerline is required for Plane1. Plane1 is a reference plane. Sketch a rectangle symmetric about the centerline. Insert the required relations and dimensions. Sketch4 is the profile for Extrude3.

11. Create the first **Extruded Cut** feature. Extrude in both directions. Select Through All for End Condition in Direction 1 and Direction 2.

12. Create **Sketch5**. Select the inside face of Extrude3 for the Sketch plane. Sketch a circle from the top midpoint. Sketch a construction circle. Construction geometry is required for future features. Complete the sketch.

13. Create the **Extruded Boss** feature. Blind is the default End Condition. Depth = 19mm.

14. Create **Sketch6**. Select the inside face for the Sketch plane. Show Sketch5. Select the construction circle in Sketch5. Apply the Convert Entities Sketch tool.

15. Create the second **Extruded Cut** feature. Select the Up To Next End Condition in Direction 1.

💡 There are numerous ways to create the models in this chapter. The goal is to display different design intents and techniques

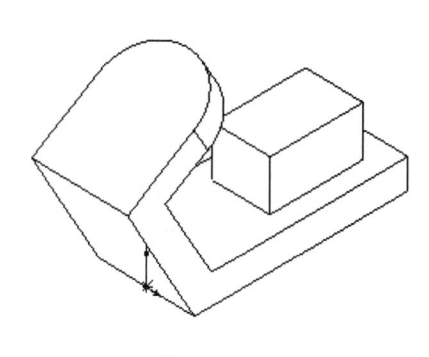

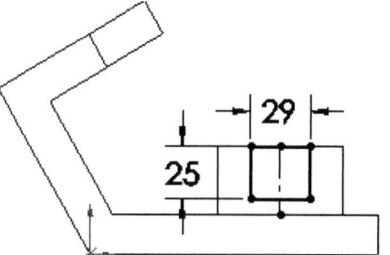

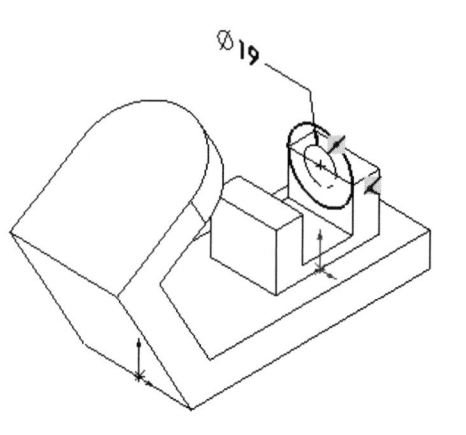

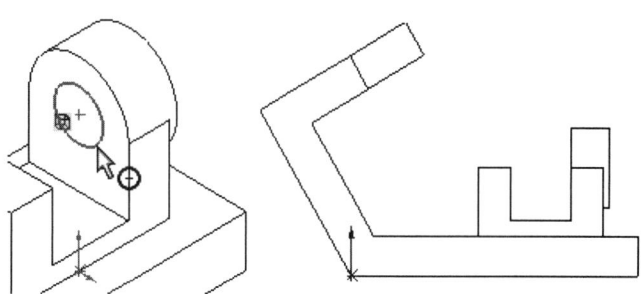

Page 4 - 29

Advanced Part Modeling

16. Create **Plane1**. Apply symmetry. Create Plane1 to mirror Extrude4 and Extrude5. Create a Parallel Plane at Point. Select the midpoint of Sketch4, and Face<1> as illustrated. Point1@Sketch4 and Face<1> is displayed in the Selections box.

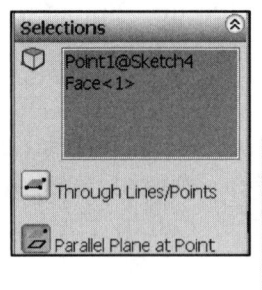

17. Create the **Mirror** feature. Mirror Extrude4, and Extrude5 about Plane1.

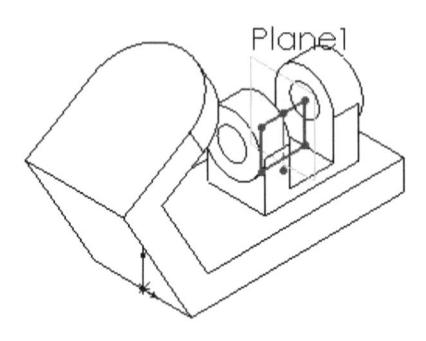

☼ The Mirror feature creates a copy of a feature, (or multiple features), mirrored about a face or a plane. You can select the feature or you can select the faces that comprise the feature.

18. Create **Sketch7**. Select the top front angled face of Extrude-Thin1 as the Sketch plane. Apply the Centerline Sketch tool. Insert the required geometric relations and dimensions.

19. Create the third **Extruded Cut** feature. Select Through All for End Condition in Direction 1. Select the angle edge for the vector to extrude as illustrated.

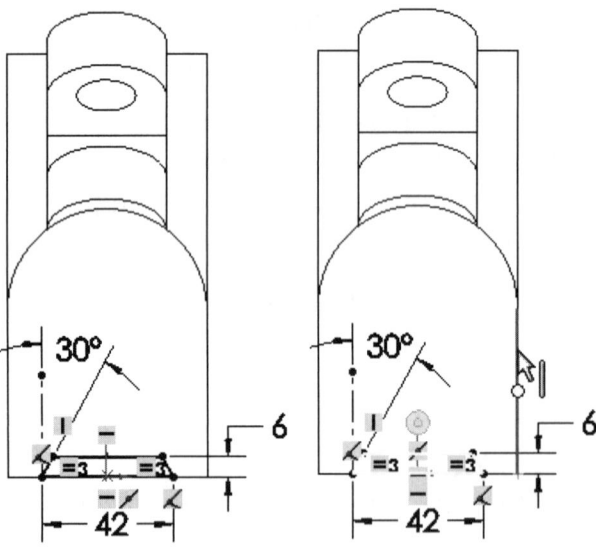

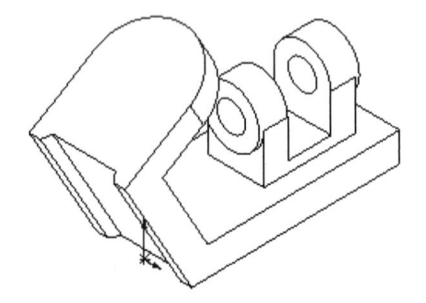

Page 4 - 30

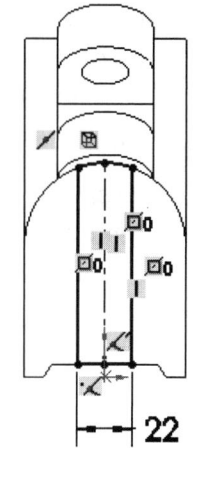

Advanced Part Modeling

20. Create **Sketch8**. Select the top front angled face of Extrude-Thin1 as the Sketch plane. Sketch a centerline. Sketch two vertical lines and a horizontal line. Select the top arc edge. Apply the Convert Entities Sketch tool. Apply the Trim Sketch tool to remove the unwanted arc geometry. Insert the required geometric relations and dimension.

21. Create the forth **Extruded Cut** feature. Blind is the default End Condition in Direction 1. Depth = 6mm.

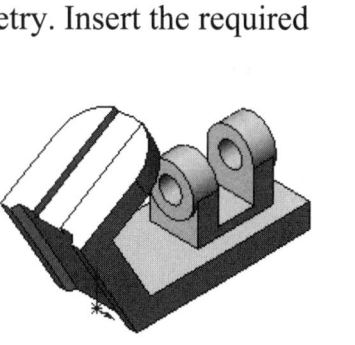

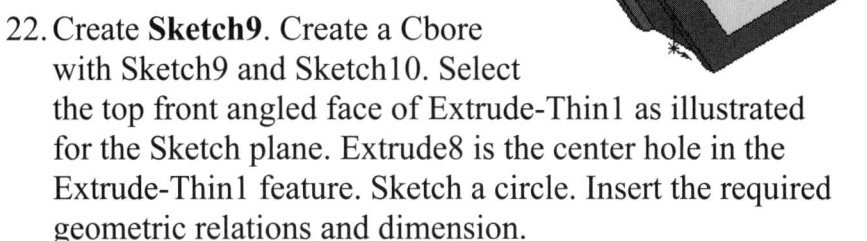

22. Create **Sketch9**. Create a Cbore with Sketch9 and Sketch10. Select the top front angled face of Extrude-Thin1 as illustrated for the Sketch plane. Extrude8 is the center hole in the Extrude-Thin1 feature. Sketch a circle. Insert the required geometric relations and dimension.

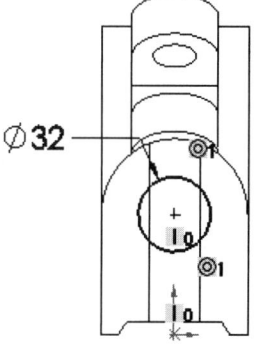

23. Create the fifth **Extrude Cut** feature. Blind is the default End Condition. Depth = 9mm. Note: This is the first feature for the Cbore.

24. Create **Sketch10**. Select the top front angled face of Extrude-Thin1 as the Sketch plane. Sketch a circle. Insert the required geometric relation and dimension. Note: A = Ø19.

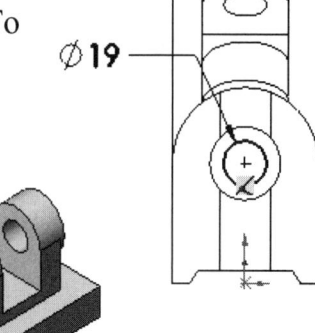

25. Create the sixth **Extruded Cut** feature. Select the Up To Next End Condition in Direction 1. The Cbore is complete.

☼ There are numerous ways to create the models in this chapter. The goal is to display different design intents and techniques

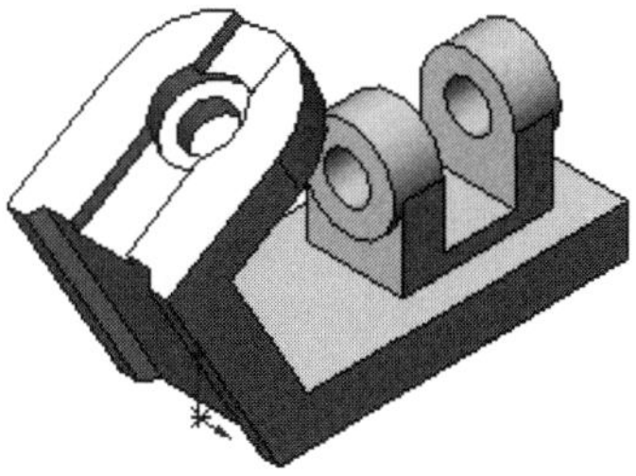

Page 4 - 31

Advanced Part Modeling

26. Create **Sketch11**. Select the front angle face of Extrude6 for the Sketch plane. Sketch two circles. Insert the required geometric relations and dimensions.

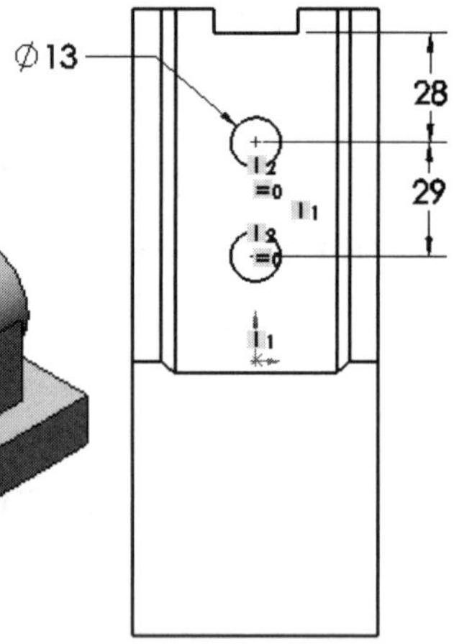

27. Create the last **Extruded Cut** feature. Select the Up To Next End Condition in Direction 1.

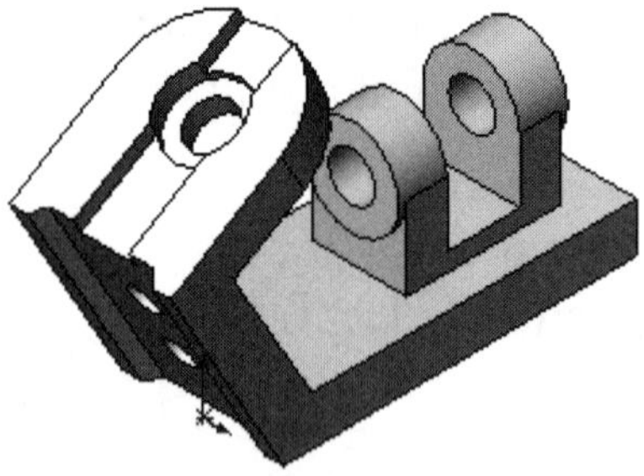

☼ The FilletXpert manages, organizes, and reorders constant radius fillets.

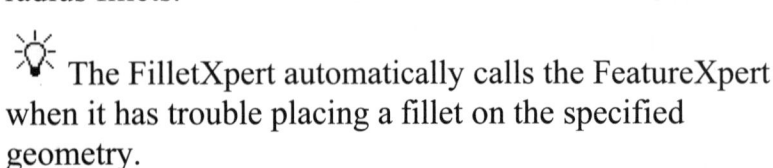

☼ The FilletXpert automatically calls the FeatureXpert when it has trouble placing a fillet on the specified geometry.

28. **Assign** the material to the part.

29. **Calculate** the overall mass of the part. The overall mass = 2536.59 grams.

30. **Locate** the Center of mass relative to the part Origin:

- X: 0.00 millimeters
- Y: 34.97 millimeters
- Z: -46.67 millimeters

31. **Save** the part and name it Advanced Part 4-6.

☼ Due to software rounding, you may view a negative -0.00 coordinate location in the Mass Properties dialog box.

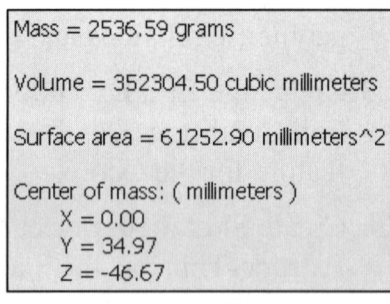

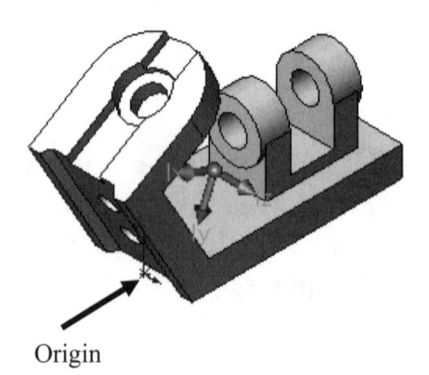

```
Mass = 2536.59 grams
Volume = 352304.50 cubic millimeters
Surface area = 61252.90 millimeters^2
Center of mass: ( millimeters )
    X = 0.00
    Y = 34.97
    Z = -46.67
```

Origin

Tutorial: Advanced part 4-6A

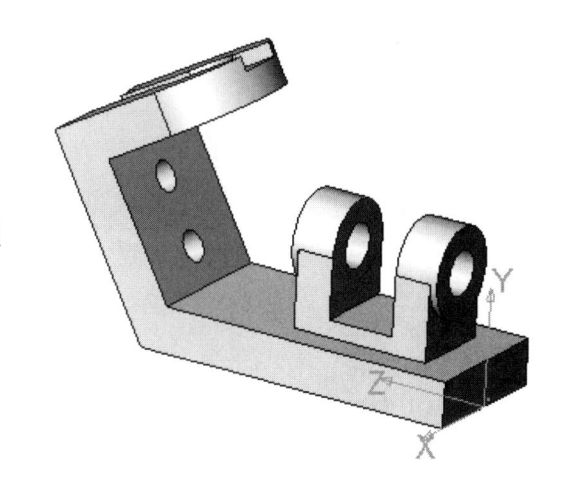

An exam question in this category could read: Build this part. Locate the Center of mass for the illustrated coordinate system.

Where do you start? Build the illustrated model as you did in the Tutorial: Advanced Part 4-6. Create Coordinate System1 to locate the Center of mass for the model.

1. **Open** Advanced Part 4-6 from your SolidWorks folder.

Create the illustrated Coordinate system.

2. Click the **Coordinate System** tool from the Consolidated Reference Geometry toolbar. The Coordinate System PropertyManager is displayed.

3. Click the **bottom midpoint** of Extrude-Thin1 as illustrated. Point<1> is displayed in the Origin box.

4. Click **OK** from the Coordinate System PropertyManager. Coordinate System1 is displayed.

5. **Locate** the Center of mass based on the location of the illustrated coordinate system. Select Coordinate System1.

- X: 0.00 millimeters
- Y: 34.97 millimeters
- Z: 93.33 millimeters

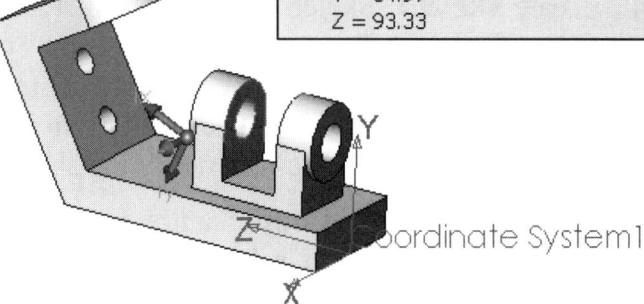

6. **Save** the part and name it Advanced Part 4-6A.

7. **View** the Center of mass with the default coordinate system.

8. **Close** the model.

Advanced Part Modeling

Tutorial: Advanced part 4-7

An exam question in this category could read: Build this part. Calculate the overall mass and locate the Center of mass of the illustrated model.

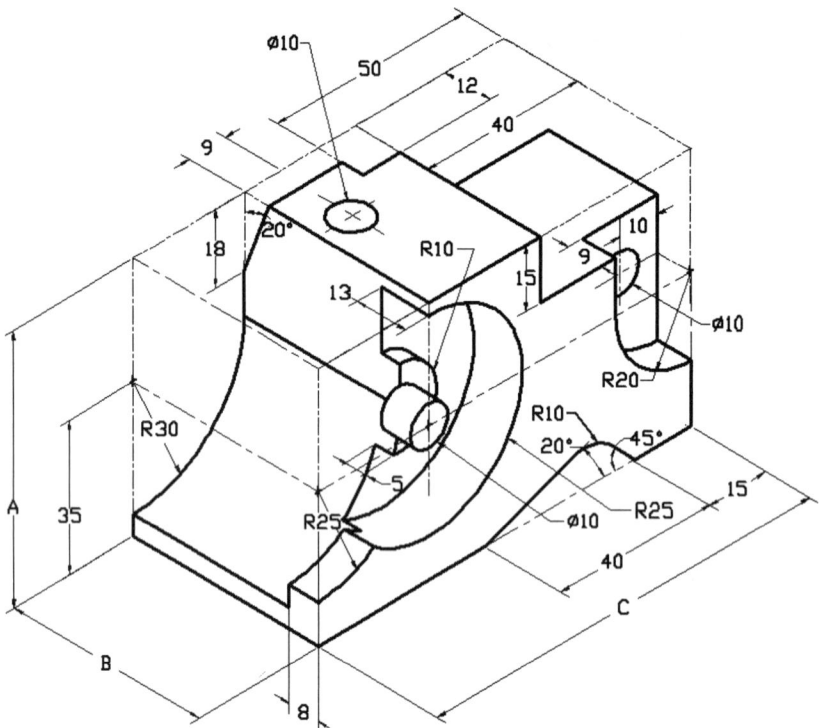

1. **Create** a New part in SolidWorks.

2. **Build** the illustrated dimensioned model. Insert thirteen features: Extruded Base, nine Extruded Cuts, two Extruded Bosses, and a Chamfer. Note: The center point of the top hole is located 30mm from the top right edge.

Given:
A = 63, B = 50, C = 100
Material: Copper
Units: MMGS
Density: .0089 g/mm^3
All HOLES THROUGH ALL

Think about the steps that you would take to build the illustrated part. Review the centerlines that outline the overall size of the part.

3. **Set** the document properties for the model.

4. Create **Sketch1**. Sketch1 is the Base sketch. Select the Right Plane as the Sketch plane. Sketch a rectangle. Insert the required geometric relations and dimensions. The part Origin is located in the bottom left corner of the sketch.

5. Create the **Extruded Base** feature. Blind is the default End Condition in Direction 1. Depth = 50mm. Extrude1 is the Base feature.

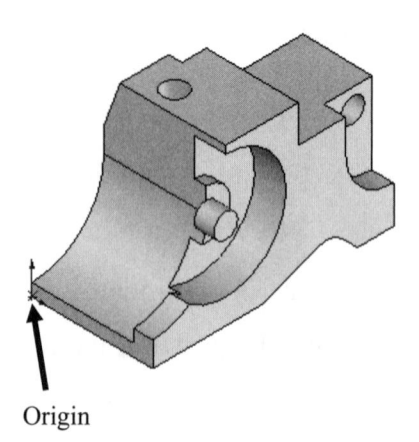

Page 4 - 34

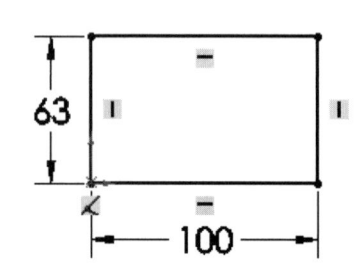

Advanced Part Modeling

6. Create **Sketch2**. Select the right face of Extrude1 as the Sketch plane. Sketch a 90deg tangent arc. Sketch three lines to complete the sketch. Insert the required geometric relations and dimensions.

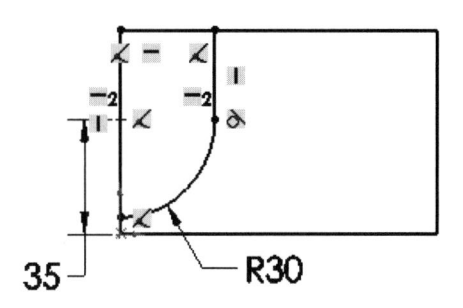

7. Create the first **Extruded Cut** feature. Offset the extrude feature. Select the Offset Start Condition. Offset value = 8.0mm. Blind is the default End Condition. Depth = 50mm.

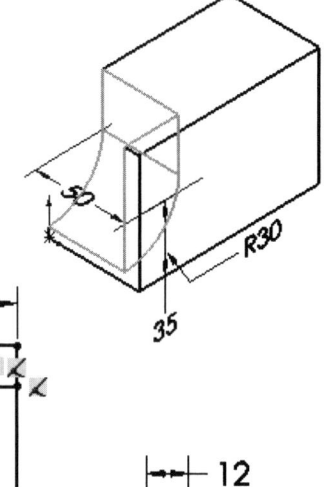

💡 The default Start Condition in the Extrude PropertyManager is Sketch Plane. The Offset start condition starts the extrude feature on a plane that is offset from the current Sketch plane.

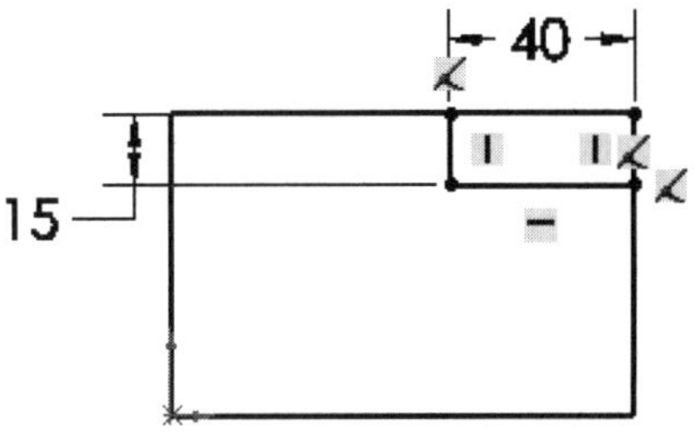

8. Create **Sketch3**. Select the right face as the Sketch plane. Create the Extrude3 profile. Insert the required geometric relations and dimensions.

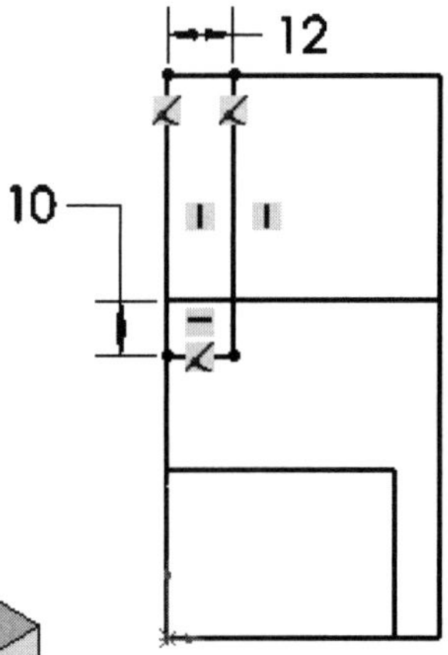

9. Create the second **Extruded Cut** feature. Select Through All for End Condition in Direction 1.

10. Create **Sketch4**. Select the top face of Extrude1 as the Sketch Plane. Select the top edge to reference the 10mm dimension. Insert the required geometric relations and dimensions.

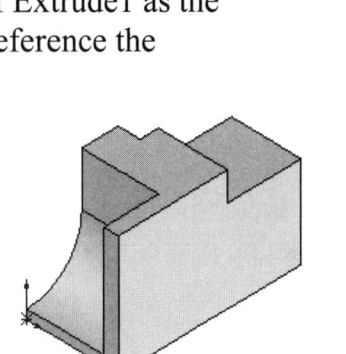

11. Create the third **Extruded Cut** feature. Select Through All for End Condition in Direction 1.

12. Create **Sketch5**. Select the right face of Extrude1 as the Sketch Plane. Apply construction geometry. Sketch a 90deg tangent arc. Sketch three lines to complete the sketch. Insert the required geometric relations and dimensions.

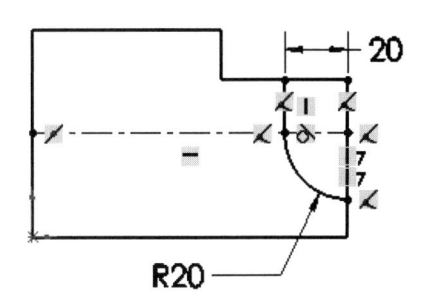

Advanced Part Modeling

13. Create the forth **Extruded Cut** feature. Blind is the default End Condition. Depth = 9mm.

14. Create **Sketch6**. Select the right face of Extrude5 as the Sketch plane. Sketch a circle. Insert the required dimensions.

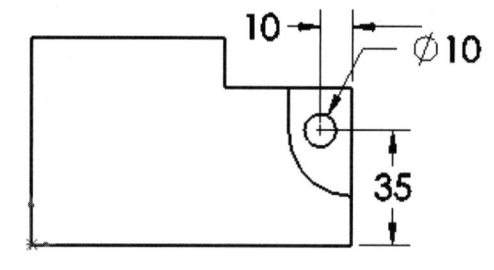

15. Create the fifth **Extruded Cut** feature. Select Through All for End Condition in Direction 1.

16. Create **Sketch7**. Select the top face of Extrude1 as the Sketch Plane. Sketch a circle. Insert the required dimensions and relations.

17. Create the sixth **Extruded Cut** feature. Select Through All for End Condition in Direction 1.

☼ There are numerous ways to create the models in this chapter. The goal is to display different design intents and techniques.

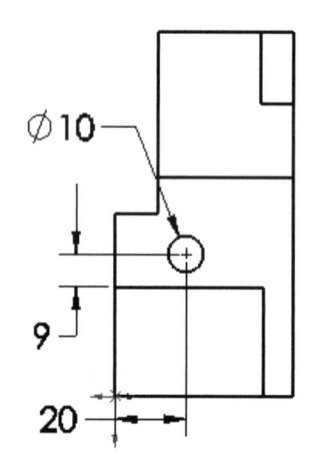

18. Create **Sketch8**. Select the right face of Extrude1 as the Sketch plane. Insert a tangent arc as illustrated. Complete the sketch. Insert the required relations and dimensions.

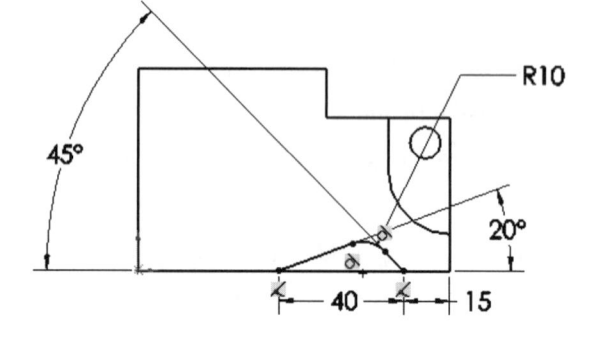

19. Create the seventh **Extruded Cut** feature. Apply symmetry. Select the Through All End Condition in Direction 1 and Direction 2.

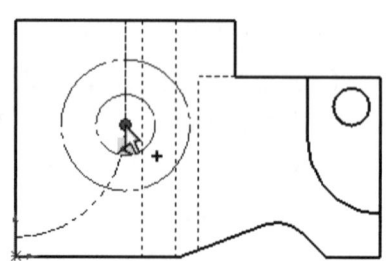

20. Create **Sketch9**. Select the right face of Extrude1 as the Sketch plane. Select Hidden Lines Visible. Sketch two construction circles centered about the end point of the arc. Apply the 3 Point Arc Sketch tool. Complete the sketch. Insert the required relations and dimensions.

Page 4 - 36

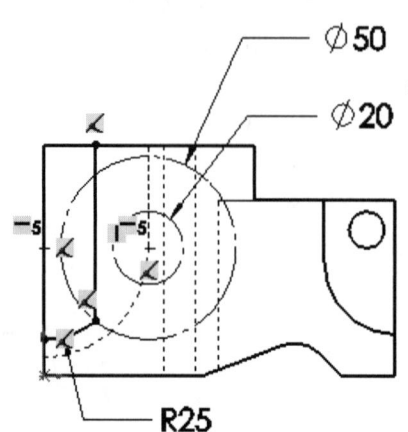

Advanced Part Modeling

21. Create the eight **Extruded Cut** feature. Select the Through All End Condition in Direction 1 and Direction 2. Note the direction of the extrude feature from the illustration.

22. Create **Sketch10**. Select the right face of Extrude1 as the Sketch plane. Sketch a circle centered at the end point of the arc as illustrated. Apply the Trim Entities Sketch tool. Display Sketch9. Complete the sketch. Insert the required geometric relations.

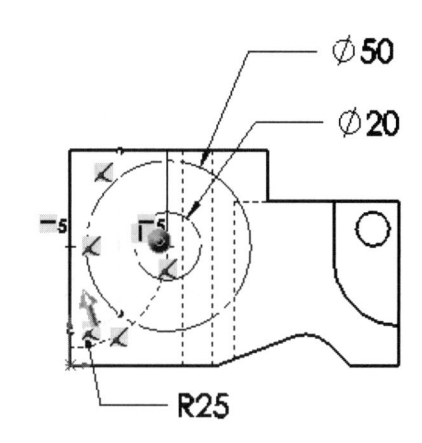

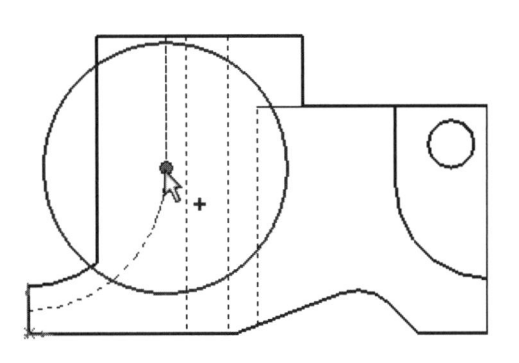

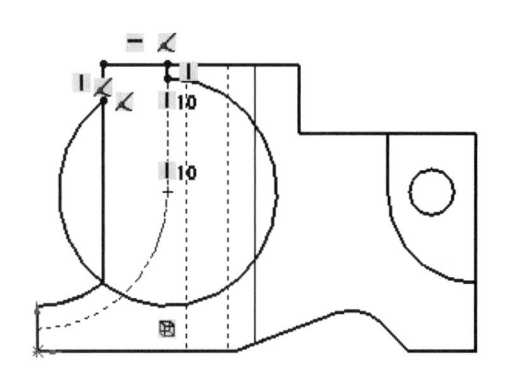

23. Create the ninth **Extruded Cut** feature. Blind is the default End Condition in Direction 1. Depth = 13mm. Extrude10 is displayed.

24. Create **Sketch11**. Select the right face of Extrude10 as the Sketch plane. Select the construction circle from Sketch9, the left arc, and the left edge as illustrated. Apply the Convert Entities Sketch tool. Apply the Trim Sketch tool. Insert the required relations.

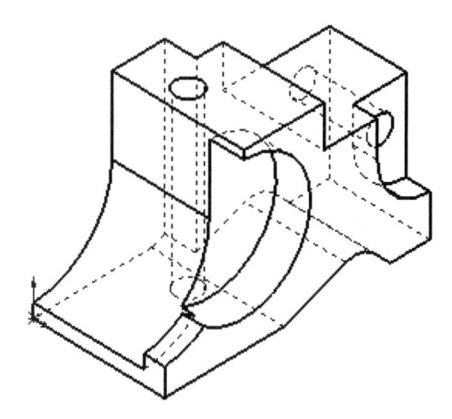

25. Create the **Extruded Boss** feature. Blind is the default End Condition in Direction 1. Depth = 5.00mm.

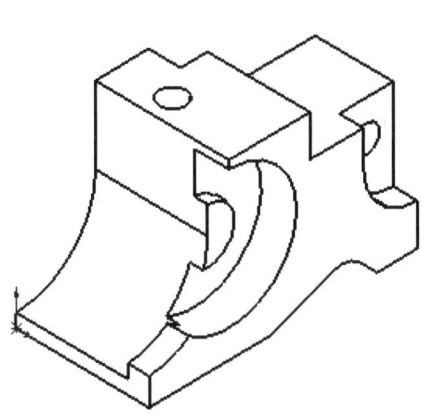

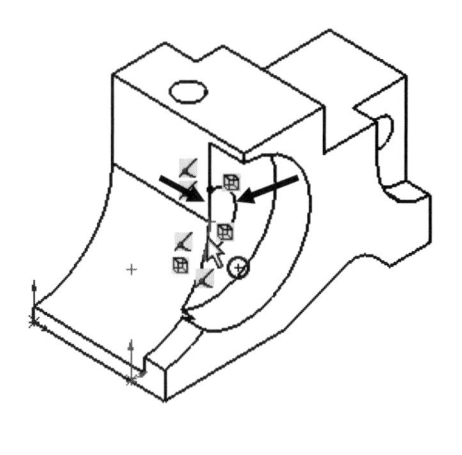

Page 4 - 37

26. Create **Sketch12**. Select the right face of Extrude11 as the Sketch plane. Sketch a circle. Insert the require relation and dimension.

27. Create the **Extruded Boss** feature. Select the Up To Surface End Condition in Direction 1. Select the right face of Extrude1 for Direction 1.

28. Create the **Chamfer** feature. Chamfer the left edge as illustrated. Distance = 18mm. Angle = 20deg.

29. **Assign** the material to the part.

30. **Calculate** the overall mass of the part. The overall mass = 1280.33 grams.

31. **Locate** the Center of mass relative to the part Origin:

    - X: 26.81 millimeters
    - Y: 25.80 millimeters
    - Z: -56.06 millimeters

32. **Save** the part and name it Advanced Part 4-7.

33. **Close** the model.

This example was taken from the SolidWorks website, www.solidworks.com/cswa as an example of an Advanced Part on the CSWA exam. This model has thirteen features and twelve sketches. As stated throughout this book, there are numerous ways to create the models in these chapters.

One of the goals in this book is to display different design intents and techniques, and to provide you with the ability to successfully address the provided models in the given time frame of the CSWA exam.

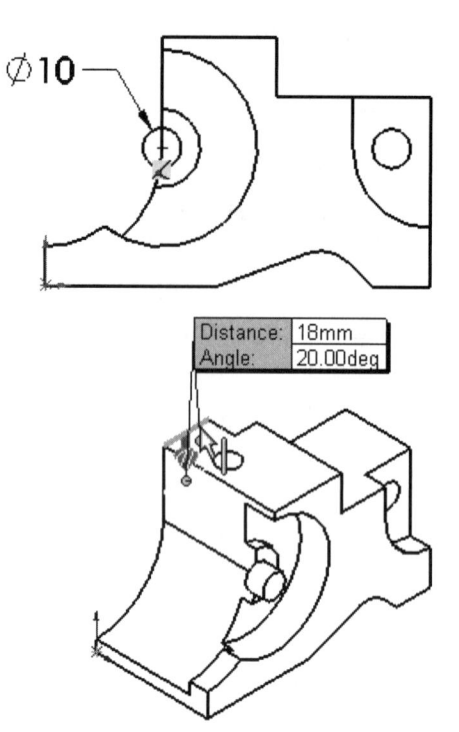

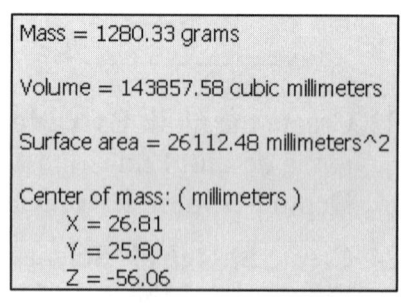

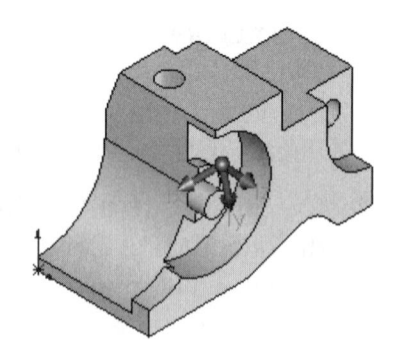

Summary

Advanced Part Modeling is one of the five categories on the CSWA exam. The main difference between the Advanced Part modeling and the Part Modeling category is the complexity of the sketches and the number of dimensions and geometric relations along with an increase in the number of features.

There is one question on the CSWA exam in this category. The question is worth twenty points and is in a multiple choice single answer or fill in the blank format. You are required to create a model, with eight or more features and to answer a question either on the location of the Center of mass relative to part Origin or to a new created coordinate system and all of the mass properties provided in the Mass Properties dialog box. Spend no more than 40 minutes on the question in this category. This is a timed exam. Manage your time.

At this time, there are no sheet metal questions, or questions on Loft, Swept, or Shell features on the exam.

Assembly Modeling is the next chapter in this book. Up to this point, a simple or advanced part was the focus. The Assembly Modeling category addresses an assembly with numerous sub-components. This chapter covers the general concepts and terminology used in Assembly Modeling and then addresses the core elements that are aligned to the exam. Knowledge of Standard mates is required in this category. There is one question on the CSWA exam in this category. The question is worth 30 points. The question is in a multiple choice single answer or fill in the blank format.

Key terms

- *Annotation.* A text note or a symbol that adds specific design intent to a part, assembly, or drawing.

- *Base feature.* The first feature of a part is called the Base feature.

- *Base sketch.* The first sketch of a part is called the Base sketch. The Base sketch is the foundation for the 3D model. Create a 2D sketch on a default plane: Front, Top, and Right in the FeatureManager design tree, or on a created plane.

- *Chamfer.* Bevels a selected edge or vertex. You can apply chamfers to both sketches and features.

- *Constraints.* Geometric relations such as Perpendicular, Horizontal, Parallel, Vertical, Coincident, Concentric, etc. Insert constraints to your model to incorporate design intent.

- *Construction geometry.* The characteristic of a sketch entity that the entity is used in creating other geometry, but is not itself used in creating features. Construction geometry is also called reference geometry.
- *Coordinate system.* A system of planes used to assign Cartesian coordinates to features, parts, and assemblies. Part and assembly documents contain default coordinate systems; other coordinate systems can be defined with reference geometry. Coordinate systems can be used with measurement tools and for exporting documents to other file formats.
- *Geometric tolerance.* A set of standard symbols that specify the geometric characteristics and dimensional requirements of a feature.
- *Hole Wizard.* Provides the ability to specify the parameters for a hole.
- *Mass Properties tool.* Displays the mass properties of a part or assembly model, or the section properties of faces or sketches.
- *Mirror Feature.* Creates a copy of a feature, (or multiple features), mirrored about a selected face or a plane. You can select the feature or you can select the faces that comprise the feature.
- *Pattern.* A pattern repeats selected sketch entities, features, or components in an array, which can be linear, circular, or sketch-driven. If the seed entity is changed, the other instances in the pattern update.
- *Reference Axis.* A reference axis is also called a construction axis. Reference geometry defines the shape or form of a surface or a solid. Reference geometry includes planes, axes, coordinate systems, and points.
- *Reference Plane.* Insert a plane as a reference to apply restraints. Reference geometry defines the shape or form of a surface or a solid. Reference geometry includes planes, axes, coordinate systems, and points.
- *Revolved.* A feature that creates a base or boss, a revolved cut, or revolved surface by revolving one or more sketched profiles around a centerline.

Check your understanding

1. In Tutorial: Advanced Part 4-1 you created the illustrated part. Modify the Base flange thickness from .40in to .50in. Modify the Chamfer feature angle from 45deg to 33deg. Modify the Fillet feature radius from .10in to .125in. Modify the material from 1060 Alloy to Nickel.

Calculate the overall mass of the part, volume, and locate the Center of mass with the provided information.

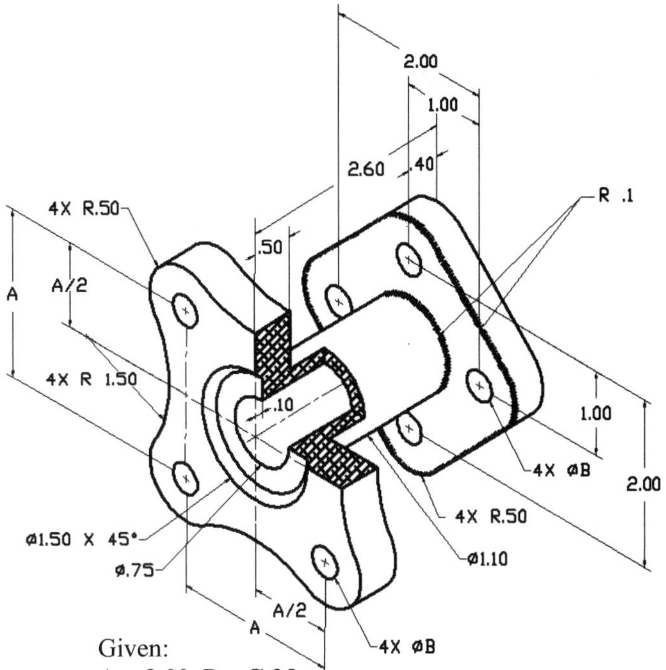

Given:
A = 2.00, B = Ø.35
Material: 1060 Alloy
Density: 0.097 lb/in^3
Units: IPS
Decimal places = 2

2. In Tutorial: Advanced Part 4-2 you created the illustrated part. Modify the CirPattern1 feature. Modify the number of instances from 6 to 8. Modify the seed feature from an 8mm diameter to a 6mm diameter.

Calculate the overall mass, volume, and the location of the Center of mass relative to the part Origin.

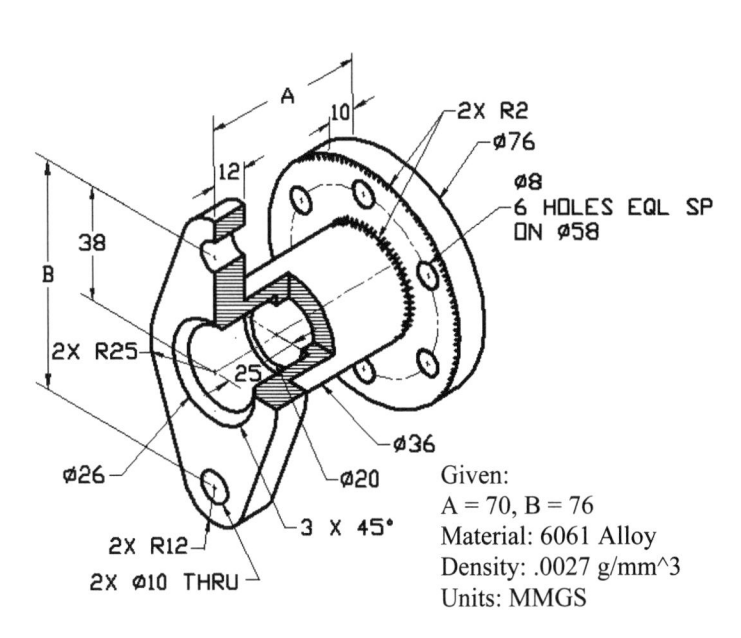

Given:
A = 70, B = 76
Material: 6061 Alloy
Density: .0027 g/mm^3
Units: MMGS

3. In Tutorial: Advanced Part 4-3 you created the illustrated part. Modify the material from 6061 Alloy to Copper. Modify the B dimension from 4.000in. to 3.500in. Modify the Fillet radius from .12in to .14in. Modify the unit system from IPS to MMGS.

Calculate the volume of the part and the location of the Center of mass. Save the part and name it Advance Part 4-3 Copper.

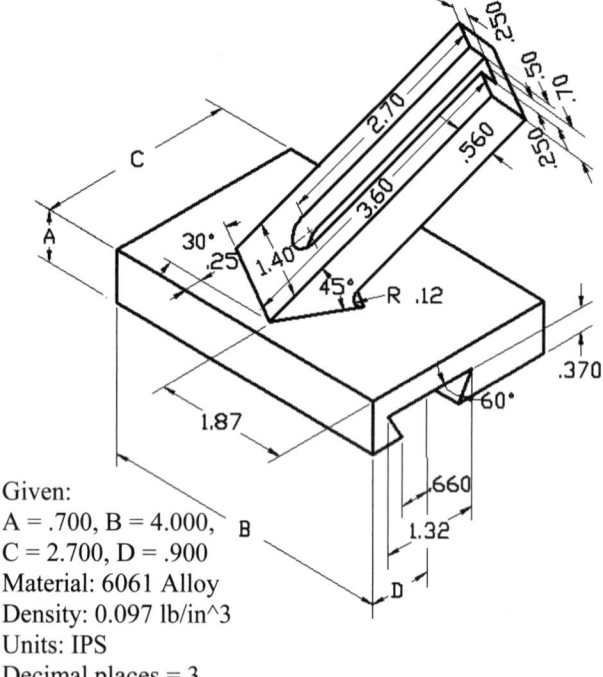

Given:
A = .700, B = 4.000,
C = 2.700, D = .900
Material: 6061 Alloy
Density: 0.097 lb/in^3
Units: IPS
Decimal places = 3

4. Build this illustrated model. Set document properties, identify the correct Sketch planes, apply the correct Sketch and Feature tools, and apply material. Calculate the overall mass of the part, volume and locate the Center of mass with the provided information.

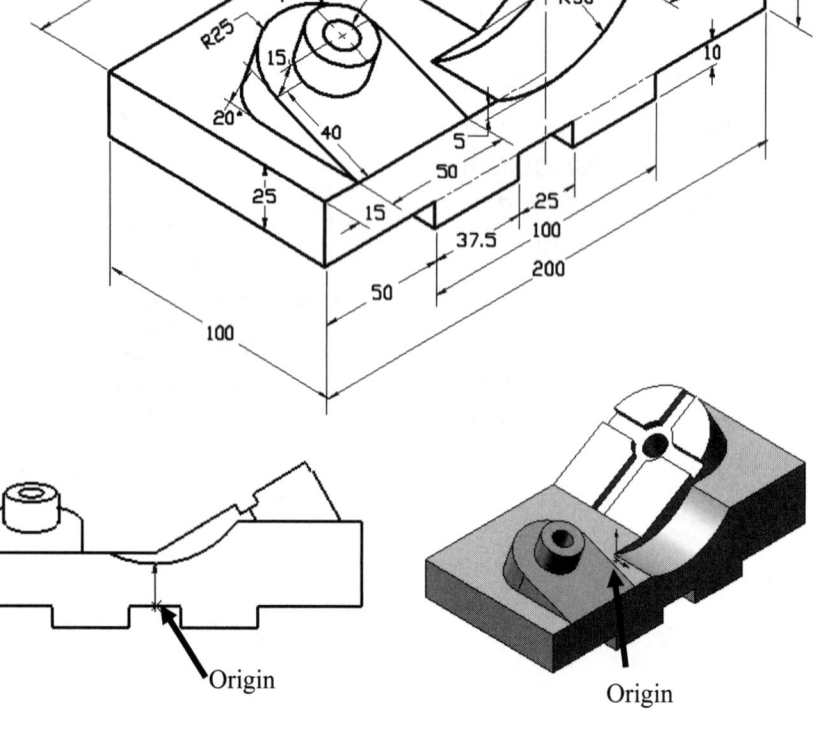

- Material: 6061 Alloy
- Units: MMGS

Origin

Origin

Page 4 - 42

Advanced Part Modeling

5 Build this illustrated model. Calculate the overall mass of the part, volume and locate the Center of mass with the provided information. Where do you start? Build the model, as you did in the above exercise. Create Coordinate System1 to locate the Center of mass for the model.

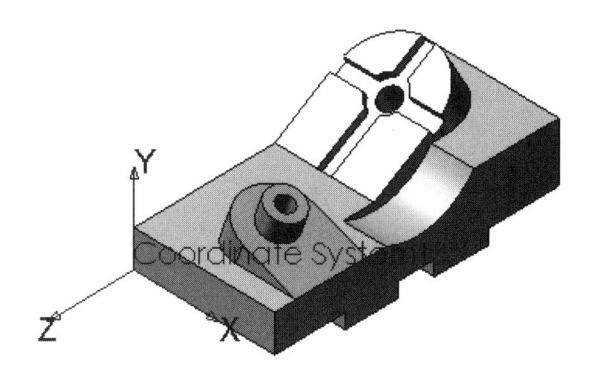

- Material: 6061 Alloy
- Units: MMGS

6. Build this illustrated model. Calculate the overall mass of the part, volume, and locate the Center of mass with the provided information.

- Material: 6061 Alloy
- Units: MMGS

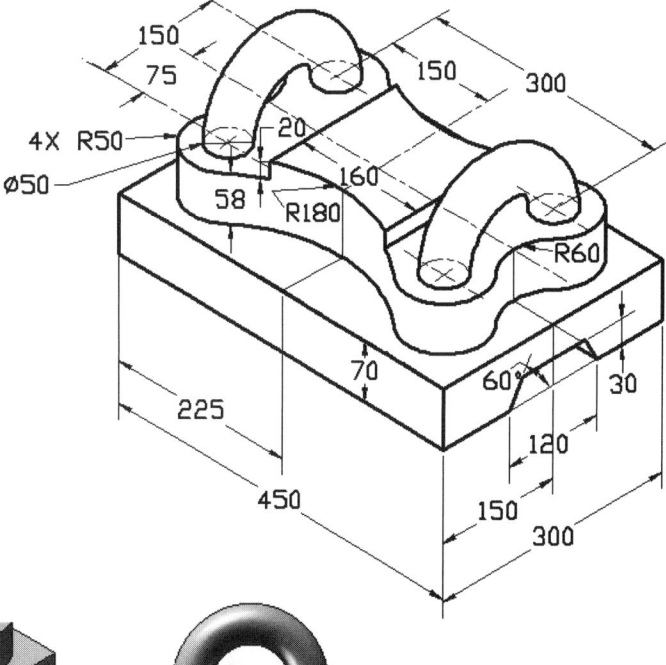

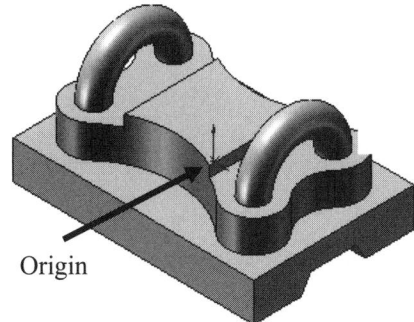

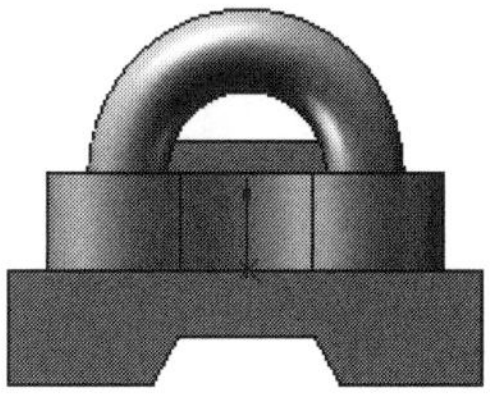

7 Build this model. Calculate the overall mass of the part, volume and locate the Center of mass with the provided information. Where do you start? Build the illustrated model, as you did in the above exercise. Create Coordinate System1 to locate the Center of mass for the model

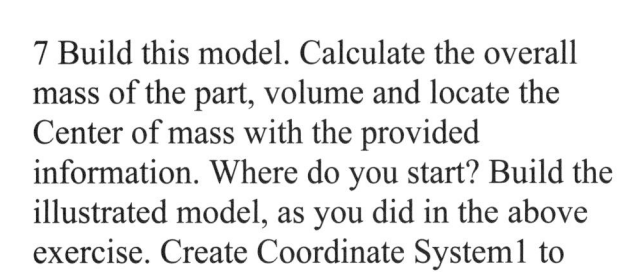

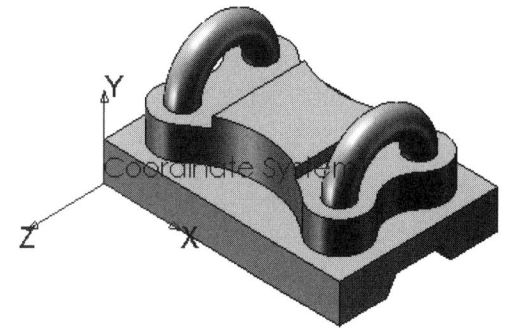

- Material: 6061 Alloy
- Units: MMGS

Notes:

Chapter 5: Assembly Modeling

Chapter Objective

Assembly Modeling is one of the five categories on the CSWA exam. In the last two chapters, a simple or advanced part was the focus. The Assembly Modeling category addresses an assembly with numerous sub-components.

This chapter covers the general concepts and terminology used in Assembly modeling and then addresses the core elements that are aligned to the CSWA exam. Knowledge to build simples parts and to insert Standard mates is required in this category.

There is one question on the CSWA exam in this category. The question is worth thirty points. The question is in a multiple choice single answer or fill in the blank format.

On the completion of the chapter, you will be able to:

- Specify Document Properties
- Identify and build the components to construct the assembly from a detailed illustration using the following features:
  - Extruded Boss/Base
  - Extruded Cut
  - Fillet
  - Mirror
  - Revolved Boss/Base
  - Revolved Cut
  - Linear Pattern
  - Chamfer
  - Hole Wizard
- Identify the first fixed component in an assembly
- Build a Bottom-up assembly with the following Standard Mates:
  - Coincident, Concentric, Parallel, Perpendicular, Tangent, Angle, and Distance

Assembly Modeling

- Aligned, Anti-Aligned option
- Apply the Mirror Component tool
- Locate the Center of mass relative to the assembly
- Create a coordinate system location
- Locate the Center of mass relative to a created coordinate system
- Calculate the overall mass and volume for the created assembly
- Mate the first component with respect to the assembly reference planes

💡 At this time, Advance Mates and Mechanical Mates are NOT addressed on the CSWA exam.

Assembly modeling techniques

There are two key assembly modeling techniques:

- Top-down, "in-context" assembly modeling
- Bottom-up assembly modeling

In top-down assembly modeling, one or more features of a part are defined by something in an assembly, such as a sketch or the geometry of another component. The design intent comes from the top, and moves down into the individual components, hence the name top-down assembly modeling.

💡 Whenever you create a part or feature using top-down assembly modeling techniques, external references are created to the geometry you referenced.

💡 At this time, sheet metal assemblies are not included on the CSWA exam.

Bottom-up assembly modeling is a traditional method that combines individual components. Based on design criteria, the components are developed independently. The three major steps in a bottom-up design approach are:

1. Create each part independent of any other component in the assembly.
2. Insert the parts into the assembly.
3. Mate the components in the assembly as they relate to the physical constraints of your design.

Assembly Modeling

☼ To modify a component in an assembly using the bottom-up assembly approach, you must edit the individually part. The modification is then made to the assembly.

The bottom-up assembly modeling approach is used in this book to address the models in the Assembly Modeling category of the CSWA exam.

Mates

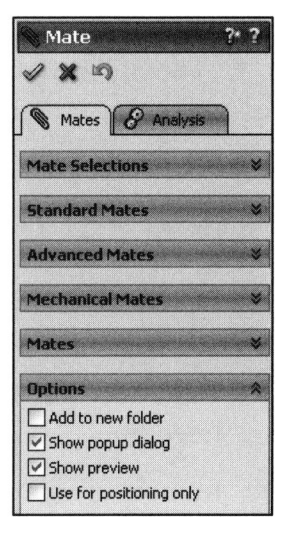

Mates create geometric relationships between assembly components. As you insert mates, you define the allowable directions of linear or rotational motion of the components in the assembly. You can move a component within its degrees of freedom, visualizing the assembly's behavior.

The Mate PropertyManager displays' Standard Mates, Advanced Mates, and Mechanical Mate. The Standard Mate types are:

- *Standard Mates*. The Standard Mates box provides the following selection:

  - **Coincident**. Locates the selected faces, edges, or planes so they use the same infinite line. A Coincident mate positions two vertices for contact.

  - **Parallel**. Locates the selected items to lie in the same direction and to remain a constant distance apart.

  - **Perpendicular**. Locates the selected items at a 90 degree angle to each other.

  - **Tangent**. Locates the selected items in a tangent mate. At least one selected item must be either a conical, cylindrical, spherical face.

  - **Concentric**. Locates the selected items so they can share the same center point.

  - **Lock**. Maintains the position and orientation between two components.

  - **Distance**. Locates the selected items with a specified distance between them. Use the drop-down arrow box or enter the distance value directly.

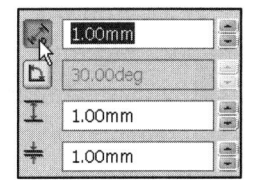

Assembly Modeling

- **Angle**. Locates the selected items at the specified angle to each other. Use the drop-down arrow box or enter the angle value directly.

- **Mate alignment**. Provides the ability to toggle the mate alignment as necessary. There are two options. They are:

  - **Aligned**. Locates the components so the normal or axis vectors for the selected faces point in the same direction.

  - **Anti-Aligned**. Locates the components so the normal or axis vectors for the selected faces point in the opposite direction.

☼ The SmartMates functionality saves time by allowing you to create commonly-used mates without using the Mate PropertyManager.

Creating mates

When creating mates, there are a few basic procedures to remember. They are:

- Click and drag components in the Graphics window to assess their degrees of freedom.

- Remove display complexity. Hide components when visibility is not required.

- Utilize Section views to select internal geometry. Utilize Transparency to see through components required for mating.

- Apply the Move Component and Rotate Component tool from the Assembly toolbar before mating. Position the component in the correct orientation.

- Use a Coincident mate when the distance value between two entities is zero. Utilize a Distance mate when the distance value between two entities in not zero.

- Apply various colors to features and components to improve visibility for the mating process.

- Resolve a mate error as soon as it occurs. Adding additional mates will not resolve the earlier mate problem.

- Use the View Mates tool, (Right-click a **component** and select **View Mates**) or expand the component in the FeatureManager design tree using **Tree Display, View Mates**, and **Dependencies** to view the mates for each component.

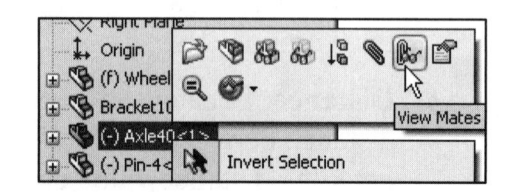

The following tutorials provide a general review on some of the Standard Mate types.

Tutorial: Standard mate 5-1

Create a Weight-Hook assembly. Insert a Concentric mate and two Coincident mates.

1. **Open** Mate 5-1 assembly from the SolidWorks CSWA Folder\Chapter5 location. View the two components: WEIGHT and HOOK. The WEIGHT component is fixed to the assembly Origin.

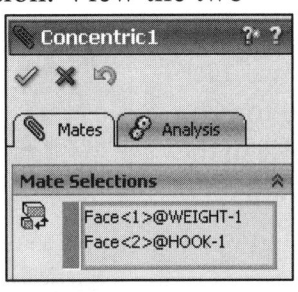

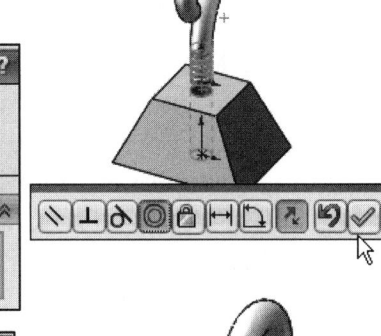

2. **Insert** a Concentric mate between the inside top cylindrical face of the WEIGHT and the cylindrical face of the thread. Concentric is the default mate.

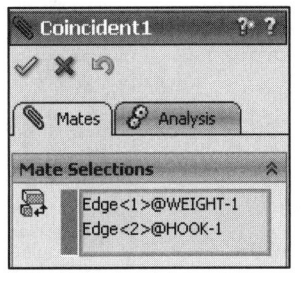

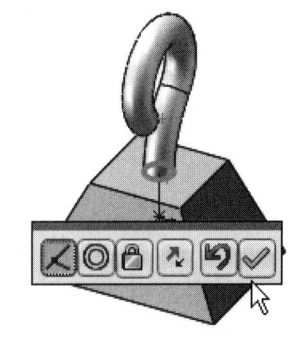

3. Click the **Green check mark**. Concentric1 is created.

4. **Insert** the first Coincident mate between the top edge of the circular hole of the WEIGHT and the top circular edge of Sweep1, above the thread. Coincident is the default mate.

5. Click the **Green check mark**. Coincident1 is created. The HOOK can rotate in the WEIGHT. Fix the position of the HOOK.

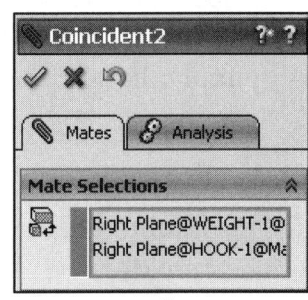

6. **Insert** the second Coincident mate between the Right Plane of the WEIGHT and the Right Plane of the HOOK. Coincident is the default mate.

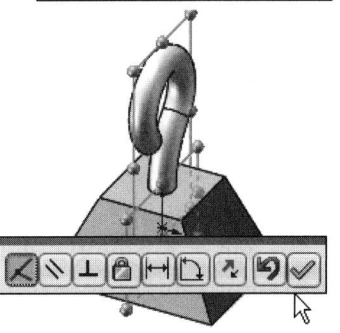

7. Click the **Green check mark**. Coincident2 is created.

8. **Expand** the Mates folder from the FeatureManager. View the three created mates.

9. **Save** the part and name it Mate 5-1.

10. **Close** the model.

Assembly Modeling

Tutorial: Standard mate 5-2

Create a Weight-Link assembly. Insert a Tangent and Coincident mate.

1. **Open** Mate 5-2 assembly from the SolidWorks CSWA Folder\Chapter5 location. View the two components and sub-assembly. The Axle component is fixed to the assembly Origin.

2. **Insert** a Tangent mate between the inside bottom cylindrical face of the FLATBAR and the top circular face of the HOOK, in the WEIGHT-AND-HOOK sub-assembly. Tangent mate is selected by default.

3. Click the **Green check mark**. Tangent1 is created.

4. **Insert** a Coincident mate between the Front Plane of the FLATBAR and the Front Plane of the WEIGHT-AND-HOOK sub-assembly. Coincident mate is selected by default. The WEIGHT-AND-HOOK sub-assembly is free to move in the bottom circular hole of the FLATBAR.

5. Click the **Green check mark**. Coincident3 is created.

6. **Expand** the Mates folder from the FeatureManager. View the created mates.

7. **Save** the part and name it Mate 5-2.

8. **Close** the model.

💡 MateXpert is a tool that allows you to identify mating problems in an assembly. You can examine the details of mates which are not satisfied, and identify groups of mates which over define the assembly.

💡 You can also use mate callouts and View Mate Errors to help identify and resolve mating problems. Right-click a **component**, of the assembly or of a sub-assembly and select **View Mates**.

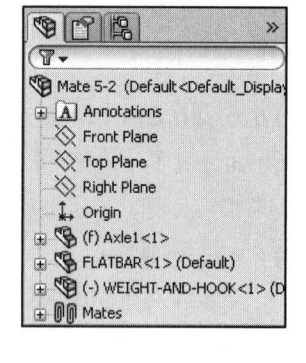

Assembly Modeling

Tutorial: Standard mate 5-3

Insert a Parallel mate.

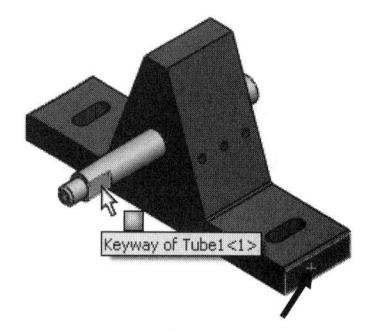

1. **Open** Mate 5-3 assembly from the SolidWorks CSWA Folder\Chapter5 location. View the two components. The Wedge-1 component is fixed to the assembly Origin.

2. Click and drag the **Tube1** component in a horizontal direction. Tube1 travels linearly in Wedge-1.

3. **Rotate** Tube1 until the Keyway feature is approximately parallel to the right face of Wedge1.

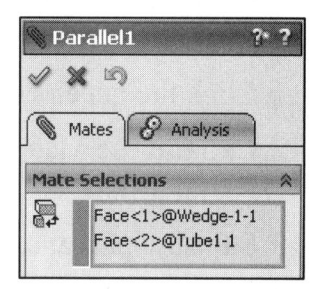

4. **Insert** a Parallel mate between the face of the Keyway feature and the flat right face of Wedge1. The selected entities are displayed in the Mate Selections box.

5. Select **Parallel** mate.

6. Click the **Green check mark**. Parallel1 is created.

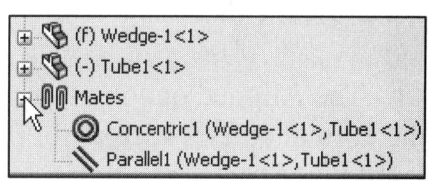

7. **Expand** the Mates folder from the FeatureManager. View the create mate.

8. **Save** the part and name it Mate 5-3.

9. **Close** the model.

Tutorial: Standard mate 5-4

Modify the component state from fixed to float. Insert two Coincident mates and a Distance mate.

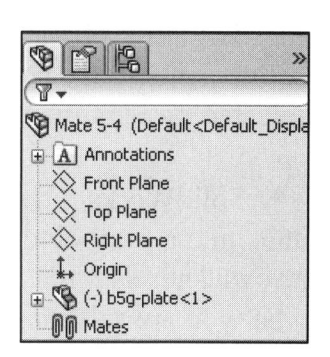

1. **Open** Mate 5-4 assembly from the SolidWorks CSWA folder.

2. Right-click **bg5-plate** from the FeatureManager.

3. Click **Float**. The bg5-plate part is not fixed to the Origin of the assembly.

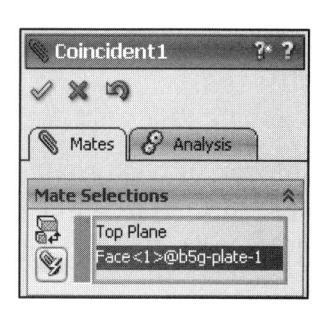

4. **Insert** a Coincident mate between the Top Plane of Mate 5-4 and the back face of bg5-plate. Top Plane and Face<1>@b5g-plate-1 are displayed in the Mate Selections box. Coincident is selected by default.

5. Click the **green check mark**. Coincident1 is created.

Page 5 - 7

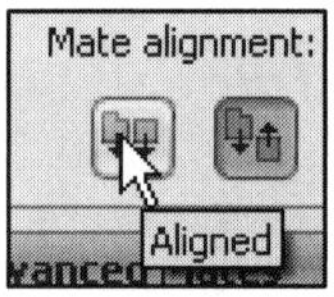

Assembly Modeling

6. **Insert** a Coincident mate between the Front Plane of Mate 5-4 and the Top Plane of bg5-plate.

7. Click **Aligned** from the Standard Mates box.

8. Click **Anti-Aligned** to review the position.

9. **Display** an Isometric view. The narrow cut faces front. Click **OK** from the Coincident2 PropertyManager.

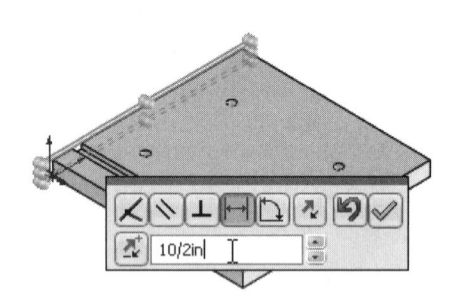

10. **Insert** a Distance mate between the Right Plane of Mate 5-4 and the Right Plane of bg5-plate. Distance = 10/2in. Note: The part is in inches and the assembly is in millimeter units. Click the Flip direction box if required.

11. Click the **green check mark**. Distance1 is created.

12. **Expand** the Mates folder from the FeatureManager. View the created mates.

13. **Save** the part and name it Mate 5-4.

14. **Close** the model.

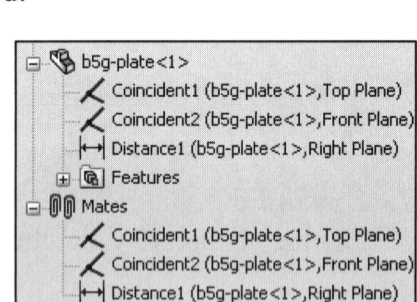

💡 The Aligned option places the components so the normal or axis vectors for the selected faces point in the same direction.

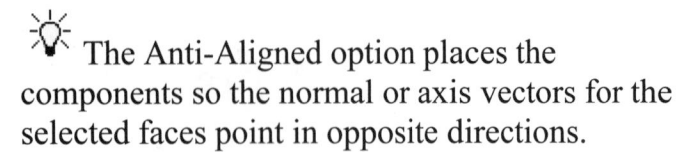

💡 The Anti-Aligned option places the components so the normal or axis vectors for the selected faces point in opposite directions.

💡 SolidWorks provides system feedback by attaching a symbol to the mouse pointer cursor arrow. The system feedback symbol indicates what you are selecting or what the system is expecting you to select. As you move the mouse pointer across the model, system feedback is provided to you in the form of symbols, riding next to the cursor arrow.

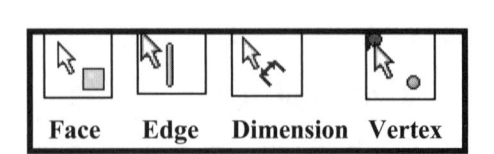

In the next section, build an assembly from the illustrated model. Information is provided on the individual parts from the illustrated assembly. You are required to build each part, and then insert the part into the assembly and insert the correct Standard mates. It is very important to select the correct face, edge, vertex, etc.

Build an assembly from a detailed dimensioned illustration

An exam question in this category could read: Build this assembly. Locate the Center of mass of the model with respect to the illustrated coordinate system.

The assembly contains the following: One Clevis component, three Axle components, two 5 Hole Link components, two 3 Hole Link components, and six Collar components. All holes Ø.190 THRU unless otherwise noted. Angle A = 150deg. Angle B = 120deg. Note: The location of the illustrated coordinate system: (+X, +Y, +Z).

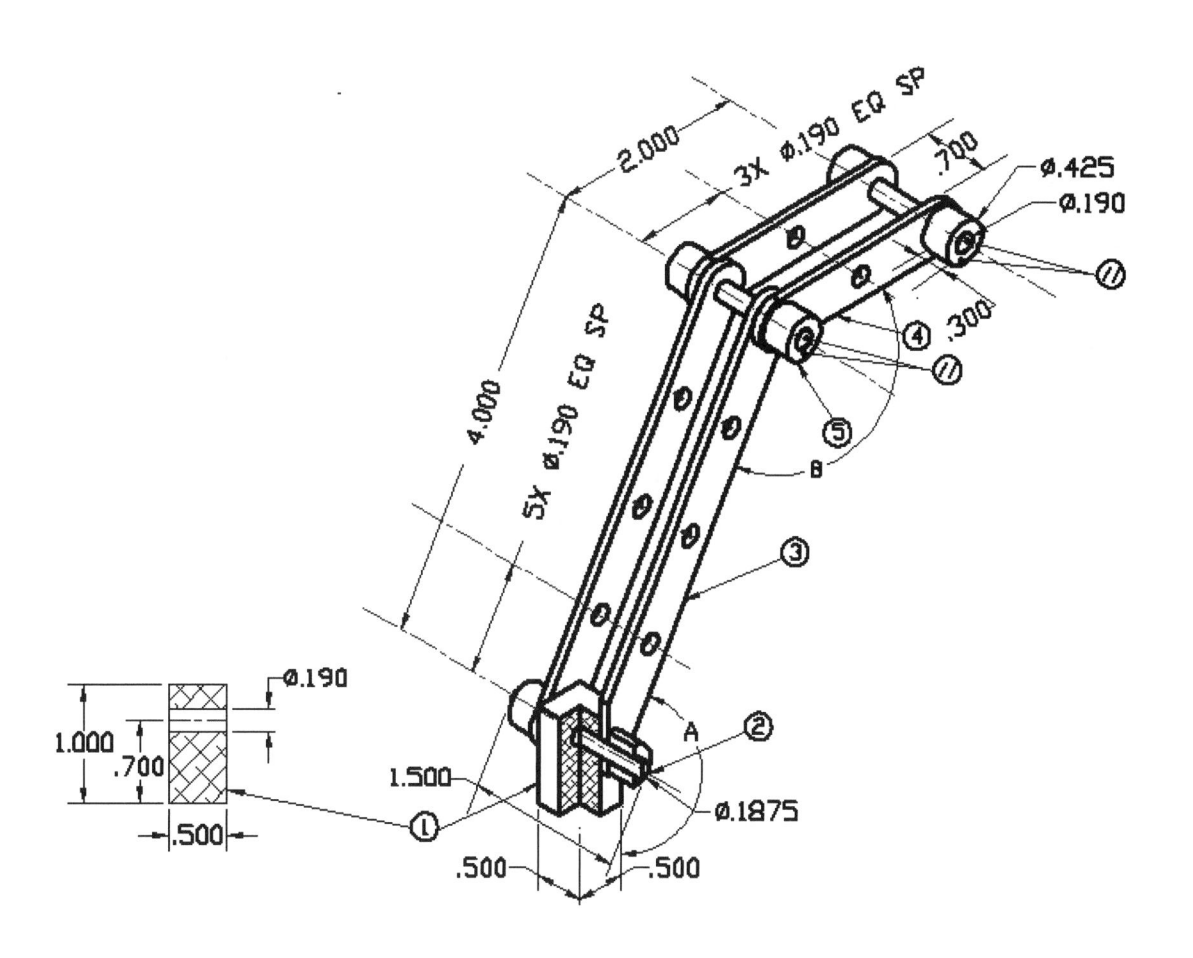

Assembly Modeling

- Clevis, (Item 1): Material: 6061 Alloy. The two 5 Hole Link components are positioned with equal Angle mates, (150deg) to the Clevis component.

- Axle, (Item 2): Material: AISI 304. The first Axle component is mated Concentric and Coincident to the Clevis. The second and third Axle components are mated Concentric and Coincident to the 5 Hole Link and the 3 Hole Link components respectively.

- 5 Hole Link, (Item 3): Material: 6061 Alloy. Material thickness = .100in. Radius = .250in. Five holes located 1in. on center. The 5 Hole Link components are position with equal Angle mates, (120deg) to the 3 Hole Link components.

- 3 Hole Link, (Item 4): Material: 6061 Alloy. Material thickness = .100in. Radius = .250in. Three holes located 1in. on center. The 3 Hole Link components are positioned with equal Angle mates, (120deg) to the 5 Hole Link components.

- Collar, (Item 5): Material: 6061 Alloy. The Collar components are mated Concentric and Coincident to the Axle and the 5 Hole Link and 3 Hole Link components respectively.

Think about the steps that you would take to build the illustrated assembly. Identify and build the required parts. Identify the first fixed component. Position the Base component features in the part so they are in the correct orientation in the assembly. Insert the required Standard mates. Locate the Center of mass of the model with respect to the illustrated coordinate system. In this example, start with the Clevis part.

Tutorial Assembly model 5-1

Build the Clevis part.

1. **Create** a New part in SolidWorks.

2. **Build** the illustrated Clevis part. Insert two features: Extruded Base, Extruded Cut. Think about the steps that you would take to build this part. Identify the location of the part Origin. Reflect back to the assembly illustration.

3. **Set** the document properties for the model.

4. Create **Sketch1**. Sketch1 is the Base sketch. Select the Top Plane as the Sketch plane. Sketch a square. Apply construction geometry. Insert the required geometric relations and dimension. Note the location of the Origin.

5. Create the **Extruded Base** feature. Extrude1 is the Base feature. Blind is the default End Condition in Direction 1. Depth = 1.000in.

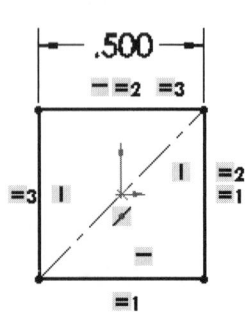

Assembly Modeling

💡 There are numerous ways to create the models in this chapter. The goal is to display different design intents and techniques.

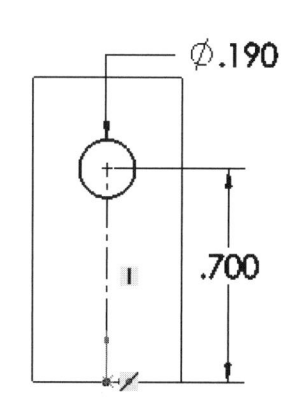

6. Create **Sketch2**. Select the right face of Extrude1 as the Sketch plane. Sketch a circle. Insert the required geometric relation and dimensions.

7. Create the **Extruded Cut** feature. Select Through All for End Condition in Direction 1.

8. **Assign** 6061 Alloy material to the part.

9. **Save** the part and name it Clevis.

💡 Leave the created models open. This will save time when you build the assembly during the exam. The open models are displayed in the Insert Components PropertyManager.

Build the Axle part.

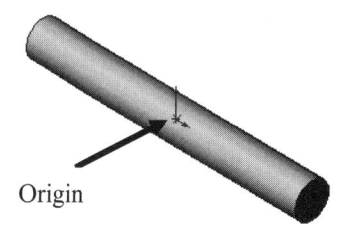

1. **Create** a New part in SolidWorks.

2. **Build** the illustrated Axle part. Think about the steps that you would take to build this part. Identify the location of the part Origin.

3. **Set** the document properties for the model.

4. Create **Sketch1**. Sketch1 is the Base sketch. Select the Right Plane as the Sketch plane. Sketch a circle. Insert the required dimension. Note the location of the Origin.

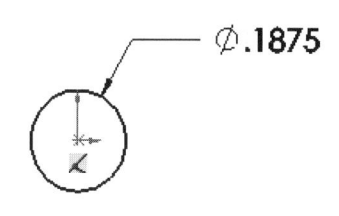

5. Create the **Extrude1** feature. Extrude1 is the Base feature. Select Mid Plane for End Condition in Direction 1. Depth = 1.500in.

6. **Assign** AISI 304 material to the part.

7. **Save** the part and name it Axle.

Create the 3 Hole Link part.

1. **Create** a New part in SolidWorks.

2. **Build** the illustrated 3 Hole Link part. Insert three features: Extruded Base, Extruded Cut, and Linear Pattern. Identify the location of the part Origin.

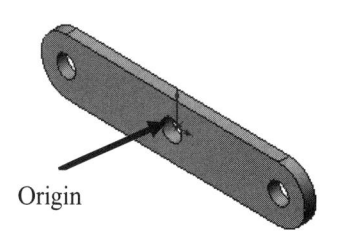

Assembly Modeling

3. **Set** the document properties for the model.

4. Create **Sketch1**. Sketch1 is the Base sketch. Select the Front Plane as the Sketch plane. Apply construction geometry. Use the Tangent Arc and Line Sketch tool. Insert the required geometric relations and dimensions.

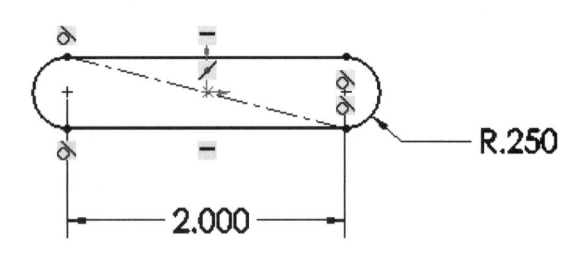

5. Create the **Extruded Base** feature. Extrude1 is the Base feature. Select Mid Plane for End Condition in Direction 1. Depth = .100in.

6. Create **Sketch2**. Select the front face of Extrude1 as the Sketch plane. Sketch a circle centered at the left tangent arc. Insert the required geometric relation and dimension.

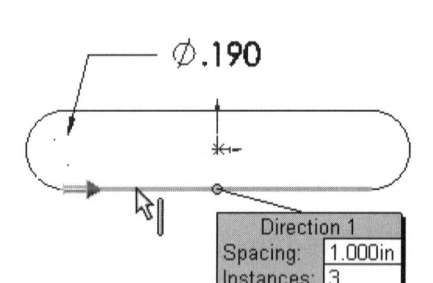

7. Create the **Extruded Cut** feature. Select Through All for End Condition in Direction 1. Extrude2 is the seed feature for LPattern1.

8. Create the **Linear Pattern** feature. Spacing = 1.000in. Instances = 3.

9. **Assign** 6061 Alloy material to the part.

10. **Save** the part and name it 3 Hole Link.

Create the 5 Hole Link part.

1. Apply the **Save as Copy** command to save the 5 Hole Link. The CSWA exam is a timed exam. Conserve design time. You can either: create a new part, apply the ConfigurationManager, or use the Save as Copy command to modify the existing 3 Hole Link to the 5 Hole Link part.

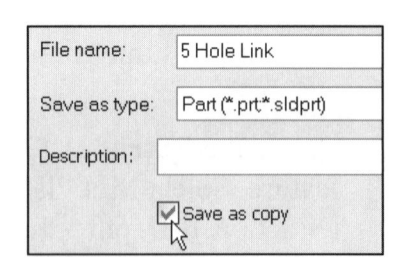

2. **Open** the 5 Hole Link part.

3. **Edit** Sketch1. Modify the dimension as illustrated for the 5 Hole Link part.

4. **Edit** LPattern1. Modify the number of Instances for the 5 Hole Link part. Instances = 5.

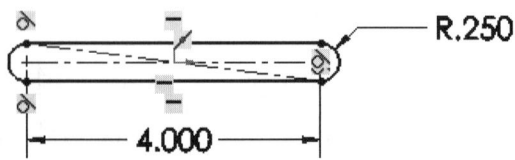

5. **Save** the 5 Hole Link part. Note: Do not overwrite the 3 Hole Link part.

Page 5 - 12

Assembly Modeling

Create the Collar part.

1. **Create** a New part in SolidWorks.

2. **Build** the illustrated Collar part. Insert an Extrude1 feature. Apply symmetry. Identify the location of the part Origin.

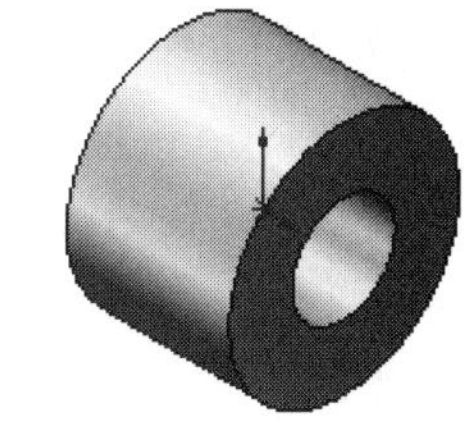

3. **Set** the document properties for the model.

4. Create **Sketch1**. Sketch1 is the Base sketch. Select the Right Plane as the Sketch plane. Sketch two circles centered about the Origin. Insert the required dimensions.

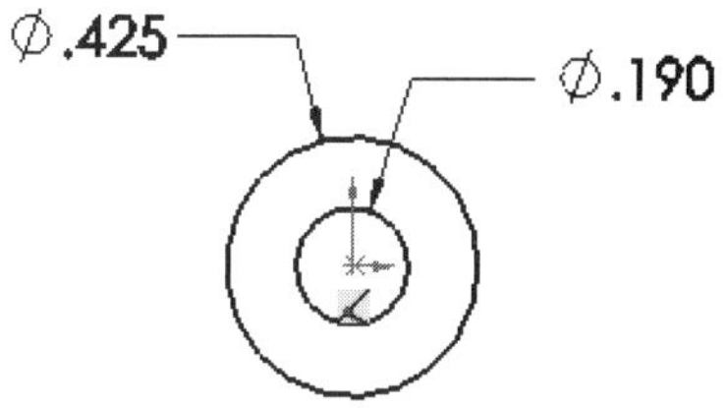

5. Create the **Extruded Base** feature. Extrude1 is the Base feature. Apply Symmetry. Select Mid Plane for End Condition in Direction 1. Depth = .300in.

6. **Assign** 6061 Alloy material to the part.

7. **Save** the part and name it Collar.

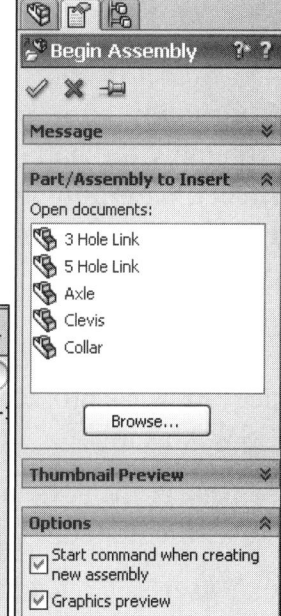

Create the assembly. The illustrated assembly contains the following: One Clevis component, three Axle components, two 5 Hole Link components, two 3 Hole Link components, and six Collar component s. All holes Ø.190 THRU unless otherwise noted. Angle A = 150deg. Angle B = 120deg.

1. **Create** a New assembly in SolidWorks. The created models are displayed in the Open documents box. Click Cancel ✖ from the Begin Assembly PropertyManager. Assem1 is the default document name. Assembly documents end with the extension; .sldasm.

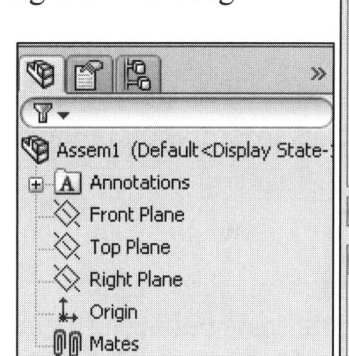

There are numerous ways to insert components to a new or existing assembly.

In Chapter 3, you addressed symbols from the Dimension PropertyManager. During the exam, various engineering symbols will be provided in the illustration of the part or assembly, such as parallelism.

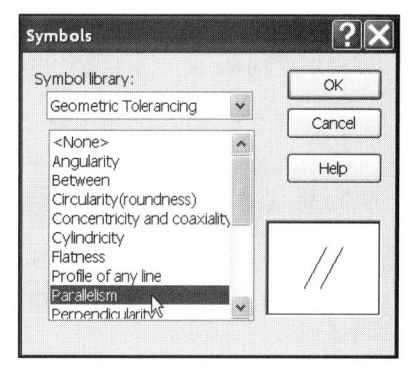

Assembly Modeling

2. **Set** the document properties for the model.
3. **Insert** the Clevis part. Display the Origin.
4. **Fix** the component to the assembly Origin. Click OK from the Insert Component PropertyManager. The Clevis is displayed in the Assembly FeatureManager and in the Graphics window.

☼ Fix the position of a component so that it cannot move with respect to the assembly Origin. By default, the first part in an assembly is fixed; however, you can float it at any time.

☼ To remove the fixed state, Right-click a **component name** in the FeatureManager. Click **Float**. The component is free to move.

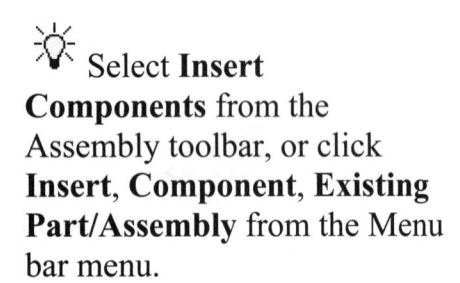

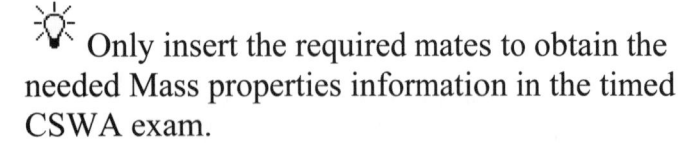

☼ Select **Insert Components** from the Assembly toolbar, or click **Insert, Component, Existing Part/Assembly** from the Menu bar menu.

☼ Only insert the required mates to obtain the needed Mass properties information in the timed CSWA exam.

5. **Insert** the Axle part above the Clevis component as illustrated.
6. **Insert** a Concentric mate between the inside cylindrical face of the Clevis and the outside cylindrical face of the Axle. The selected face entities are displayed in the Mate Selections box. Concentric1 is created.
7. **Insert** a Coincident mate between the Right Plane of the Clevis and the Right Plane of the Axle. Coincident1 mate is created.
8. **Insert** the 5 Hole Link part. Locate and rotate the component as illustrated. Note the location of the Origin.

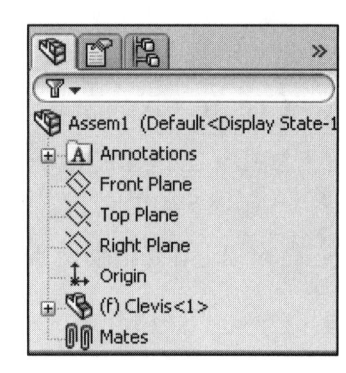

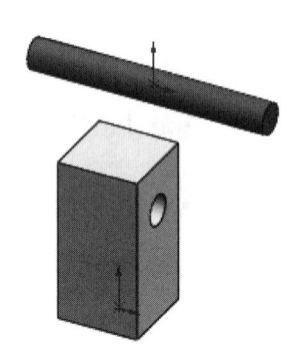

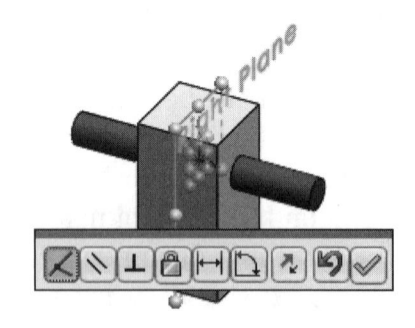

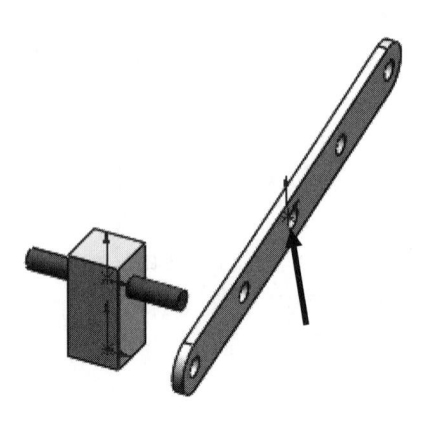

Page 5 - 14

Assembly Modeling

9. **Insert** a Concentric mate between the outside cylindrical face of the Axle and the inside cylindrical face of the 5 Hole Link. Concentric2 is created.

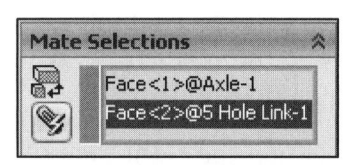

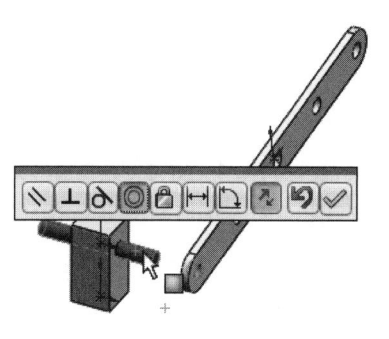

10. **Insert** a Coincident mate between the right face of the Clevis and the left face of the 5 Hole Link. Coincident2 is created.

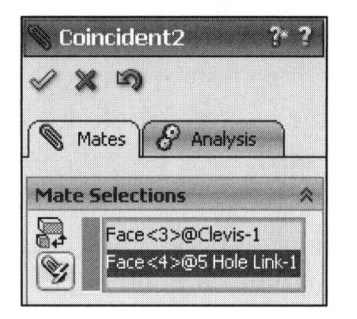

11. **Insert** an Angle mate between the bottom face of the 5 Hole Link and the back face of the Clevis. Angle = 30deg. The selected faces are displayed in the Mate Selections box. Angle1 is created. Flip direction if needed.

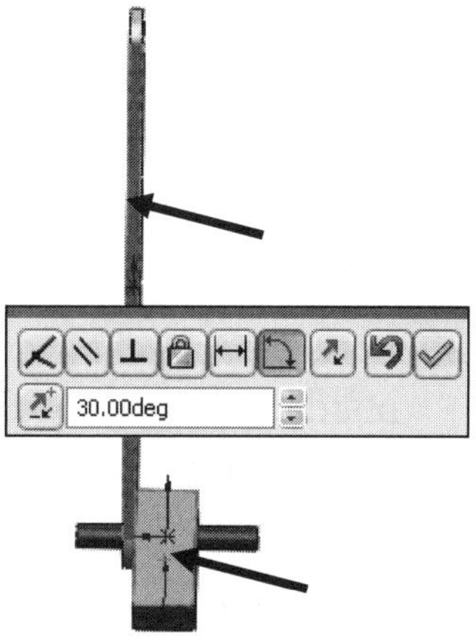

☼ Depending on the component orientation, select the Flip Dimension option and or enter the supplement of the angle.

12. **Insert** the second Axle part. Locate the second Axle component near the end of the 5 Hole Link as illustrated.

13. **Insert** a Concentric mate between the inside cylindrical face of the 5 Hole Link and the outside cylindrical face of the Axle. Concentric3 is created.

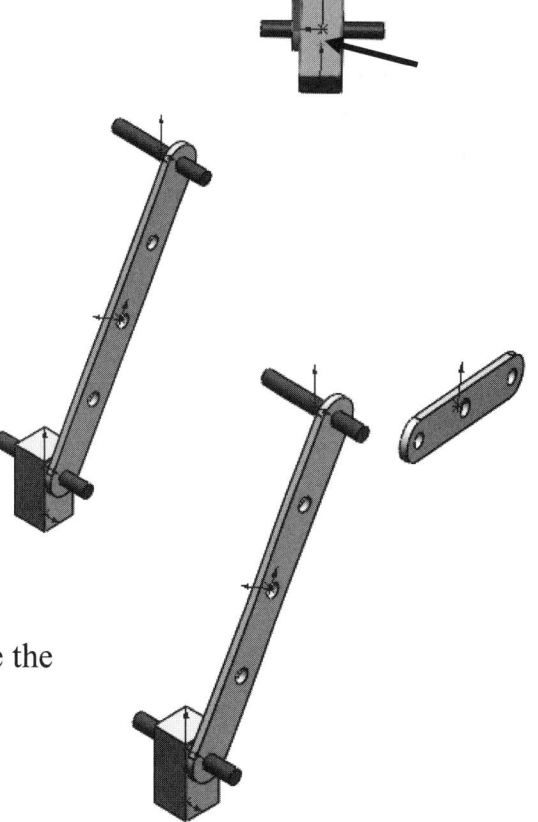

14. **Insert** a Coincident mate between the Right Plane of the assembly and the Right Plane of the Axle. Coincident3 is created.

15. **Insert** the 3 Hole Link part. Locate and rotate the component as illustrated. Note the location of the Origin.

Assembly Modeling

16. **Insert** a Concentric mate between the outside cylindrical face of the Axle and the inside cylindrical face of the 3 Hole Link. Concentric4 is created.

17. **Insert** a Coincident mate between the right face of the 5 Hole Link and the left face of the 3 Hole Link.

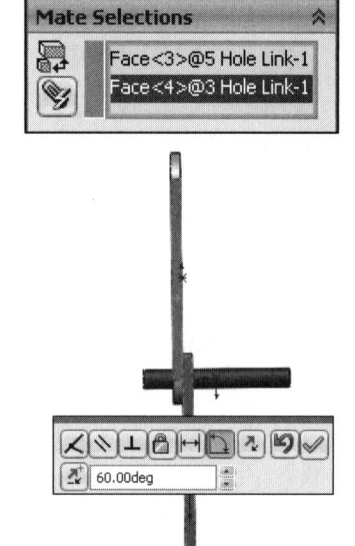

18. **Insert** an Angle mate between the bottom face of the 5 Hole Link and the bottom face of the 3 Hole Link. Angle = 60deg. Angle2 is created.

☼ Depending on the component orientation, select the Flip Dimension option and or enter the supplement of the angle when needed.

☼ Apply the Measure tool to check the angle. In this case, the Measure tool provides the supplemental angle.

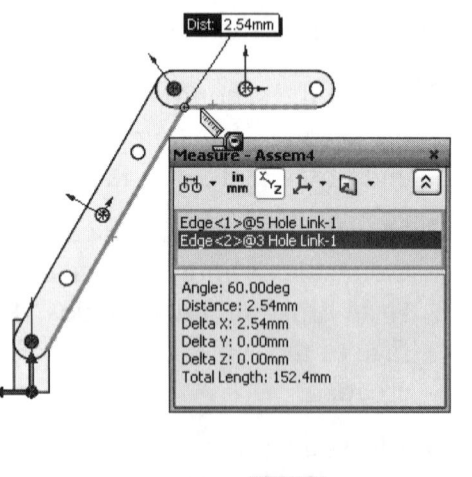

19. **Insert** the third Axle part.

20. **Insert** a Concentric mate between the inside cylindrical face of the 3 Hole Link and the outside cylindrical face of the Axle.

21. **Insert** a Coincident mate between the Right Plane of the assembly and the Right Plane of the Axle.

22. **Insert** the Collar part. Locate the Collar near the first Axle component.

23. **Insert** a Concentric mate between the inside cylindrical face of the Collar and the outside cylindrical face of the first Axle.

Assembly Modeling

24. **Insert** a Coincident mate between the right face of the 5 Hole Link and the left face of the Collar.

25. **Insert** the second Collar part. Locate the Collar near the second Axle component

26. **Insert** a Concentric mate between the inside circular face of the second Collar and the outside circular face of the second Axle.

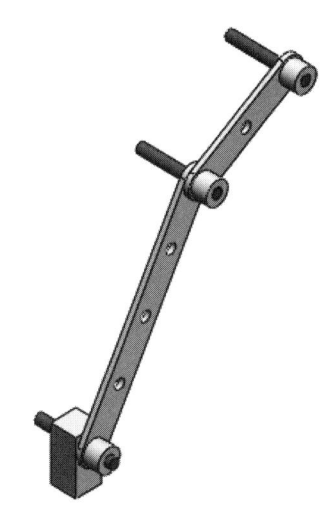

27. **Insert** a Coincident mate between the right face of the 3 Hole Link and the left face of the second Collar.

28. **Insert** the third Collar part. Locate the Collar near the third Axle component.

29. **Insert** a Concentric mate between the inside cylindrical face of the Collar and the outside cylindrical face of the third Axle.

30. **Insert** a Coincident mate between the right face of the 3 Hole Link and the left face of the third Collar.

31. **Mirror** the components. Mirror the three Collars, 5 Hole Link and 3 Hole Link about the Right Plane. Do not check any components in the Components to Mirror box. Check the Recreate mates to new components box. Click Next in the Mirror Components PropertyManager. Check the Preview instanced components box.

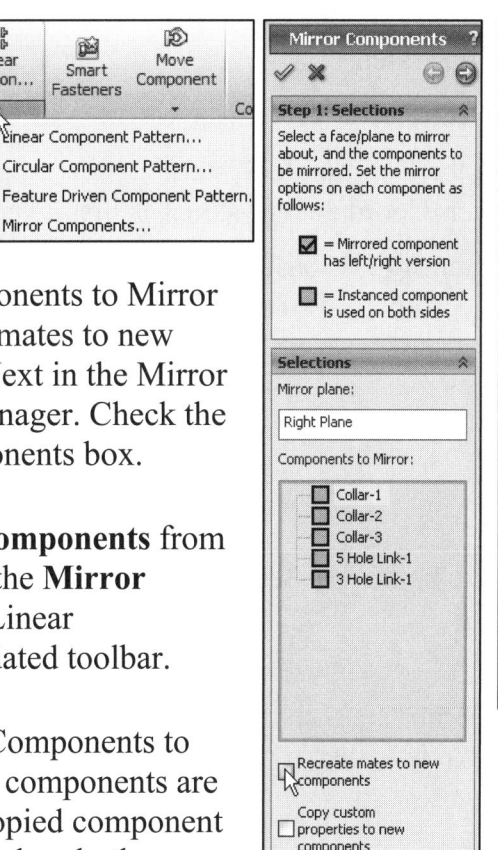

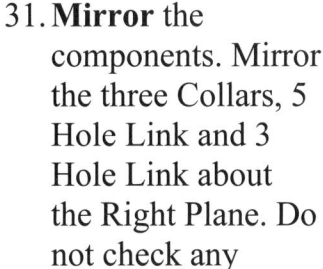

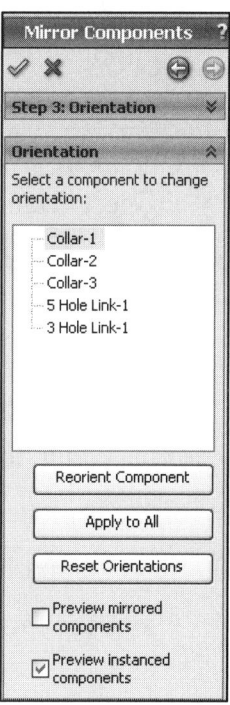

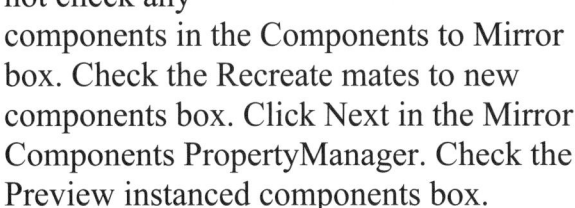

Click **Insert**, **Mirror Components** from the Menu bar menu or click the **Mirror Components** tool from the Linear Component Pattern Consolidated toolbar.

No check marks in the Components to Mirror box indicates that the components are copied. The geometry of a copied component is unchanged from the original, only the orientation of the component is different.

Page 5 - 17

Assembly Modeling

💡 Check marks in the Components to Mirror box indicates that the selected is mirrored. The geometry of the mirrored component changes to create a truly mirrored component.

💡 To preserve any mates between the selected components when you mirror more than one component, select **Recreate mates to new components**.

Create the coordinate system location for the assembly.

32. Select the front right **vertex** of the Clevis component as illustrated.

33. Click the **Coordinate System** tool from the Reference Geometry Consolidated toolbar. The Coordinate System PropertyManager is displayed.

34. Click the **right bottom edge** of the Clevis component.

35. Click the **front bottom edge** of the Clevis component as illustrated.

36. Address the **direction** for X, Y, Z as illustrated.

37. Click **OK** from the Coordinate System PropertyManager. Coordinate System1 is displayed

38. **Locate** the Center of mass based on the location of the illustrated coordinate system. Select Coordinate System1.

- X: 1.79 inches
- Y: 0.25 inches
- Z: 2.61 inches

39. **Save** the part and name it Assembly Modeling 5-1.

40. **Close** the model.

💡 There are numerous ways to create the models in this chapter. The goal is to display different design intents and techniques.

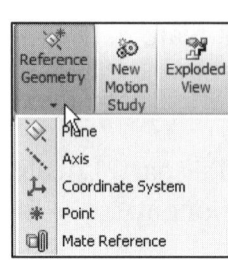

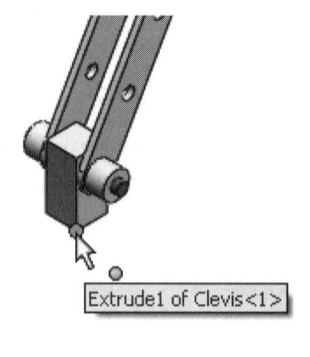

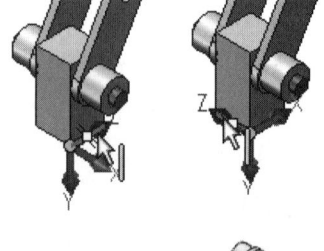

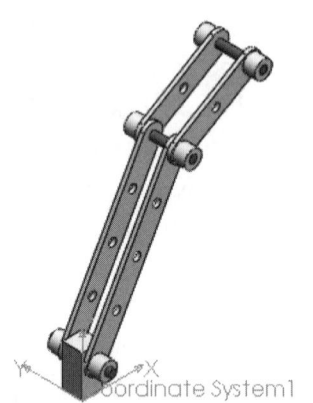

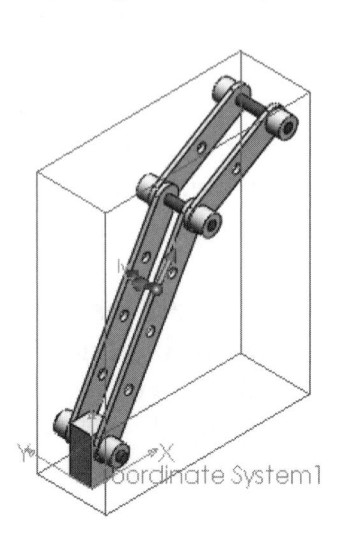

Assembly Modeling

Tutorial: Assembly model 5-2

An exam question in this category could read: Build this assembly. Locate the Center of mass of the model with the illustrated coordinate system.

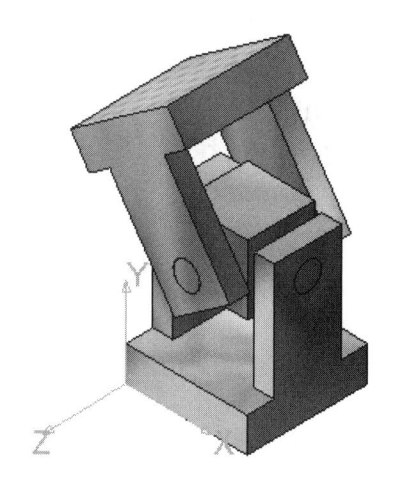

The assembly contains the following: two U-Bracket components, four Pin components, and one Square block component.

- U-Bracket, (Item 1): Material: AISI 304. Two U-Bracket components are combined together Concentric to opposite holes of the Square block component. The second U-Bracket component is positioned with an Angle mate, to the right face of the first U-Bracket and a Parallel mate between the top face of the first U-Bracket and the top face of the Square block component. Angle A = 125deg.

- Square block, (Item 2): Material: AISI 304. The Pin components are mated Concentric and Coincident to the 4 holes in the Square block, (no clearance). The depth of each hole = 10mm.

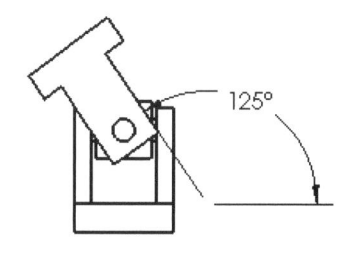

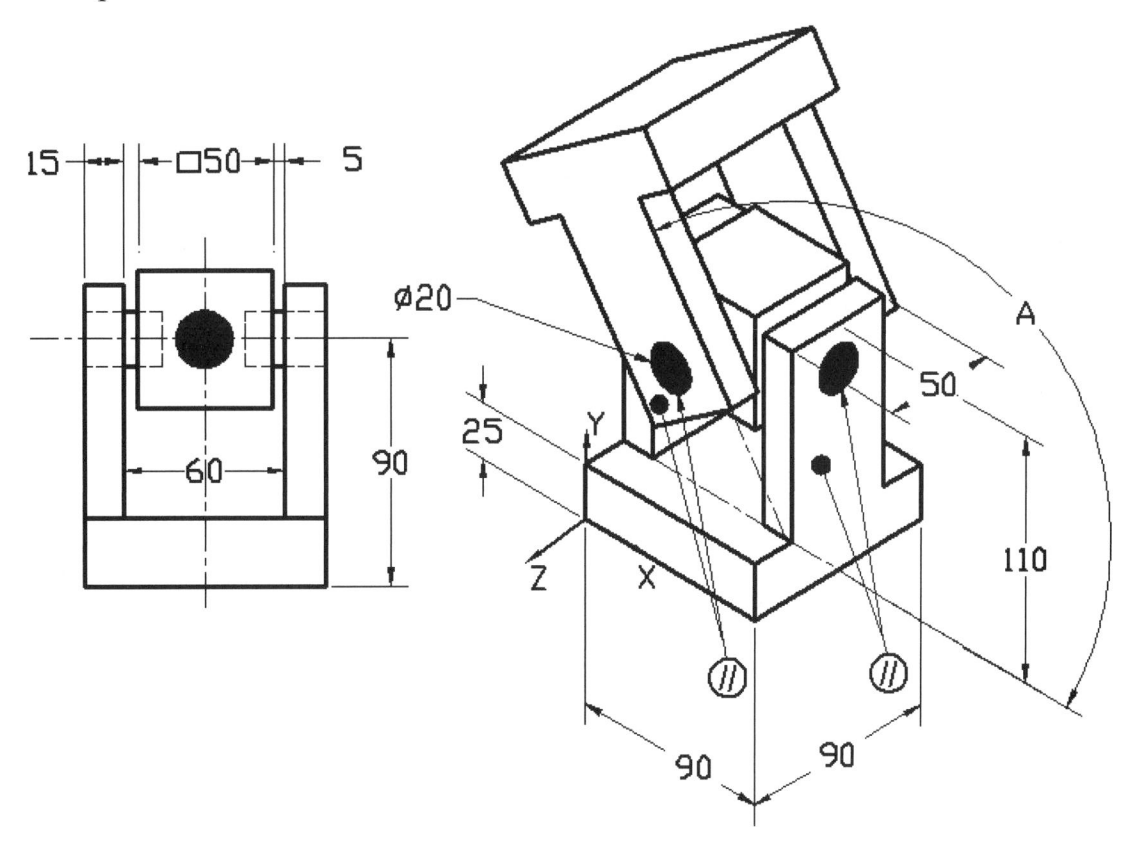

Page 5 - 19

Assembly Modeling

- Pin, (Item 3): Material: AISI 304. The Pin components are mated Concentric to the hole, (no clearance). The end face of the Pin components are Coincident to the outer face of the U-Bracket components. The Pin component has a 5mm spacing between the Square block component and the two U-Bracket components.

Think about the steps that you would take to build the illustrated assembly. Identify and build the required parts. Identify the first fixed component. This is the Base component of the assembly. Position the Base component features in the part so they are in the correct orientation in the assembly. Insert the required Standard mates. Locate the Center of mass of the model with respect to the illustrated coordinate system. In this example, start with the U-Bracket part.

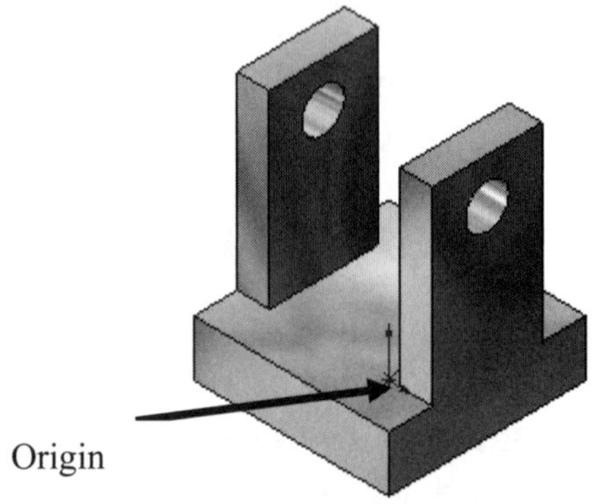

Build the U-Bracket part.

1. **Create** a New part in SolidWorks.

2. **Build** the illustrated U-Bracket part. Insert three features: Extruded Base, and two Extruded Cuts. Think about the steps that you would take to build this part. Identify the location of the part Origin.

3. **Set** the document properties for the model.

4. Create **Sketch1**. Sketch1 is the Base sketch. Select the Right Plane as the Sketch plane. Apply symmetry. Insert the required geometry relations and dimensions.

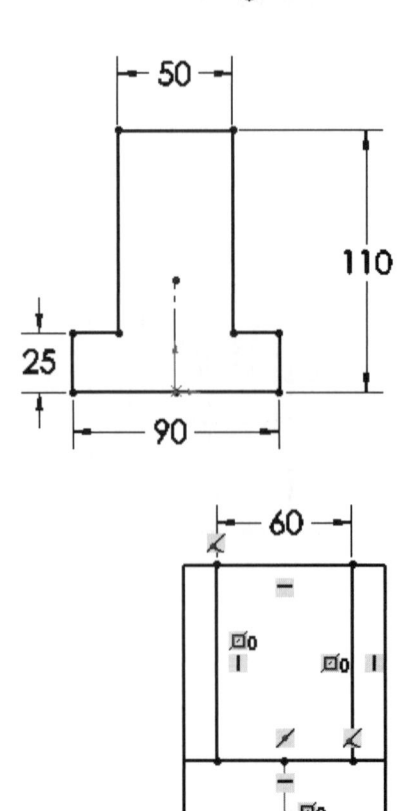

5. Create the **Extruded Base** feature. Extrude1 is the Base feature. Apply Symmetry. Select Mid Plane for End Condition. Depth = 90mm.

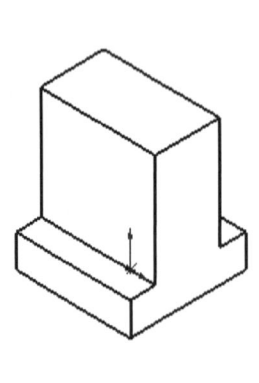

6. Create **Sketch2**. Select the Front Plane as the Sketch plane. Sketch a rectangle. Apply construction reference geometry. Insert the required geometry relations and dimension.

Assembly Modeling

7. Create the first **Extruded Cut** feature. Select Through All for the End Condition in Direction 1 and Direction 2.

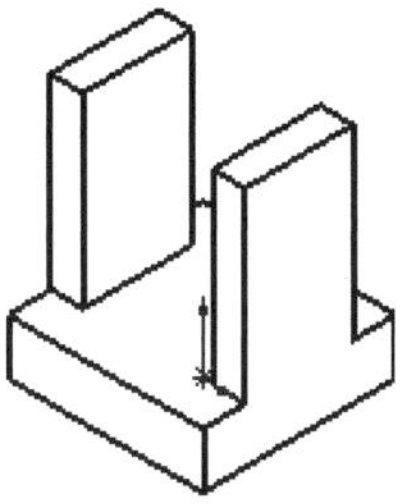

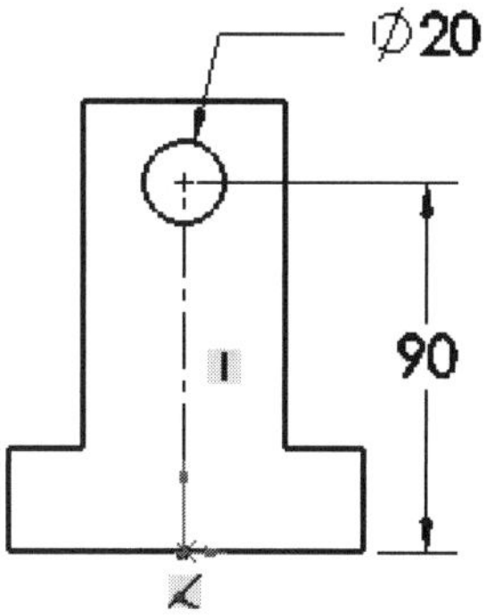

8. Create **Sketch3**. Select the right flat face of Extrude1 as the Sketch plane. Sketch a circle. Apply construction geometry. Insert the required relation and dimensions.

9. Create the second **Extruded Cut** feature. Select Through All for the End Condition in Direction 1. Two holes are created.

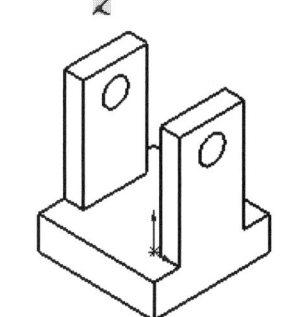

10. **Assign** AISI 304 material to the part.

11. **Save** the part and name it U-Bracket.

Create the Pin part.

1. **Create** a New part in SolidWorks.

2. **Build** the illustrated Pin part. Insert an Extrude1 feature. Identify the location of the part Origin.

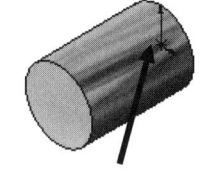

3. **Set** the document properties for the model.

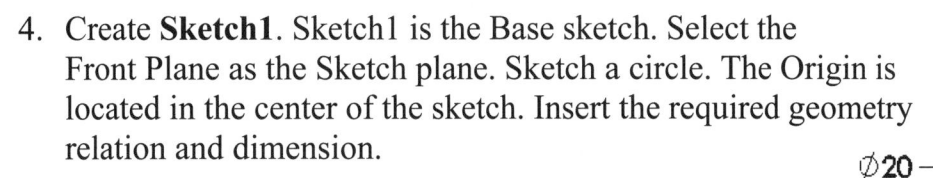

4. Create **Sketch1**. Sketch1 is the Base sketch. Select the Front Plane as the Sketch plane. Sketch a circle. The Origin is located in the center of the sketch. Insert the required geometry relation and dimension.

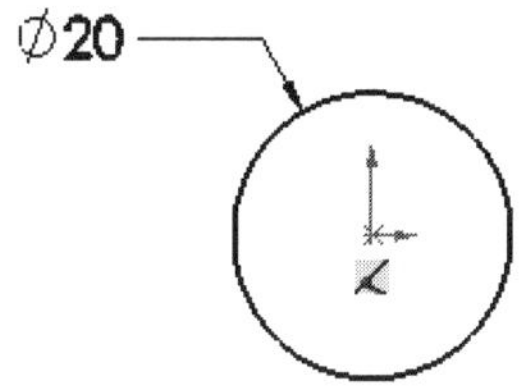

5. Create the **Extruded Base** feature. Extrude1 is the Base feature. Blind is the default End Condition. Depth = 30mm. Note: The direction of the extrude feature.

6. **Assign** AISI 304 material to the part.

7. **Save** the part and name it Pin.

Create the Square block part.

1. **Create** a New part in SolidWorks.

2. **Build** the illustrated Square block part. Insert five features: Extruded Base, two Extruded Cuts, and two Mirror. Identify the location of the part Origin.

3. **Set** the document properties for the model.

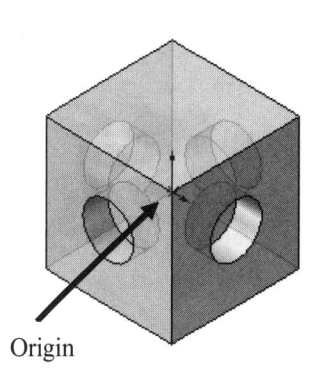

Assembly Modeling

4. Create **Sketch1**. Sketch1 is the Base sketch. Select the Front Plane as the Sketch plane. Sketch a square. Apply construction reference geometry. Insert the required geometry relation and dimensions.

5. Create the **Extruded Base** feature. Extrude1 is the Base feature. Apply Symmetry. Select Mid Plane for End Condition. Depth = 50mm.

6. Create **Sketch2**. Select the front face as the Sketch plane. Sketch a circle centered about the Origin. Insert the required dimension.

7. Create the first **Extruded Cut** feature. Blind is the default End Condition in Direction 1.
Depth = 10mm.

8. Create **Sketch3**. Select the right face of Extrude1 as the Sketch plane. Sketch a circle centered about the Origin. Insert the required dimension.

9. Create the second **Extruded Cut** feature. Blind is the default End Condition in Direction 1.
Depth = 10mm.

10. Create the first **Mirror** feature. Mirror the second Extruded Cut feature about the Right Plane.

11. Create the second **Mirror** feature. Mirror the first Extruded Cut feature about the Front Plane. Four holes are displayed in the Extrude1 feature.

12. **Assign** AISI 304 material to the part.

13. **Save** the part and name it Square block.

Create the assembly. The illustrated assembly contains the following: Two U-Block components, four Pin components, and one Square block component.

1. **Create** a New assembly in SolidWorks. The created models are displayed in the Open documents box.

2. Click **Cancel** ✖ from the Begin Assembly PropertyManager.

3. **Set** the document properties for the model.

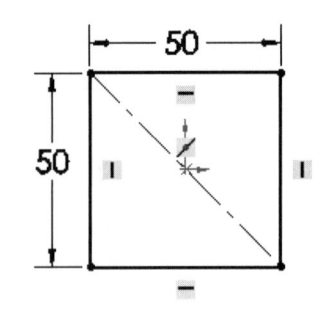

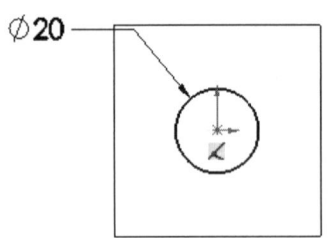

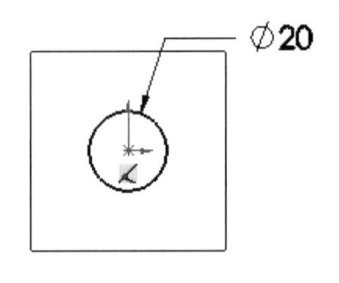

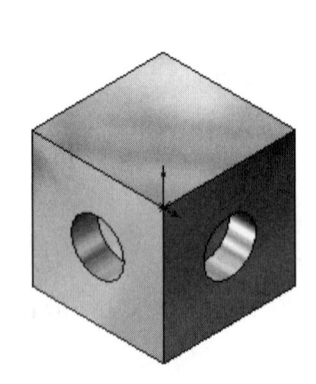

Assembly Modeling

4. **Insert** the first U-Bracket.

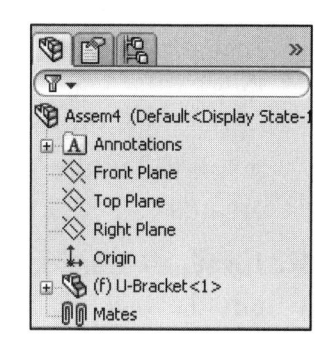

5. **Fix** the component to the assembly Origin. Click OK from the PropertyManager. The U-Bracket is displayed in the Assembly FeatureManager and in the Graphics window.

6. **Insert** the Square block above the U-Bracket component as illustrated.

7. **Insert** the first Pin part. Locate the first Pin to the front of the Square block.

8. **Insert** the second Pin part. Locate the second Pin to the back of the Square block.

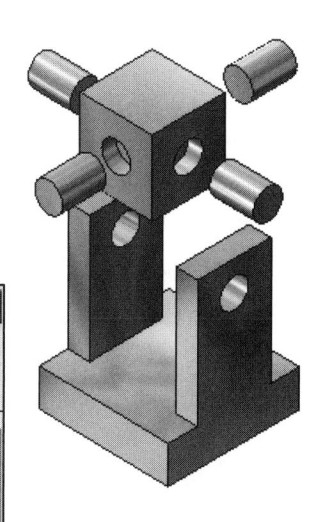

9. **Insert** the third Pin part. Locate the third Pin to the left side of the Square block. Rotate the Pin.

10. **Insert** the fourth Pin part. Locate the fourth Pin to the right side of the Square block. Rotate the Pin.

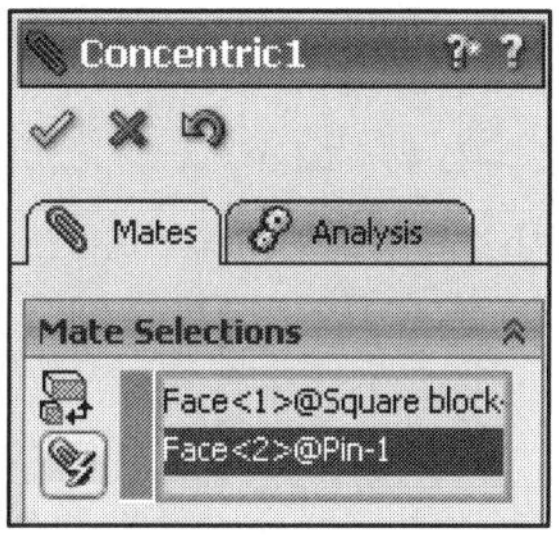

11. **Insert** a Concentric mate between the inside cylindrical face of the Square block and the outside cylindrical face of the first Pin. The selected face entities are displayed in the Mate Selections box. Concentric1 is created.

12. **Insert** a Coincident mate between the inside back circular face of the Square block and the flat back face of the first Pin. Coincident1 mate is created.

13. **Insert** a Concentric mate between the inside cylindrical face of the Square block and the outside cylindrical face of the second Pin. The selected face sketch entities are displayed in the Mate Selections box. Concentric2 is created.

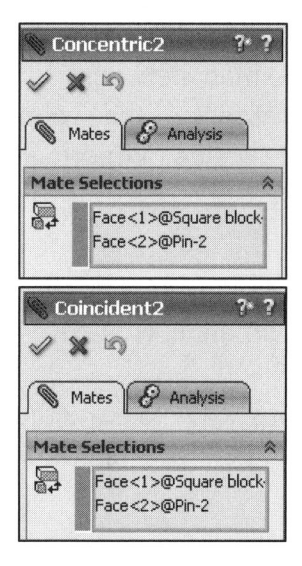

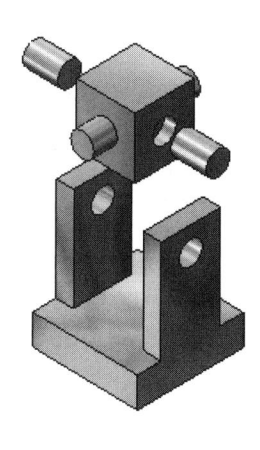

14. **Insert** a Coincident mate between the inside back circular face of the Square block and the front flat face of the second Pin. Coincident2 mate is created.

Page 5 - 23

Assembly Modeling

15. **Insert** a Concentric mate between the inside cylindrical face of the Square block and the outside cylindrical face of the third Pin. The selected face sketch entities are displayed in the Mate Selections box. Concentric3 is created.

16. **Insert** a Coincident mate between the inside back circular face of the Square block and the right flat face of the third Pin. Coincident3 mate is created.

17. **Insert** a Concentric mate between the inside circular face of the Square block and the outside cylindrical face of the fourth Pin. The selected face entities are displayed in the Mate Selections box. Concentric4 is created.

18. **Insert** a Coincident mate between the inside back circular face of the Square block and the left flat face of the fourth Pin. Coincident4 mate is created.

19. **Insert** a Concentric mate between the inside left cylindrical face of the Extrude3 feature on U-Bracket and the outside cylindrical face of the left Pin. Coincident5 is created.

20. **Insert** a Coincident mate between the Right Plane of the Square block and the Right Plane of the assembly. Coincident5 is created.

21. **Insert** the second U-Bracket part above the assembly. Position the U-Bracket as illustrated.

22. **Insert** a Concentric mate between the inside cylindrical face of the second U-Bracket component and the outside cylindrical face of the second Pin. The mate is created.

23. **Insert** a Coincident mate between the outside circular edge of the second U-Bracket and the back flat face of the second Pin. The mate is created.

💡 There are numerous ways to mate the models in this chapter. The goal is to display different design intents and techniques.

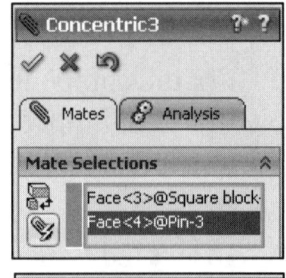

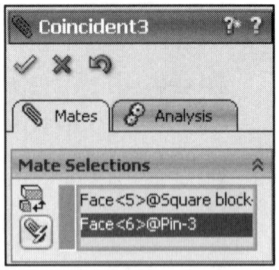

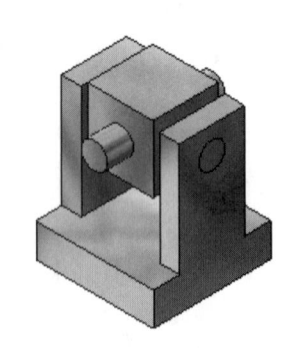

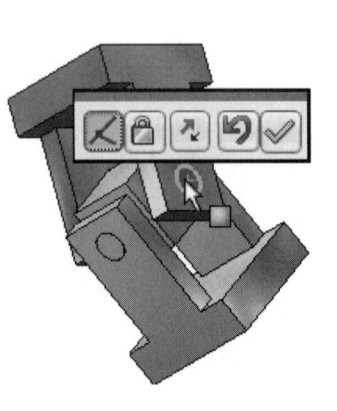

Assembly Modeling

24. **Insert** an Angle mate between the top flat face of the first U-Bracket component and the right narrow face of the second U-Bracket component as illustrated. Angle1 is created. An Angle mate is required to obtain the correct Center of mass.

25. **Insert** a Parallel mate between the top flat face of the first U-Bracket and the top flat face of the Square block component.

26. **Expand** the Mates folder and the components from the FeatureManager. View the created mates.

Create the coordinate location for the assembly.

27. Select the front bottom left **vertex** of the first U-Bracket component as illustrated.

28. Click the **Coordinate System** tool from the Reference Geometry Consolidated toolbar. The Coordinate System PropertyManager is displayed.

29. Click **OK** from the Coordinate System PropertyManager. Coordinate System1 is displayed.

30. **Locate** the Center of mass based on the location of the illustrated coordinate system. Select Coordinate System1.

- X: 31.54 millimeters
- Y: 85.76 millimeters
- Z: -45.00 millimeters

31. **Save** the part and name it Assembly Modeling 5-2.

32. **Close** the model.

At this time, there are no assembly configuration questions on the CSWA exam.

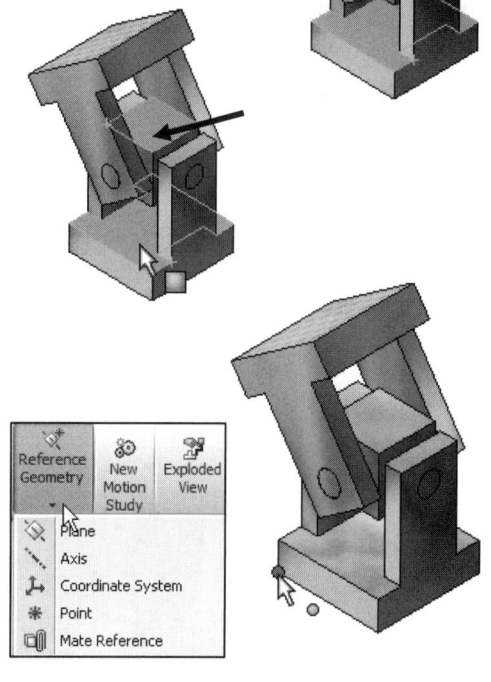

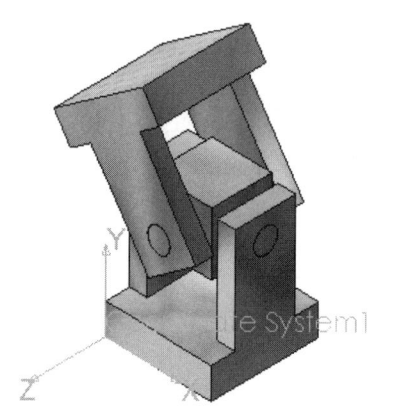

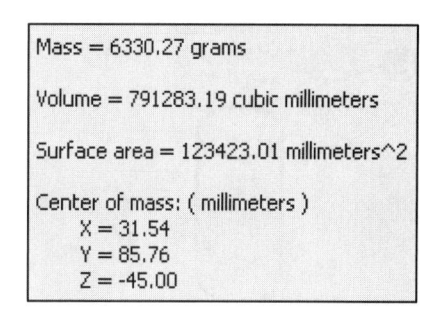

Mass = 6330.27 grams

Volume = 791283.19 cubic millimeters

Surface area = 123423.01 millimeters^2

Center of mass: (millimeters)
 X = 31.54
 Y = 85.76
 Z = -45.00

Assembly Modeling

Tutorial: Assembly model 5-3

An exam question in this category could read: Build this assembly. Locate the Center of mass using the illustrated coordinate system.

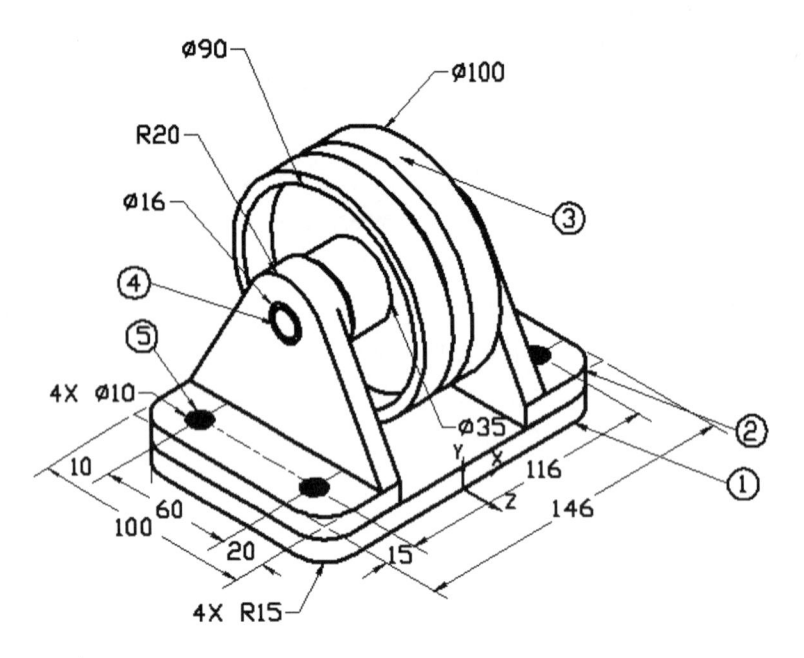

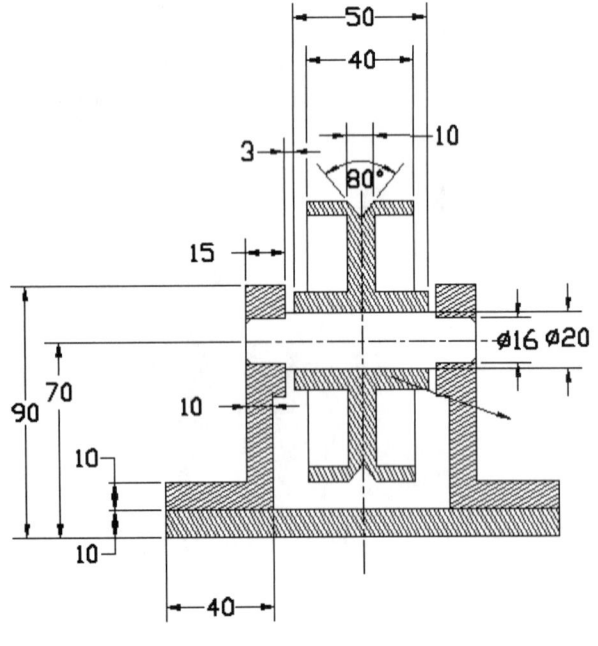

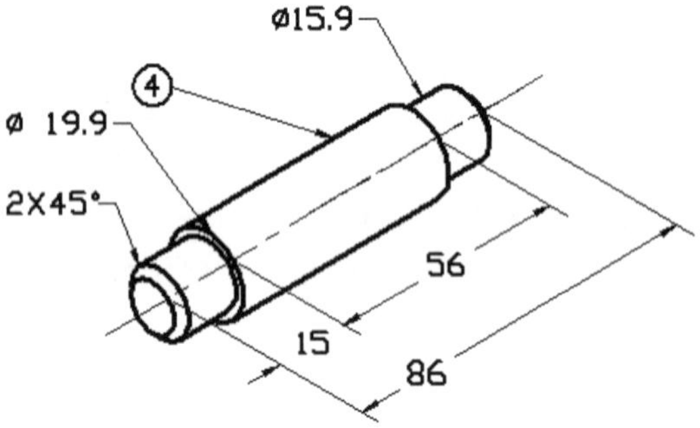

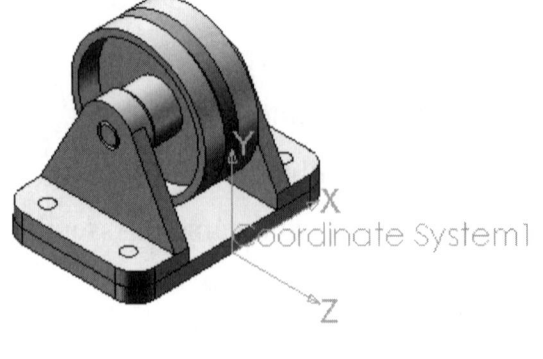

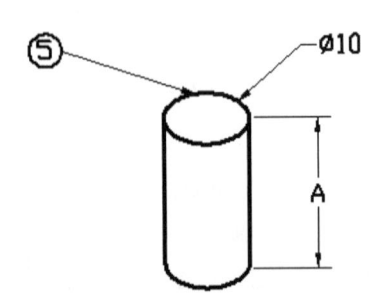

Assembly Modeling

The assembly contains the following: One WheelPlate component, two Bracket100 components, one Axle40 component, one Wheel1 component, and four Pin-4 components.

- WheelPlate, (Item 1): Material: AISI 304. The WheelPlate contains 4-Ø10 holes. The holes are aligned to the left Bracket100 and the right Bracket100 components. All holes are through-all. The thickness of the WheelPlate = 10 mm.

- Bracket100, (Item 2): Material: AISI 304. The Bracket100 component contains 2-Ø10 holes and 1- Ø16 hole. All holes are through-all.

- Wheel1, (Item 3): Material AISI 304: The center hole of the Wheel1 component is Concentric with the Axle40 component. There is a 3mm gap between the inside faces of the Bracket100 components and the end faces of the Wheel hub.

- Axle40, (Item 4): Material AISI 304: The end faces of the Axle40 are Coincident with the outside faces of the Bracket100 components.

- Pin-4, (Item 5): Material AISI 304: The Pin-4 components are mated Concentric to the holes of the Bracket100 components, (no clearance). The end faces are Coincident to the WheelPlate bottom face and the Bracket100 top face.

Identify and build the required parts. Identify the first fixed component. This is the Base component of the assembly. Position the Base component features in the part so they are in the correct orientation in the assembly. Insert the required Standard mates. Locate the Center of mass of the illustrated model with respect to the referenced coordinate system. The referenced coordinate system is located at the bottom, right, midpoint of the Wheelplate. In this example, start with the WheelPlate part.

Build the WheelPlate part.

1. **Create** a New part in SolidWorks.

2. **Build** the illustrated WheelPlate part. Insert four features: Extruded Base, Fillet, Extruded Cut, and a Linear Pattern. Think about the steps that you would take to build this part. Identify the location of the part Origin.

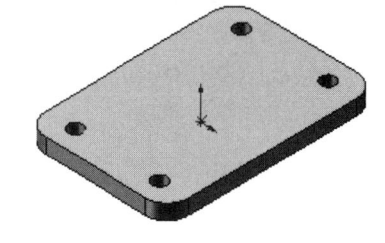

3. **Set** the document properties for the model.

4. Create **Sketch1**. Sketch1 is the Base sketch. Select the Top Plane as the Sketch plane. Sketch a rectangle. Insert the required geometry relations and dimensions. The Origin is located in the middle of the sketch.

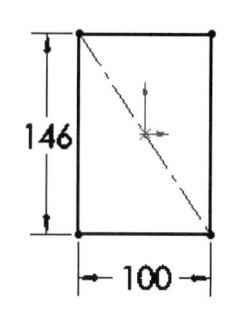

Assembly Modeling

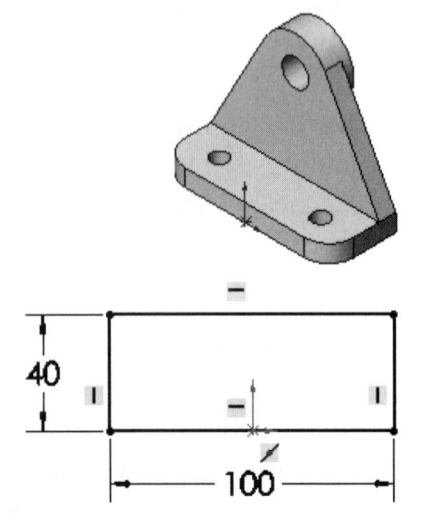

5. Create the **Extruded Base** feature. Extrude1 is the Base feature. Blind is the default End Condition in Direction 1. Depth = 10mm.

6. Create the **Fillet** feature. Fillet the four outside edges. Radius = 15mm.

7. Create **Sketch2**. Select the top face of Extrude1 as the Sketch plane. Sketch a circle. Insert the required geometric relations and dimensions.

☼ In a timed exam, do not insert note annotation. It will not affect your answer for the exam.

8. Create the **Extruded Cut** feature. Select Through All for End Condition in Direction 1. Extrude2 is the seed feature.

9. Create the **Linear Pattern** feature. Create a vertical and horizontal hole pattern. Extrude2 is the seed feature. Spacing = 116mm in Direction 1. Spacing = 60mm in Direction 2.

10. **Assign** AISI 304 material to the part.

11. **Save** the part and name it WheelPlate.

Create the Bracket100 part for the assembly. There are two Bracket100 components in the assembly.

1. **Create** a New part in SolidWorks.

2. **Build** the illustrated Bracket100 part. Insert seven features: Extruded Base, two Extruded Bosses, Extruded Cut, Hole Wizard, Mirror, and Fillet. Think about the steps that you would take to build this part. Identify the location of the part Origin.

3. **Set** the document properties for the model.

4. Create **Sketch1**. Sketch1 is the Base sketch. Select the Top Plane as the Sketch plane. Sketch a rectangle. Insert the required geometric relations and dimensions. Note the location of the Origin.

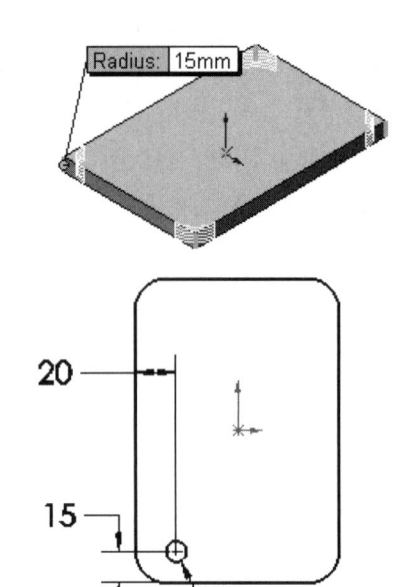

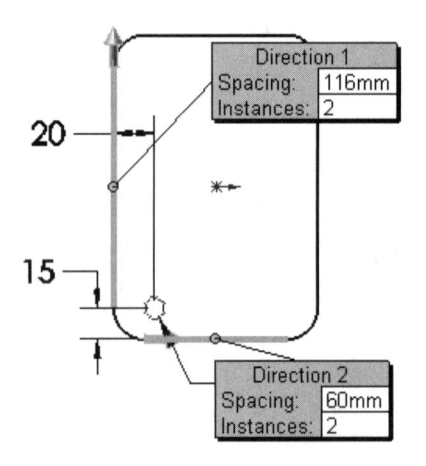

Page 5 - 28

Assembly Modeling

5. Create the **Extruded Base** feature. Extrude1 is the Base feature. Blind is the default End Condition. Depth = 10mm.

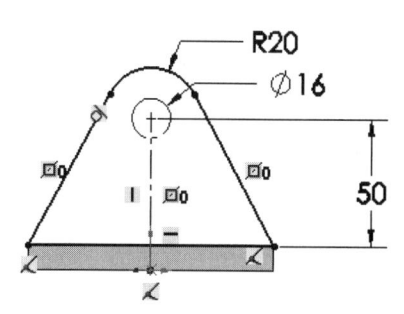

6. Create **Sketch2**. Select the back face of Extrude1 as the Sketch plane. Apply the Tangent Arc Sketch tool. Complete the sketch. Insert the required relations and dimensions.

7. Create the first **Extruded Boss** feature. Blind is the default End Condition in Direction 1. Depth = 10mm.

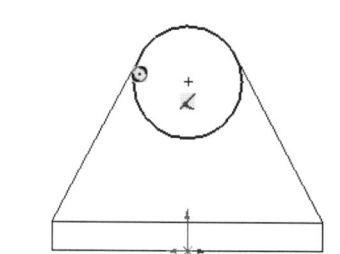

8. Create **Sketch3**. Select the front face of Extrude2 as the Sketch plane. Wake-up the centerpoint of the tangent arc. Sketch a circle Coradial and Coincident to the tangent arc.

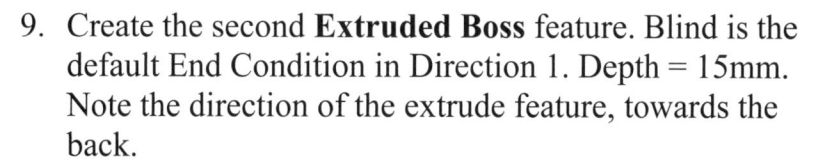

9. Create the second **Extruded Boss** feature. Blind is the default End Condition in Direction 1. Depth = 15mm. Note the direction of the extrude feature, towards the back.

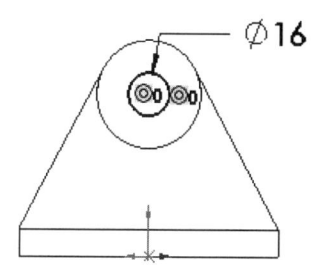

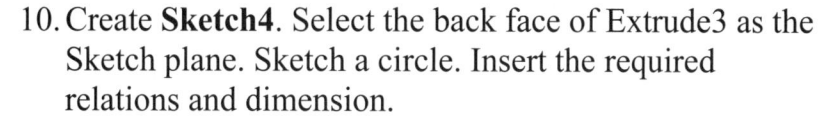

10. Create **Sketch4**. Select the back face of Extrude3 as the Sketch plane. Sketch a circle. Insert the required relations and dimension.

11. Create the **Extrude Cut** feature. Select Through All for End Condition in Direction 1.

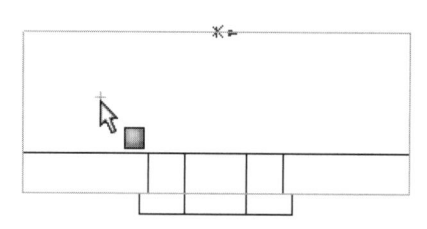

12. Create **Sketch5**. Select the bottom face of Extrude1 for the Sketch plane. Select a location as illustrated for the first hole. Sketch5 is the profile for the Hole Wizard feature.

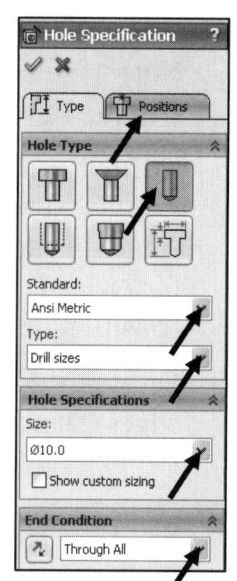

13. Create the **Hole Wizard** feature. Select Hole for Hole Specification type. Select Ansi Metric for Standard. Select Drill sizes for Type. Select Ø10.0 for Size. Select Through All for End Condition. Click the Positions tab. Insert the required dimensions.

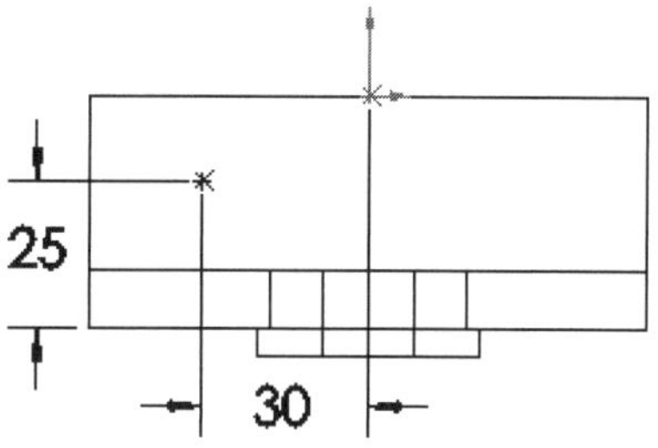

Page 5 - 29

Assembly Modeling

☼ The Hole Wizard PropertyManager is displayed when you create a Hole Wizard hole. Two tabs are displayed in the Hole Wizard PropertyManager:

- Type (default). Sets the hole type parameters.
- Positions. Locates the Hole Wizard holes on planar or non-planar faces. Use the dimension and other sketch tools to position the center of the holes.

14. Create the **Mirror** feature. Mirror the Ø10.0mm hole about the Right Plane.

15. Create the **Fillet** feature. Fillet the two front edges. Radius = 15m.

16. **Assign** AISI 304 material to the part.

17. **Save** the part and name it Bracket100.

Create the Axle40 part for the assembly.

1. **Create** a New part in SolidWorks.

2. **Build** the illustrated Axle40 part. Insert two features: Revolved Base and Chamfer. Identify the location of the part Origin.

3. **Set** the document properties for the model.

4. Create **Sketch1**. Sketch1 is the Base sketch. Select the Right Plane as the Sketch plane. Apply construction reference geometry. Insert the required geometric relations and dimensions. Note the location of the Origin.

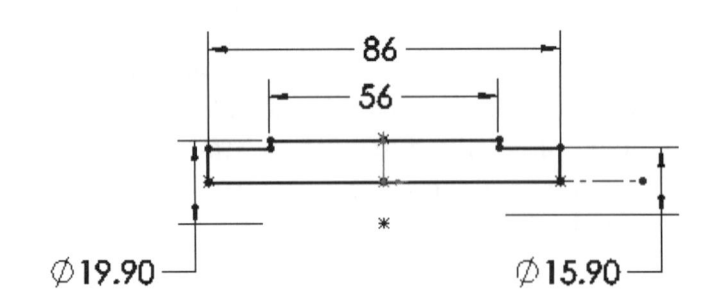

5. Create the **Revolved Base** feature. Revolve1 is an Angel-distance feature. Revolve1 is the Base feature for Axle40. 360deg is the default angle. Apply construction geometry for Axis of revolution.

6. Create the **Chamfer** feature. Chamfer the two outside edges. Distance = 2mm. Angle = 45deg.

7. **Assign** AISI 304 material to the part.

8. **Save** the part and name it Axle40.

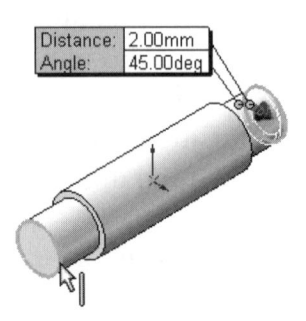

Assembly Modeling

💡 There are numerous ways to create the models in this chapter. The goal is to display different design intents and techniques.

💡 A Revolve feature adds or removes material by revolving one or more profiles about a centerline. You can create Revolved Boss/Bases, Revolved Cuts, or Revolved Surfaces. The Revolve feature can be a solid, a thin feature, or a surface.

Create the Wheel1 part for the assembly.

1. **Create** a New part in SolidWorks.

2. **Build** the illustrated Wheel1 part. Insert two features: Revolved Base and Revolved Cut. Think about the steps that you would take to build this part. Identify the location of the part Origin.

3. **Set** the document properties for the model.

4. Create **Sketch1**. Sketch1 is the Base sketch. Select the Right Plane as the Sketch plane. Apply construction reference geometry. Apply symmetry. Insert the required geometric relations and dimensions. Note the location of the Origin.

5. Create the **Revolved Base** feature. Revolve1 is the Base feature for the Axle. 360deg is the default angle. Select the horizontal construction geometry line for the Axis of Revolution.

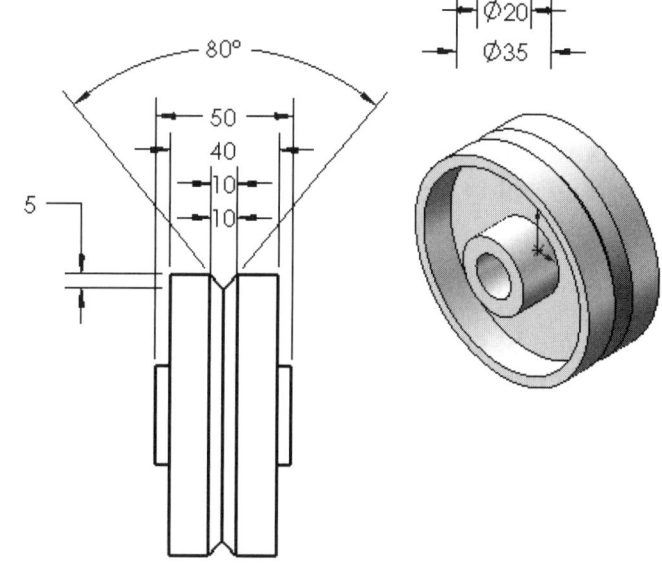

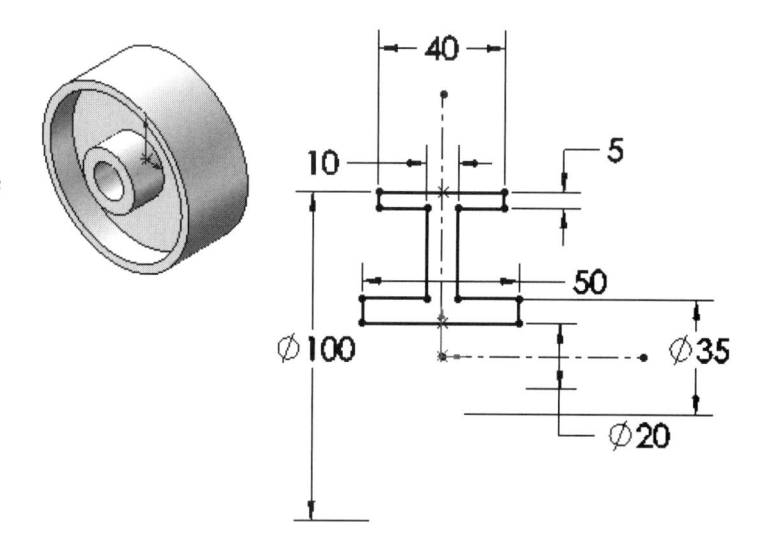

Page 5 - 31

Assembly Modeling

6. Create **Sketch2**. Select the Right Plane as the Sketch plane. Apply construction reference geometry. Sketch a triangle for the groove in the wheel. Insert the required geometric relations and dimensions.

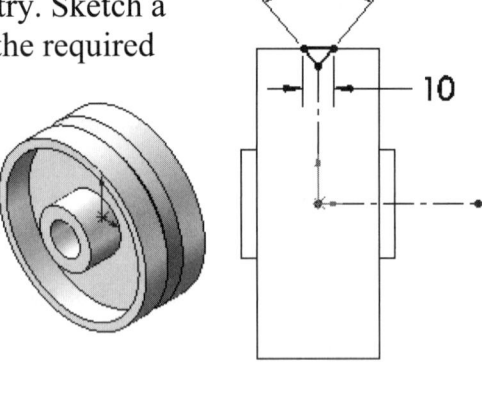

7. Create the **Revolved Cut** feature. 360deg is the default angle. Select the horizontal reference construction geometry line for the Axis of Revolution.

8. **Assign** AISI 304 material to the part.

9. **Save** the part and name it Wheel1.

Create the Pin-4 part for the assembly.

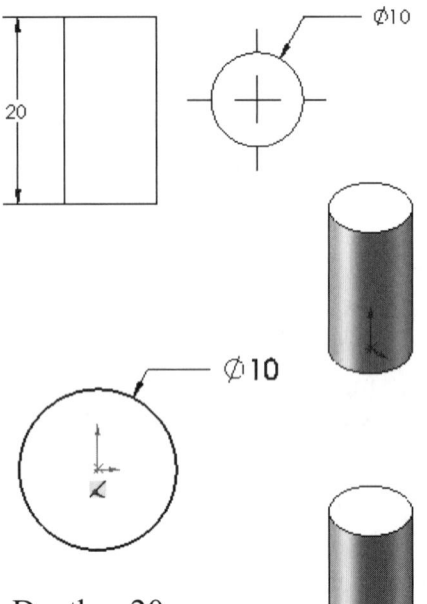

1. **Create** a New part in SolidWorks.

2. **Build** the illustrated Pin-4 part. Insert an Extrude1 feature. Think about the steps that you would take to build this part. Identify the location of the part Origin.

3. **Set** the document properties for the model.

4. Create **Sketch1**. Sketch1 is the Base sketch. Select the Top Plane as the Sketch plane. Sketch a circle. Insert the required geometric relation and dimension. Note the location of the Origin.

5. Create the **Extrude1** feature. Extrude1 is the Base feature. Blind is the default End Condition. Depth = 20mm.

6. **Assign** AISI 304 material to the part.

7. **Save** the part and name it Pin-4.

Create the assembly. The assembly contains the following: one WheelPlate component, two Bracket100 components, one Axle40 component, one Wheel1 component, and four Pin-4 components.

1. **Create** a New assembly in SolidWorks. The created models are displayed in the Open documents box.

2. Click **Cancel** ✖ from the Begin Assembly PropertyManager.

3. **Set** the document properties for the model.

Page 5 - 32

Assembly Modeling

4. **Insert** the first component. Insert the WheelPlate. Fix the component to the assembly Origin. The WheelPlate is displayed in the Assembly FeatureManager and in the Graphics window. The WheelPlate component is fixed.

5. **Insert** the first Bracket100 part above the WheelPlate component as illustrated.

6. **Insert** a Concentric mate between the inside front left cylindrical face of the Bracket100 component and the inside front left cylindrical face of the WheelPlate. Concentric1 is created.

7. **Insert** a Concentric mate between the inside front right cylindrical face of the Bracket100 component and the inside front right cylindrical face of the WheelPlate. Concentric2 is created.

8. **Insert** a Coincident mate between the bottom flat face of the Bracket100 component and the top flat face of the WheelPlate component. Coincident1 is created.

9. **Insert** the Axle40 part above the first Bracket100 component as illustrated.

10. **Insert** a Concentric mate between the outside cylindrical face of the Axle40 component and the inside cylindrical face of the Bracket100 component. Concentric3 is created.

11. **Insert** a Coincident mate between the flat face of the Axle40 component and the front outside edge of the first Bracket100 component. Coincident2 is created.

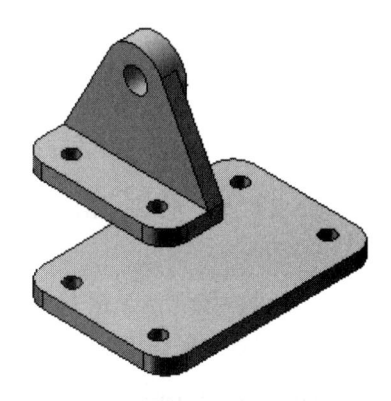

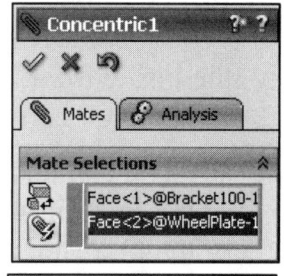

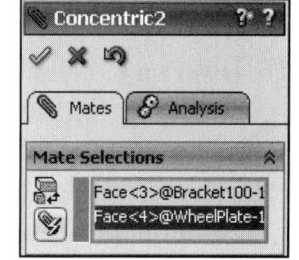

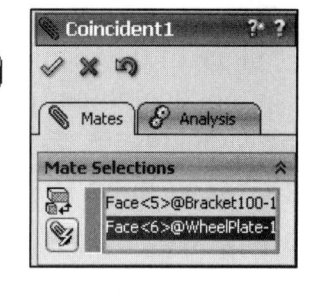

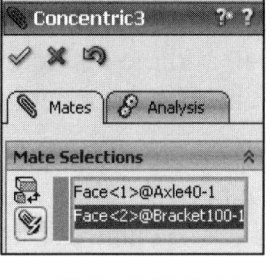

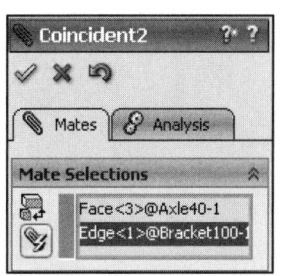

Assembly Modeling

💡 To verify that the distance between holes of mating components is equal, utilize Concentric mates between pairs of cylindrical hole faces.

12. **Insert** the first Pin-4 part above the Bracket100 component.

13. **Insert** the second Pin-4 part above the Bracket100 component.

14. **Insert** a Concentric mate between the outside cylindrical face of the first Pin-4 component and the inside front left cylindrical face of the Bracket100 component. Concentric4 is created.

15. **Insert** a Coincident mate between the flat top face of the first Pin-4 component and the top face of the first Bracket100 component. Coincident3 is created.

16. **Insert** a Concentric mate between the outside cylindrical face of the second Pin-4 component and the inside front right cylindrical face of the Bracket100 component. Concentric5 is created.

17. **Insert** a Coincident mate between the flat top face of the second Pin-4 component and the top face of the first Bracket100 component. Coincident4 is created.

18. **Insert** the Wheel1 part as illustrated.

19. **Insert** a Concentric mate between the outside cylindrical face of Axle40 and the inside front cylindrical face of the Wheel1 component. Concentric6 is created.

20. **Insert** a Coincident mate between the Front Plane of Axle40 and the Front Plane of Wheel1. Coincident5 is created.

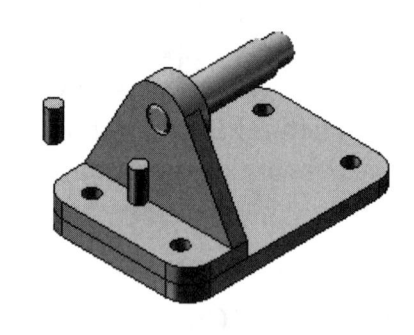

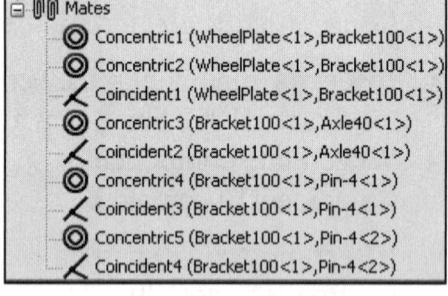

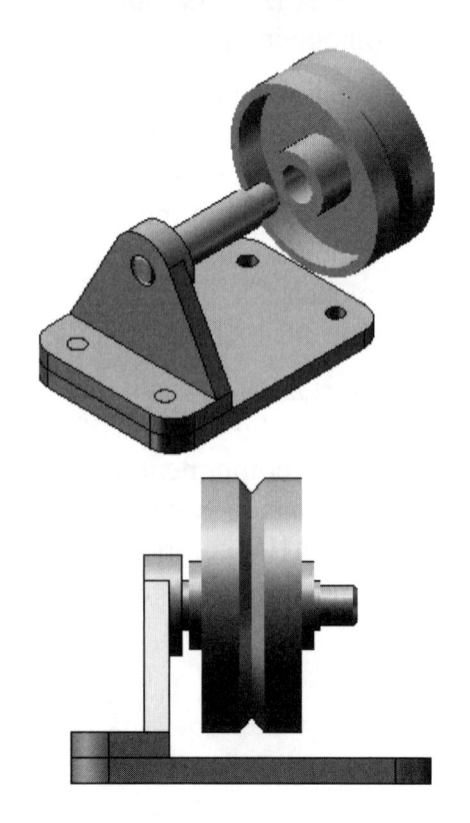

Page 5 - 34

Assembly Modeling

21. **Mirror** the components. Mirror the Bracket100, and the two Pin-4 components about the Front Plane. Do not check any components in the Components to Mirror box. Check the Recreate mates to new components box. Click Next in the PropertyManager. Check the Preview instanced components box.

☼ Click the **Mirror Component** tool from the Linear Component Pattern Consolidated toolbar to activate the Mirror Components PropertyManager.

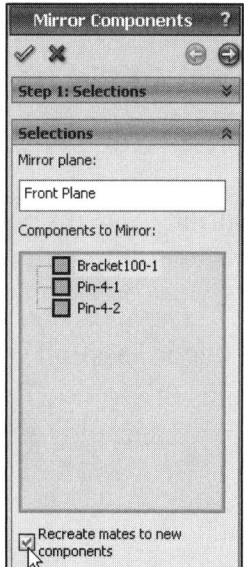

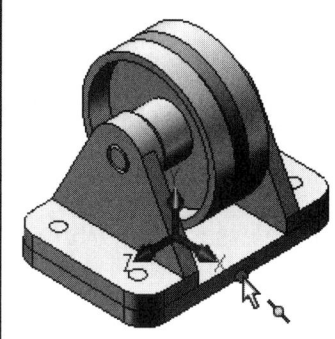

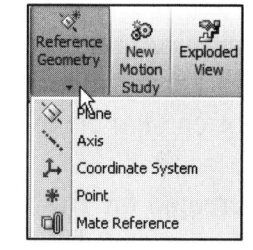

Create the coordinate location for the assembly.

22. Click the **Coordinate System** tool from the Reference Geometry Consolidated toolbar. The Coordinate System PropertyManager is displayed.

23. **Select** the right bottom midpoint as the Origin location as illustrated.

24. **Select** the bottom right edge as the X axis direction reference as illustrated.

25. Click **OK** from the Coordinate System PropertyManager. Coordinate System1 is displayed.

26. **Locate** the Center of mass based on the location of the illustrated coordinate system. Select Coordinate System1.

- X: = 0.00 millimeters
- Y: = 37.14 millimeters
- Z: = -50.00 millimeters

27. **Save** the part and name it Assembly Modeling 5-3.

28. **Close** the model.

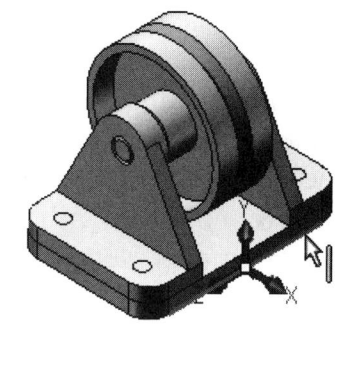

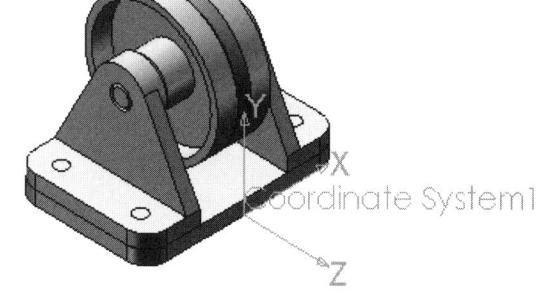

Page 5 - 35

Assembly Modeling

Mate the first component with respect to the assembly reference planes

You can fix the position of a component so that it cannot move with respect to the assembly Origin. By default, the first part in an assembly is fixed; however, you can float it at any time.

It is recommended that at least one assembly component is either fixed, or mated to the assembly planes or Origin. This provides a frame of reference for all other mates, and prevents unexpected movement of components when mates are added.

Up to this point, you identified the first fixed component, and built the required Base component of the assembly. The component features were orientated correctly to the illustrated assembly. In the exam, what if you created the Base component where the component features were not orientated correctly to the illustrated assembly.

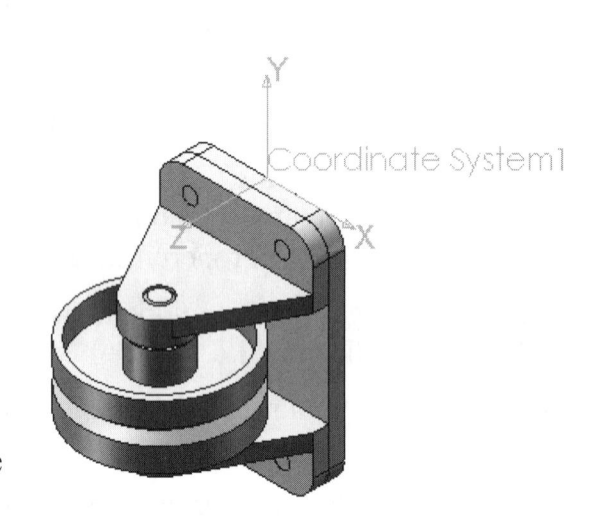

In the next tutorial, build the illustrated assembly. Insert the Base component, float the component, then mate the first component with respect to the assembly reference planes. Complete the assembly with the components from the Tutorial: Assembly model 5-3.

Tutorial: Assembly model 5-4

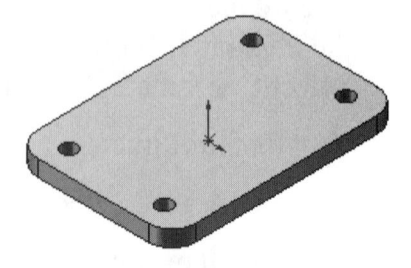

1. **Create** a New assembly in SolidWorks.

2. **Insert** the illustrated WheelPlate part that you built in Tutorial: Assembly model 5-3.

3. **Float** the WheelPlate component from the FeatureManager.

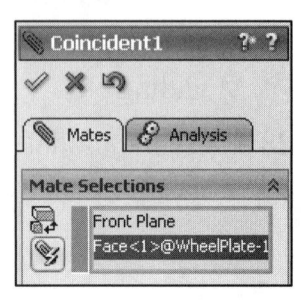

4. **Insert** a Coincident mate between the Front Plane of the assembly and the bottom flat face of the WheelPlate. Coincident1 is created.

5. **Insert** a Coincident mate between the Right Plane of the assembly and the Right Plane of the WheelPlate. Coincident2 is created.

Page 5 - 36

6. **Insert** a Coincident mate between the Top Plane of the assembly and the Front Plane of the WheelPlate. Coincident3 is created.

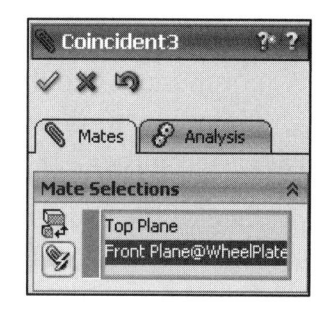

☼ When the Base component is mated to three assembly reference planes, no component status symbol is displayed in the Assembly FeatureManager.

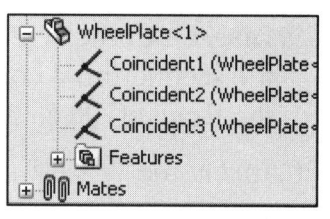

7. **Insert** the first Bracket100 part as illustrated. Rotate the component if required.

8. **Insert** a Concentric mate between the inside back left circular face of the Bracket100 component and the inside top left circular face of the WheelPlate. Concentric1 is created.

9. **Insert** a Concentric mate between the inside back right cylindrical face of the Bracket100 component and the inside top right cylindrical face of the WheelPlate. Concentric2 is created.

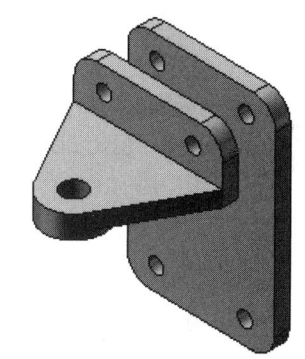

10. **Insert** a Coincident mate between the flat back face of the Bracket100 component and the front flat face of the WheelPlate component. Coincident4 is created.

11. **Insert** the Axle40 part as illustrated. Rotate the component if required.

12. **Insert** a Concentric mate between the outside cylindrical face of the Axle40 component and the inside cylindrical face of the Bracket100 component. Concentric3 is created.

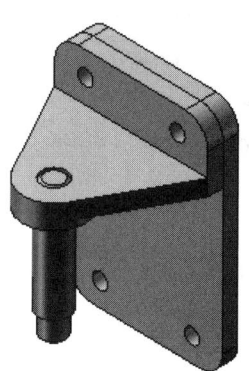

13. **Insert** a Coincident mate between the top flat face of the Axle40 component and the top outside circular edge of the Bracket100 component. Coincident5 is created.

14. **Insert** the first Pin-4 part above the Bracket100 component. Rotate the component.

15. **Insert** the second Pin-4 part above the Bracket100 component. Rotate the component.

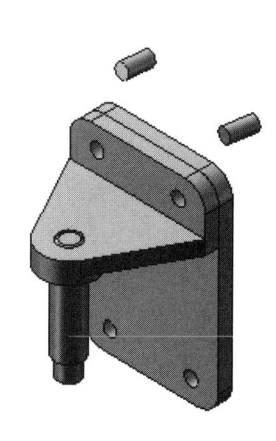

16. **Insert** a Concentric mate between the outside cylindrical face of the first Pin-4 component and the inside front left cylindrical face of the Bracket100 component. Concentric4 is created.

Assembly Modeling

17. **Insert** a Coincident mate between the flat front face of the first Pin-4 component and the top flat front face of the Bracket100 component. Coincident6 is created.

18. **Insert** a Concentric mate between the outside cylindrical face of the second Pin-4 component and the inside front right cylindrical face of the Bracket100 component. Concentric5 is created.

19. **Insert** a Coincident mate between the flat front face of the second Pin-4 component and the top flat front face of the Bracket100 component. Coincident7 is created.

20. **Insert** the Wheel1 part as illustrated.

21. **Insert** a Concentric mate between the outside cylindrical face of Axle40 and the inside top cylindrical face of the Wheel1 component. Concentric6 is created.

22. **Insert** a Coincident mate between the Right Plane of Axle40 and the Right Plane of the Wheel1 component. Coincident8 is created.

23. **Insert** a Coincident mate between the Front Plane of Axle40 and the Front Plane of the Wheel1 component. Coincident9 is created.

29. **Mirror** the components. Mirror the Bracket100, and the two Pin-4 components about the Top Plane. Do not check any components in the Components to Mirror box. Check the Recreate mates to new components box. Click Next in the PropertyManager. Check the Preview instanced components box.

Create the coordinate location for the assembly.

30. Click the **Coordinate System** tool from the Reference Geometry Consolidated toolbar. The Coordinate System PropertyManager is displayed.

31. **Select** the top back midpoint for the Origin location as illustrated.

32. Click **OK** from the Coordinate System PropertyManager. Coordinate System1 is displayed.

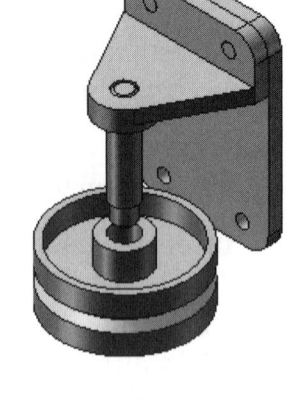

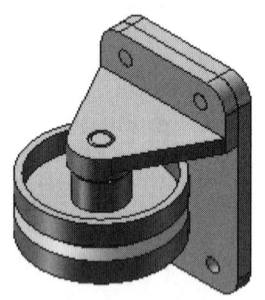

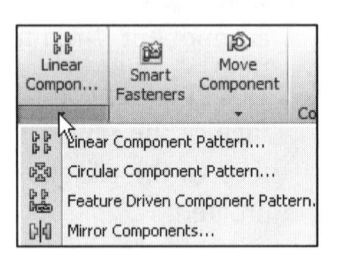

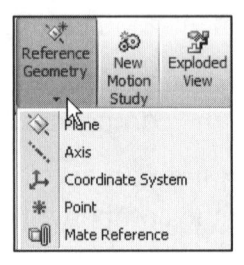

33. **Locate** the Center of mass based on the location of the illustrated coordinate system. Select Coordinate System1.

   - X: = 0.00 millimeters
   - Y: = -73.00 millimeters
   - Z: = 37.14 millimeters

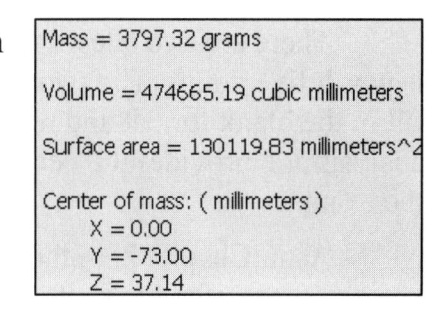

34. **Save** the part and name it Assembly Modeling 5-4.

35. **Close** the model.

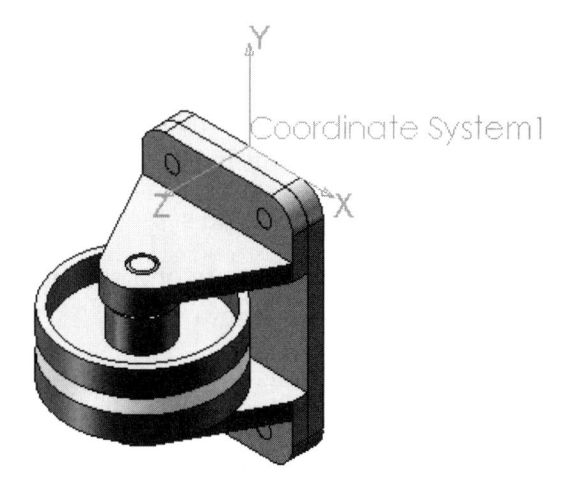

```
Mass = 3797.32 grams
Volume = 474665.19 cubic millimeters
Surface area = 130119.83 millimeters^2
Center of mass: ( millimeters )
    X = 0.00
    Y = -73.00
    Z = 37.14
```

Summary

Assembly Modeling is one of the five categories on the CSWA exam. The Assembly Modeling category addresses an assembly with numerous sub-components. Each sub-component is a simple part. Knowledge is required to know and understand how to insert Standard mates between the components to build an assembly. This chapter covered the general concepts and terminology used in Assembly modeling and then addressed the core elements that are aligned to the exam. Only insert the required mates to obtain the needed Mass properties information in the timed CSWA exam.

There is one question on the CSWA exam in this category. The question is worth 30 points. The question is in a multiple choice single answer or fill in the blank format. Spend no more than 40 minutes on the question in this category. This is a timed exam. Manage your time. At this time, there are no sheet metal assembly questions or questions on Advanced or Mechanical mates on the CSWA exam.

Advanced Modeling Theory and Analysis is the next chapter in this book. The chapter covers the general concepts and terminology used in COSMOSXpress and then addresses the core elements that are aligned to the exam.

Assembly Modeling

There are two questions on the CSWA exam in this category. Each question is worth five points. The two questions are in a multiple choice single answer or fill in the blank format and requires general terminology used in Engineering analysis and provides the needed knowledge of COSMOSXpress that is aligned to the exam.

As in Chapter 2, in the Advanced Modeling Theory and Analysis category, you are not required to perform an analysis on a part or assembly.

Key terms

- *Aligned mate*. Places the components so the normal or axis vectors for the selected faces point in the same direction.

- *Angle mate*. Places the selected items at the specified angle to each other.

- *Annotation*. A text note or a symbol that adds specific design intent to a part, assembly, or drawing.

- *Anti-Aligned mate*. Places the components so the normal or axis vectors for the selected faces point in opposite directions.

- *Assembly*. A document in which parts, features, and other assemblies, (sub-assemblies) are mated together. The parts and sub-assemblies exist in documents separate from the assembly. The extension for a SolidWorks assembly file name is .SLDASM.

- *Base sketch*. The first sketch of a part is called the Base sketch. The Base sketch is the foundation for the 3D model. Create a 2D sketch on a default plane: Front, Top, and Right in the FeatureManager design tree, or on a created plane.

- *Bottom-up design*. An assembly modeling technique where you create parts and then insert them into an assembly.

- *Coincident mate*. Forces two planar faces to become coplanar. The faces can move along one another, but cannot be pulled apart.

- *Concentric mate*. Forces two cylindrical faces to become concentric. The faces can move along the common axis, but cannot be moved away from the axis.

- *Constraints*. Geometric relations such as Perpendicular, Horizontal, Parallel, Vertical, Coincident, Concentric, etc. Insert constraints to your model to incorporate design intent.

- *Degrees of freedom.* Geometry that is not defined by dimensions or relations is free to move. In 2D sketches, there are three degrees of freedom: movement along the X and Y axes, and rotation about the Z axis. In 3D sketches and in assemblies, there are six degrees of freedom: movement along the X, Y, and Z axes, and rotation about the X, Y, and Z axes.

- *Distance mate.* Places the selected items with the specified distance between them.

- *In-context feature.* A feature with an external reference to the geometry of another component; the in-context feature changes automatically if the geometry of the referenced model or feature changes.

- *Mate.* Mates create geometric relationships between assembly components. As you add mates, you define the allowable directions of linear or rotational motion of the components. You can move a component within its degrees of freedom, visualizing the assembly's behavior.

- *Mate Alignment.* Sets the alignment condition of standard and advanced mates.

- *Parallel mate.* Places the selected items so they remain a constant distance apart from each other.

- *Perpendicular mate.* Places the selected items at a 90° angle to each other.

- *Tangent mate.* Places the selected items tangent to each other, at least one selection must be a cylindrical, conical, or spherical face.

- *Top-down design.* An assembly modeling technique where you create parts in the context of an assembly by referencing the geometry of other components. Changes to the referenced components propagate to the parts that you create in context.

Check your understanding

1: Build this assembly. Calculate the overall mass and volume of the assembly. Locate the Center of mass using the illustrated coordinate system. The assembly contains the following: one Base100 component, one Yoke component, and one AdjustingPin component. Apply the MMGS unit system.

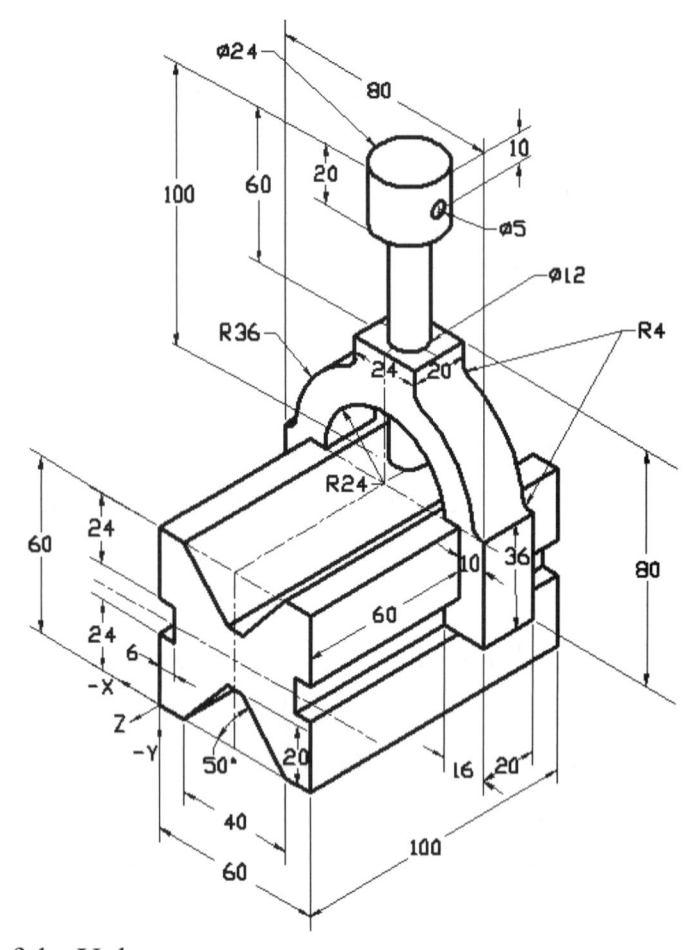

- Base100, (Item 1): Material 1060 Alloy. The distance between the front face of the Base100 component and the front face of the Yoke = 60mm.
- Yoke, (Item 2): Material 1060 Alloy. The Yoke fits inside the left and right square channels of the Base100 component, (no clearance). The top face of the Yoke contains a Ø12mm through all hole.
- AdjustingPin, (Item 3): Material 1060 alloy. The bottom face of the AdjustingPin head is located 40mm from the top face of the Yoke component. The AdjustingPin component contains a Ø5mm though all hole.

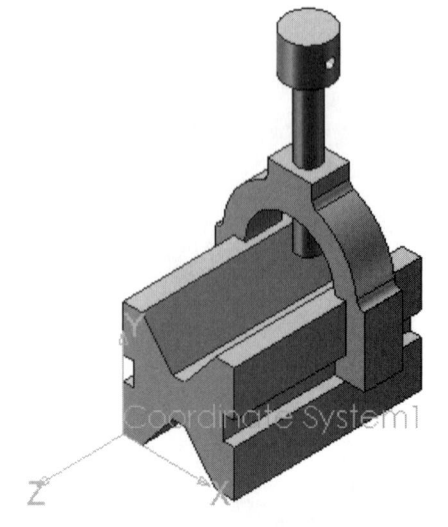

The coordinate system is located in the lower left corner of the Base100 component. The X axis points to the right.

2. Build this assembly. Calculate the overall mass and volume of the assembly. Locate the Center of mass using the illustrated coordinate system. The assembly contains the following: three MachinedBracket components, and two Pin-5 components. Apply the MMGS unit system.

☼ Insert the Base component, float the component, then mate the first component with respect to the assembly reference planes.

- MachinedBracket, (Item 1): Material 6061 Alloy. The MachineBracket component contains two Ø10mm through all holes. Each MachinedBracket component is mated with two Angle mates. The Angle mate = 45deg. The top edge of the notch is located 20mm from the top edge of the MachinedBracket.
- Pin-5, (Item 2): Material Titanium. The Pin-5 component is 5mms in length and equal in diameter. The Pin-5 component is mated Concentric to the MachinedBracket, (no clearance). The end faces of the Pin-5 component is Coincident with the outer faces of the MachinedBracket. There is a 1mm gap between the Machined Bracket components.

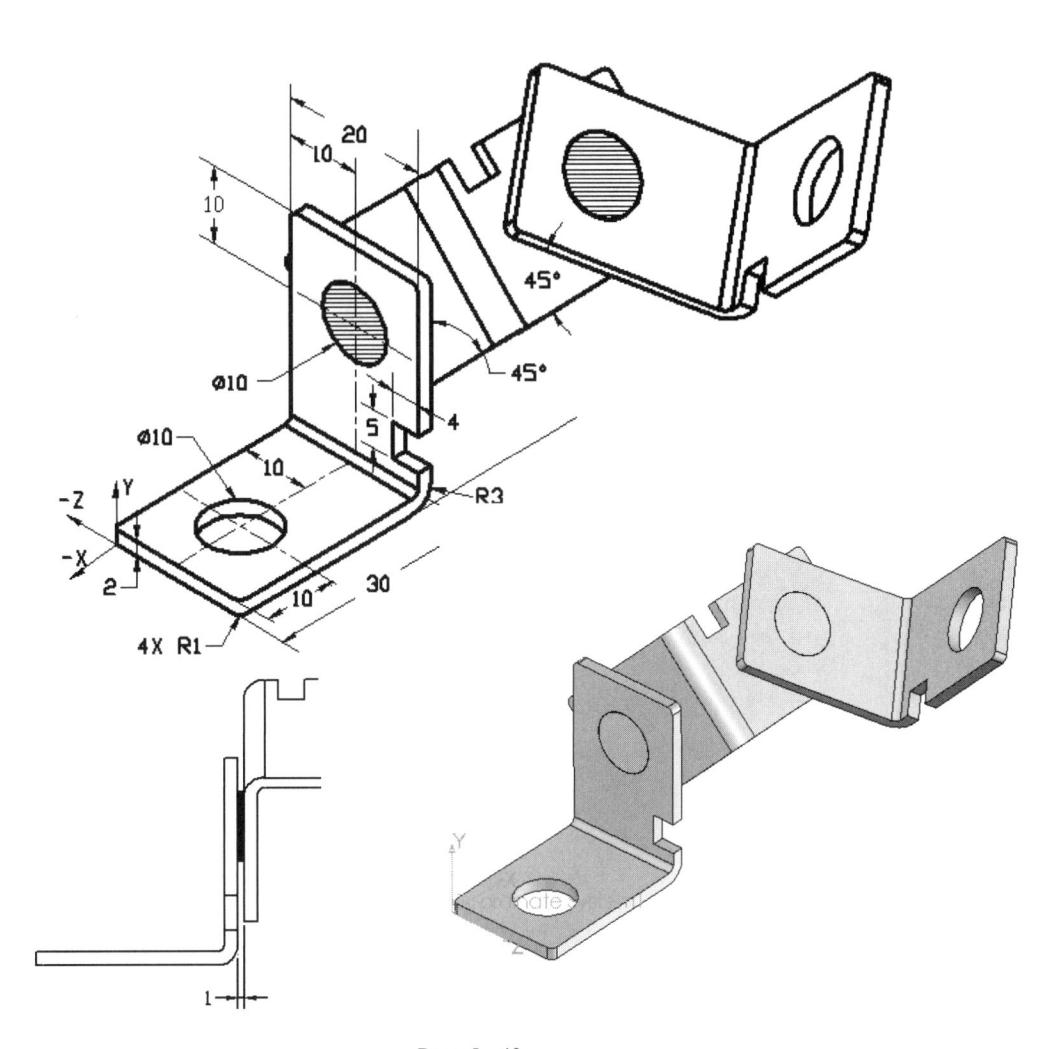

Page 5 - 43

Assembly Modeling

3. Build this assembly. Use the dimensions from the second Check your understanding problem in this chapter. Calculate the overall mass and volume of the assembly. Locate the Center of mass using the illustrated coordinate system. The illustrated assembly contains the following components: three Machined-Bracket components, and two Pin-6 components. Apply the MMGS unit system.

☼ Insert the Base component, float the component, then mate the first component with respect to the assembly reference planes.

- Machined-Bracket, (Item 1): Material 6061 Alloy. The Machine-Bracket component contains two Ø10mm through all holes. Each Machined-Bracket component is mated with two Angle mates. The Angle mate = 45deg. The top edge of the notch is located 20mm from the top edge of the MachinedBracket.
- Pin-6, (Item 2): Material Titanium. The Pin-6 component is 5mms in length and equal in diameter. The Pin-5 component is mated Concentric to the Machined-Bracket, (no clearance). The end faces of the Pin-6 component is Coincident with the outer faces of the Machined-Bracket. There is a 1mm gap between the Machined-Bracket components.

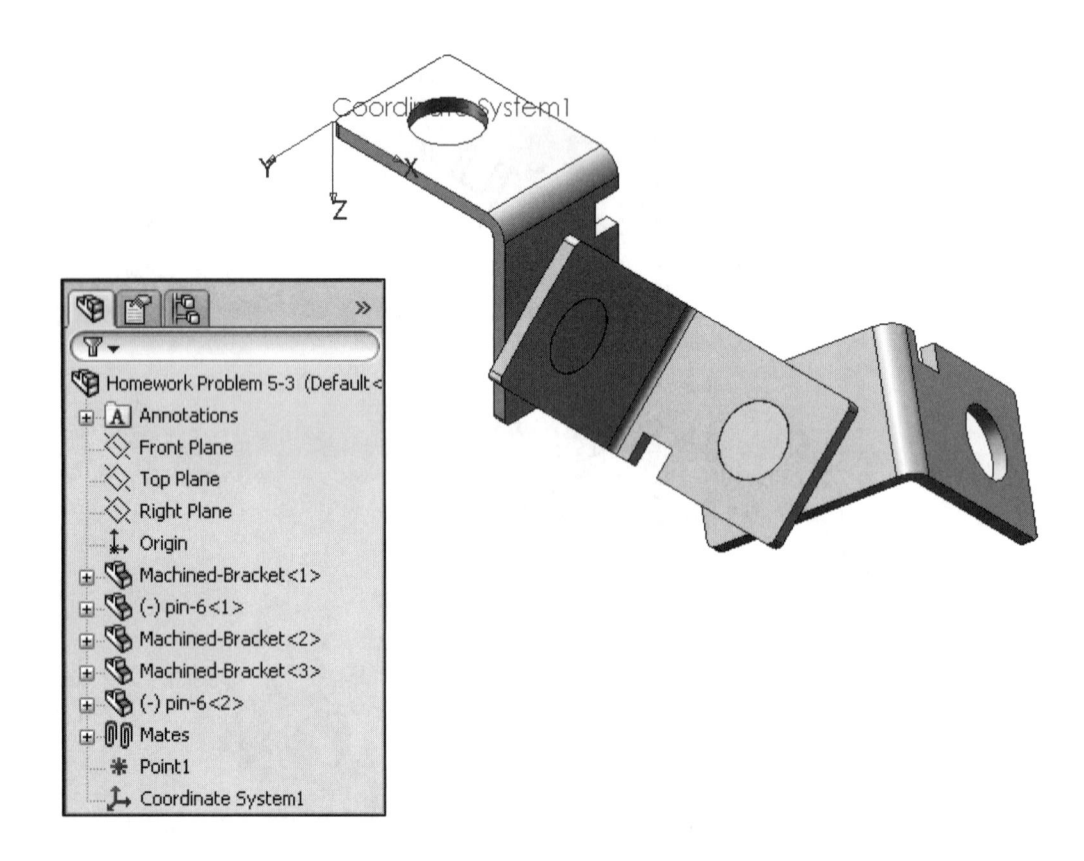

Notes:

CHAPTER 6: ADVANCED MODELING THEORY AND ANALYSIS

Chapter Objective

Advanced Modeling Theory and Analysis is one of the five categories on the CSWA exam. This chapter covers general terminology used in Engineering analysis and provides the needed knowledge of COSMOSXpress that is aligned to the exam.

There are two questions on the CSWA exam in this category. Each question is worth five points. The two questions are in a multiple choice single answer or fill in the blank format.

As in Chapter 2, in this category of the exam, you are not required to perform an analysis on a part or assembly, but are required to understand general engineering analysis terminology and how COSMOSXpress works.

On the completion of the chapter, you will be able to:

- Understand basic Engineering analysis definitions
- Know the COSMOSXpress Wizard interface
- Apply COSMOSXpress to a simple part

In SolidWorks 2009, COSMOSXpress is called SimulationXpress.

Definition Review

Buckling:

Is a failure mode characterized by a sudden failure of a structural member subjected to high compressive stresses, where the actual compressive stresses at failure are smaller than the ultimate compressive stresses that the material is capable of withstanding. This mode of failure is also described as failure due to elastic instability.

Coefficient of Thermal Expansion:

Is defined as the change in length per unit length per one degree change in temperature (change in normal strain per unit temperature).

Creep:

A term used to describe the tendency of a solid material to slowly move or deform permanently under the influence of stresses. It occurs as a result of long term exposure to levels of stress that are below the yield strength or ultimate strength of the material. Creep is more severe in materials that are subjected to heat for long periods, and near the melting point.

Degrees of Freedom:

Are the set of independent displacements and/or rotations that specify completely the displaced or deformed position and orientation of the body or system. This is a fundamental concept relating to systems of moving bodies in mechanical engineering, aeronautical engineering, robotics, structural engineering, etc. There are six degrees of freedom: Three translations and three rotations.

Density:

Is mass per unit volume. Density units are lb/in^3 in the English system, and kg/m^3 in the SI system.

Density is used in static, nonlinear, frequency, dynamic, buckling, and thermal analyses. Static and buckling analyses use this property only if you define body forces (gravity and/or centrifugal).

Ductile:

A mechanical property which describes how able the material lends itself to be formed into rod-like shapes before fracture occurs. Examples of highly ductile metals are silver, gold, copper, and aluminum. The ductility of steel varies depending on the alloying constituents. Increasing levels of carbon decreases ductility, i.e. the steel becomes more brittle.

Elastic Modulus:

For a linear elastic material, the elastic modulus is the stress required to cause a unit strain in the material. In other words stress divided by the associated strain. The modulus of elasticity was first introduced by Young and is often called the Young's Modulus.

Fatigue:

Is the progressive and localized structural damage that occurs when a material is subjected to cyclic loading. The maximum stress values are less than the ultimate tensile stress limit, and may be below the yield stress limit of the material.

Fixed Restraint:

For solids this restraint type sets all translational degrees of freedom to zero. For shells and beams, it sets the translational and the rotational degrees of freedom to zero. For truss joints, it sets the translational degrees of freedom to zero. When using this restraint type, no reference geometry is needed.

Force:

Is a push or pull upon an object resulting from the object's interaction with another object. Whenever there is an interaction between two objects, there is a force upon each of the objects. When the interaction ceases, the two objects no longer experience the force. Forces only exist as a result of an interaction.

For example, if you select 3 faces and specify a 50 lb force, COSMOSXpress applies a total force of 150 lbs (50 lbs on each face).

☼ Knowing how a design will perform under different conditions allows engineers to make changes prior to physical prototyping, thus saving both time and money.

Linear Static Analysis:

Linear static analysis allows engineers to test different load conditions and their resulting stresses and deformation. What is stress? Stress is a measure of the average amount of force exerted per unit area. It is a measure of the intensity of the total internal forces acting within a body across imaginary internal surfaces, as a reaction to external applied forces and body forces.

Deformation is a change in shape due to an applied force. This can be a result of tensile (pulling) forces, compressive (pushing) forces, shear, bending or torsion (twisting). Deformation is often described in terms of strain.

When loads are applied to a body, the body deforms and the effect of loads is transmitted throughout the body. The external loads induce internal forces and reactions to render the body into a state of equilibrium.

Linear Static analysis calculates *displacements*, *strains*, *stresses*, and *reaction forces* under the effect of applied loads. Linear static analysis makes the following assumptions:

1. The induced response is directly proportional to the applied loads.

2. The highest stress is in the linear range of the stress-strain curve characterized by a straight line starting from the origin.

3. The maximum calculated displacement is considerably smaller than the characteristic dimension of the part. For example, the maximum displacement of a plate must be considerably smaller than its thickness and the maximum displacement of a beam must be considerably smaller than the smallest dimension of its cross-section. Inertia is neglected.

4. Loads are applied slowly and gradually until they reach their full magnitudes. Suddenly applied loads cause additional displacements, strains, and stresses.

Below are three simple graphics of Stress vs. Strain:

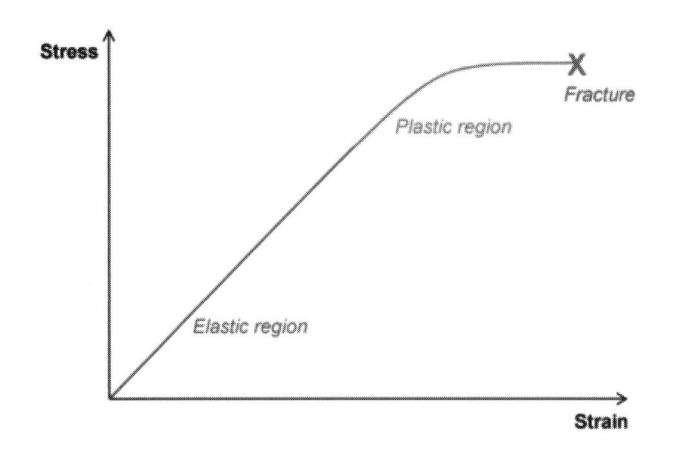

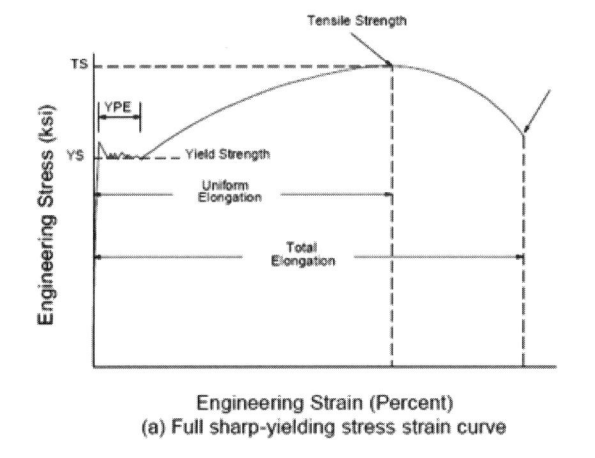

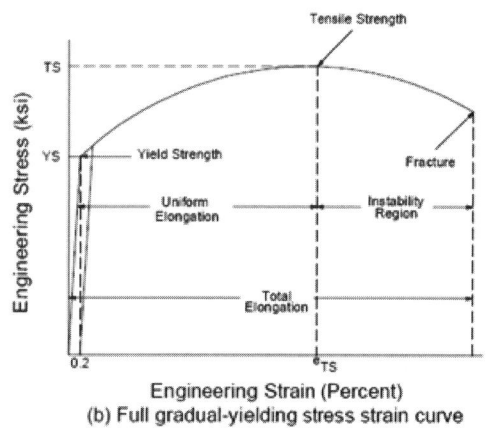

Inertia:

Describes the motion of matter and how it is affected by applied forces. The principle of inertia as described by Newton in Newton's First Law of Motion states: "An object that is not subject to any outside forces moves at a constant velocity, covering equal distances in equal times along a straight-line path." In even simpler terms, inertia means "A body in motion tends to remain in motion, a body at rest tends to remain at rest."

Material Strength:

In materials science, the strength of a material refers to the material's ability to resist an applied force.

Mohr-Columb Stress Criterion:

The Mohr-Columb stress criterion is based on the Mohr-Columb theory also known as the Internal Friction theory. This criterion is used for brittle materials with different tensile and compressive properties. Brittle materials do not have a specific yield point and hence it is not recommended to use the yield strength to define the limit stress for this criterion.

Orthotropic Material:

A material is orthotropic if its mechanical or thermal properties are unique and independent in three mutually perpendicular directions. Examples of orthotropic materials are wood, many crystals, and rolled metals. For example, the mechanical properties of wood at a point are described in the longitudinal, radial, and tangential directions. The longitudinal axis (1) is parallel to the grain (fiber) direction; the radial axis (2) is normal to the growth rings; and the tangential axis (3) is tangent to the growth rings.

Poisson's Ratio:

Extension of the material in the longitudinal direction is accompanied by shrinking in the lateral directions. For example, if a body is subjected to a tensile stress in the X-direction, then Poisson's Ratio NUXY is defined as the ratio of lateral strain in the Y-direction divided by the longitudinal strain in the X-direction. Poisson's ratios are dimensionless quantities. If not defined, the program assumes a default value of 0.

Shear Modulus:

Also called modulus of rigidity, is the ratio between the shearing stress in a plane divided by the associated shearing strain. Shear Moduli are used in static, nonlinear, frequency, dynamic and buckling analyses.

Thermal Conductivity:

Indicates the effectiveness of a material in transferring heat energy by conduction. It is defined as the rate of heat transfer through a unit thickness of the material per unit temperature difference. The units of thermal conductivity are Btu/in sec $°F$ in the English system and W/m $°K$ in the SI system.

Thermal conductivity is used in steady state and transient thermal analyses.

Tensile Strength:

Tensile strength is the maximum load sustained by the specimen in the tension test, divided by the original cross sectional area.

von Mises yield Criterion:

A scalar stress value that can be computed from the stress. In this case, a material is said to start yielding when its von Mises stress reaches a critical value known as the yield strength. The von Mises stress is used to predict yielding of materials under any loading condition from results of simple uniaxial tensile tests. The von Mises stress satisfies the property that two stress states with equal distortion energy have equal von Mises stress.

Advanced Modeling Theory and Analysis

Yield Strength:

The stress at which the metal yields or becomes permanently deformed is an important design parameter. This stress is the elastic limit below which no permanent shape changes will occur.

The elastic limit is approximated by the yield strength of the material, and the strain that occurs before the elastic limit is reached is called the elastic strain. The yield strength is defined in three ways, depending on the stress-strain characteristics of the steel as it begins to yield. The procedures in SAE J416, ASTM E8, and ASTM A370.

COSMOSXpress uses this material property to calculate the factor of safety distribution. COSMOSXpress assumes that the material starts yielding when the equivalent (von Mises) stress reaches this value.

All questions on the exam are in a multiple choice single answer format. In this category, Advanced Modeling Theory and Analysis an exam question could read:

Questions 1: Yield strength is typically determined at _____ strain.
- A = 0.1%
- B = 0.2%
- C = 0.02%
- D = 0.002%

The correct answer is B.

☼ The stress–strain curve is a graphical representation of the relationship between stress, derived from measuring the load applied on the sample, and strain, derived from measuring the deformation of the sample, i.e. elongation, compression, or distortion. The nature of the curve varies from material to material.

Questions 2: There are four key assumptions made in Linear Static Analysis: 1: Effects of inertia and damping is neglected, 2. The response of the system is directly proportional to the applied loads, 3: Loads are applied slowly and gradually, and_____ .

- A = Displacements are very small. The highest stress is in the linear range of the stress-strain curve.
- B = There are no loads.
- C = Material is not elastic.

- D = Loads are applied quickly.

The correct answer is A.

Questions 3: How many degrees of freedom does a physical structure have?

- A = Zero.
- B = Three – Rotations only.
- C = Three – Translations only.
- D = Six – Three translations and three rotations.

The correct answer is D.

Questions 4: Brittle materials has little tendency to deform (or strain) before fracture and does not have a specific yield point. It is not recommended to apply the yield strength analysis as a failure criterion on brittle material. Which of the following failure theories is appropriate for brittle materials?

- A = Mohr-Columb stress criterion.
- B = Maximum shear stress criterion.
- C = Maximum von Mises stress criterion.
- D = Minimum shear stress criterion.

The correct answer is A.

Questions 5: You are performing an analysis on your model. You select three faces and apply a 40lb load. What is the total force applied to the model?

- A = 40lbs.
- B = 20lbs.
- C = 20lbs.
- D: Additional information is required.

The correct answer is C.

Questions 6: In an engineering analysis, you select a face to restrain. What is the affect?

- A = The face will not translate but can rotate.
- B = The face will rotate but can not translate.
- C = You can not apply a restraint to a face.
- D = The face will not rotate and will not translate.

The correction answer is D.

Questions 7: A material is orthotropic if its mechanical or thermal properties are not unique and independent in three mutually perpendicular directions.

- A = True.
- B = False.

The correction answer is B.

☼ During the exam, use SolidWorks Help and COSMOSXpress help in the COSMOSXpress dialog box if needed to review and understand various engineering terms.

☼ In SolidWorks 2009, COSMOSXpress is called SimulationXpress.

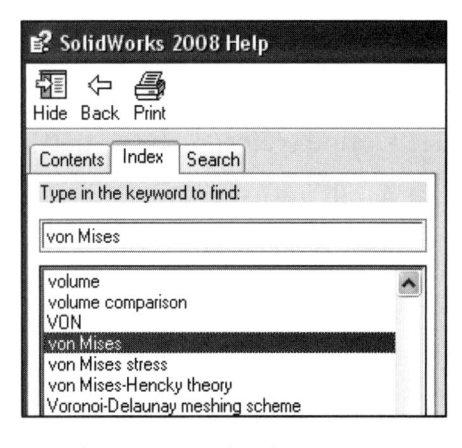

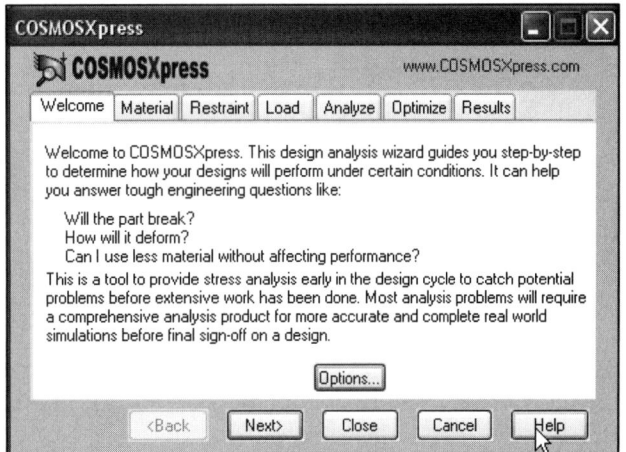

COSMOSXpress

COSMOSXpress offers an easy-to-use first pass stress analysis tool for SolidWorks users. This tool displays the effects of a force applied to a part, and simulates the design cycle and provides stress results. It also displays critical areas and safety levels at various regions in the selected part.

Based on these results, you can strengthen unsafe regions and remove material from over designed areas.

In this section, you will apply the COSMOSXpress tool to analyze a Flatbar part. There are only five required steps to analyze a part using COSMOSXpress:

1. Define material on the part.
2. Apply restraints.
3. Apply loads.
4. Analyze the part.
5. Optimize the part, (Optional).
6. View the results.

Perform a first-pass analysis on the Flatbar, its Factor of Safety, and view the applied stresses. You will then modify the Flatbar thickness and material and rerun the analysis for comparison.

💡 COSMOSXpress calculates displacements, strains, and stresses, but it only displays stresses and displacements.

💡 Only faces can be selected in COSMOSXpress to apply loads and restraints.

💡 COSMOSXpress supports the analysis of solid, single-bodied parts only. It does not support the analysis of assemblies, surface models, or multi-body parts.

💡 The accuracy of the results of the analysis depends on selected material properties, restraints, and loads. For results to be valid, the specified material properties must accurately represent the part material, and the restraints and loads must accurately represent the part working conditions.

COSMOSXpress User Interface

COSMOSXpress guides you through six steps to define material properties, restraints, loads, analyze the model, view the results, and the optional Optimization. The COSMOSXpress interface consists of the following tabs:

- **Welcome tab**: Allows you to set the default units and to specify a folder for saving the analysis results.
- **Material tab**: Applies material properties to the part. The material can be assigned from the material library or you can input the material properties.
- **Restraint tab**: Applies restraints to faces of the part.

Advanced Modeling Theory and Analysis

- **Load tab**: Applies forces and pressures to faces of the part.
- **Analyze tab**: Provides the ability to either display the analysis with the default settings or to change the settings.
- **Optimize tab**: Optimizes a model dimension based on a specified criterion.
- **Results tab**: Displays the analysis results in the following ways:
  - Shows critical areas where the factor of safety is less than a specified value.
  - Displays the stress distribution in the model with or without annotation for the maximum and minimum stress values.
  - Displays resultant displacement distribution in the model with or without annotation for the maximum and minimum displacement values.
  - Shows deformed shape of the model.
  - Generates an HTML report.
  - Generates eDrawings files for the analysis results.
- **Start Over button**: Deletes existing analysis data and results and starts a new analysis session.
- **Update button**: Runs COSMOSXpress analysis if the restraints and loads are resolved. Otherwise, it displays a message and you need to resolve the invalid restraints or loads. The Update button is displayed if you modify geometry after applying loads or restraints. It also is displayed if you modify material properties, restraints, loads, or geometry after completing the analysis. Once any of these values are changed, an exclamation mark is displayed on the Analyze and Results tabs. An exclamation mark on the Restraint or Load tab indicates that a restraint or load became invalid after a change in geometry.

☼ You can use COSMOSXpress only on an active part document. If you create a new part document or open an existing document with COSMOSXpress active, COSMOSXpress automatically saves the analysis information and closes the current analysis session.

Tutorial: COSMOSXpress 6-1

Close all parts, assemblies, and drawings.

1. **Open** the COSMOSXpress-Flatbar part from the SolidWorks CSWA Folder\Chapter6 location.
2. **Activate** COSMOSXpress from the Menu bar menu. The Welcome box is displayed.
3. Click the **Options** button to select the system units and to specify a save in folder. Select **IPS** for unit system.

4. Click **Next>**.

5. Apply a material to the part. Select **6061 Alloy**.

💡 Brittle materials do not have a specific yield point and hence it is not recommended to use the yield strength to define the limit stress for the criterion.

6. Click **Apply**.

7. Click **Next>**. Select the fixed points for the part. Fixed points are called restraints.

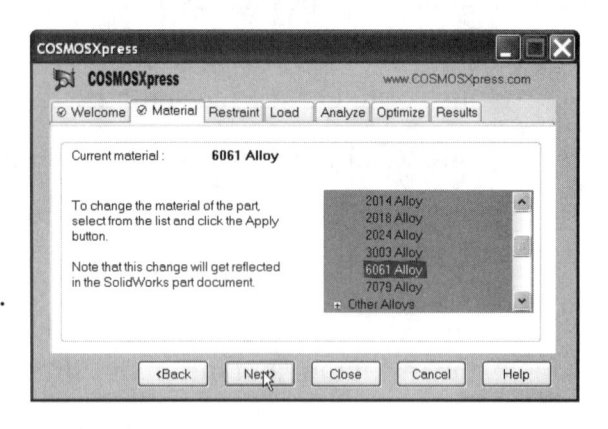

💡 You can specify multiples sets of restraints for a part. Each set of restraints can have multiple faces.

8. Click **Next>**. The Restraint tab is highlighted. Restraint1 is the default restraint name for the first set of restraints.

9. Rename Restraint1 to **Flatbar**.

10. **Select** the two illustrated faces of the Flatbar to be restrained in the Graphics window. Face1 and Face2 are displayed.

11. Click **Next>**. This box provides the ability to add, delete, or edit a restraint to the part.

12. Click **Next>**. The Load tab is highlighted. The load section provides the ability to input information of the loads acting on the part. You can specify multiple loads either in forces or pressures.

💡 You can apply multiple forces to a single face or to multiple faces.

💡 You can apply multiple pressures to a single face or to multiple faces. COSMOSXpress applies pressure loads normal to each face.

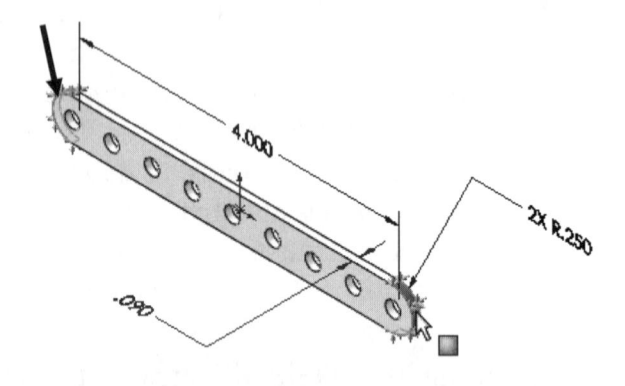

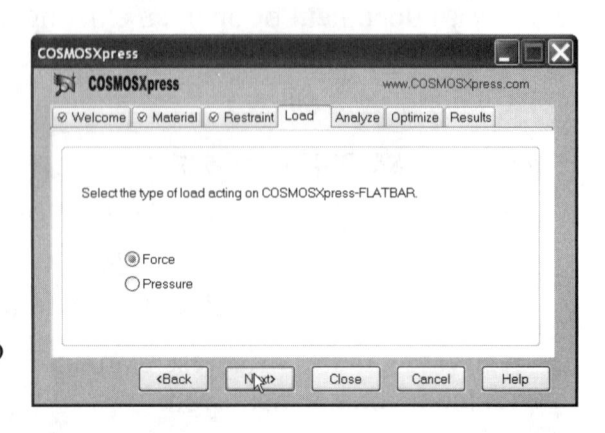

Advanced Modeling Theory and Analysis

13. Click **Next>**. Select the load type. The default is Force. Accept the default Force load type.

14. Click **Next>**. The Load tab is highlighted.

15. Enter **Force** for the load set name.

16. Select the **flat face** of the Flatbar as illustrated. The edges of the Flatbar are fixed.

17. Click **Next>**. The direction of the applied force is displayed. The direction is downward and is normal to the selected face.

18. Enter the applied **100 lb** force value.

19. Click **Next>**. This box provides the ability to add, edit, or delete a load set. Click **Next>**. The Analyze tab is displayed. The default is Yes.

20. Click **No, I want to change the settings**. Click **Next>**. The box provides the ability to modify the element size and element tolerance during the mesh period to analyze the part. The larger the element size and element tolerance, the longer it will take to calculate. Accept the default conditions.

    ☼ Specifying a smaller element size provides a more accurate result.

21. Click **Next>**. Click the **Run** button to run the analysis.

22. **View** the results. The Factor of safety for the specified parameters is approximately 0.054. What does this mean? The part will fail with the specified parameters. What can you do? You can increase the part thickness, modify the part material, modify the applied load, select different fixed points, or perform a variation of all of the above.

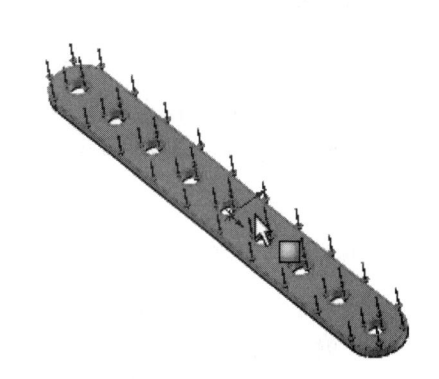

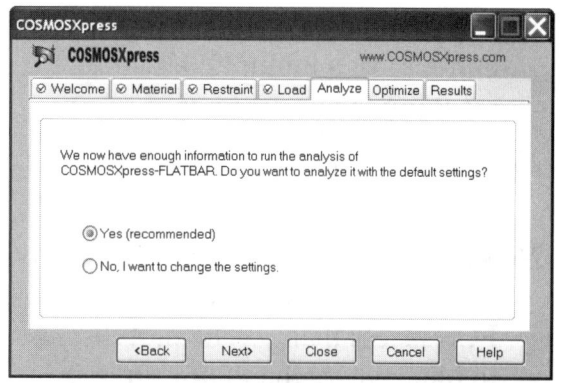

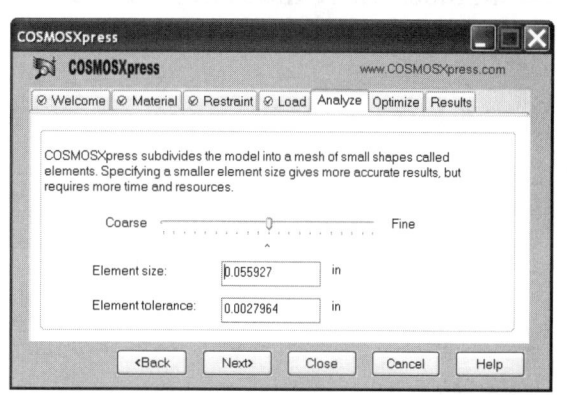

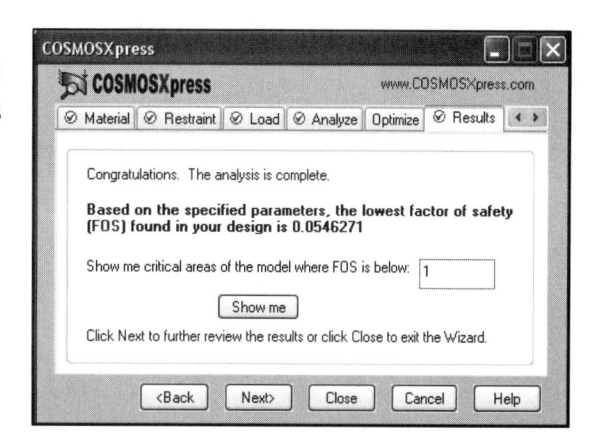

Page 6 - 12

💡 The Factor of Safety is a ratio between the material strength and the calculated stress.

Interpretation of factor of safety values:

- A factor of safety less than 1.0 at a location indicates that the material at that location has yielded and that the design is not safe.

- A factor of safety of 1.0 at a location indicates that the material at that location has just started to yield.

- A factor of safety greater than 1.0 at a location indicates that the material at that location has not yielded.

- The material at a location will start to yield if you apply new loads equal to the current loads multiplied by the resulting factor of safety.

25. Accept the default for the plot. Click the **Show me** button.

26. **View** the Max von Mises Stress plot. The plot displays in red, where the FOS is less than 1. The plot displays in blue, where the FOS is greater than one.

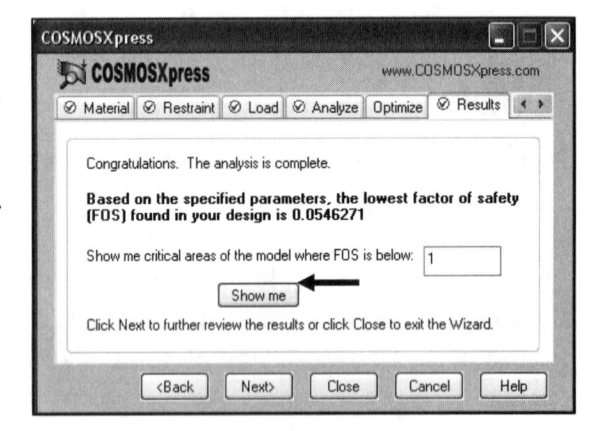

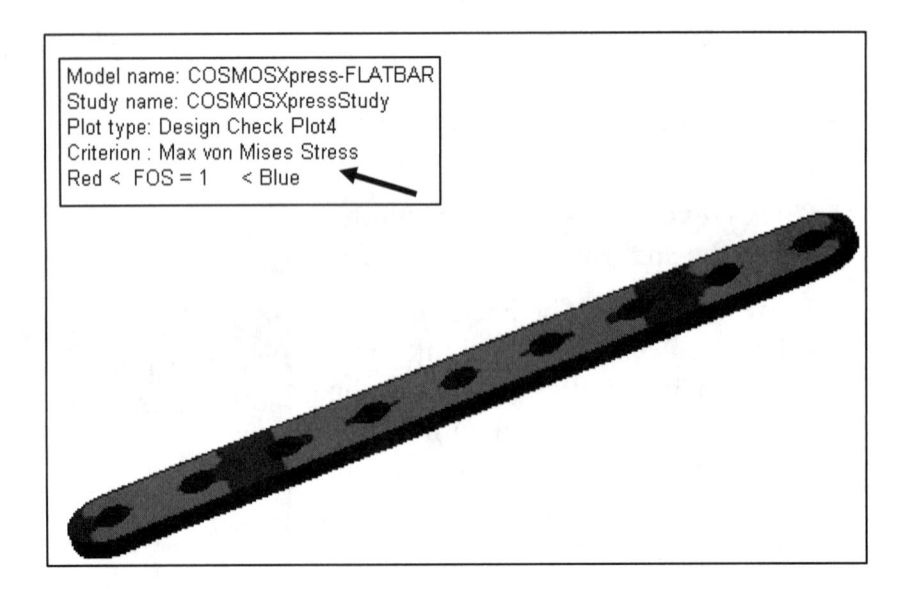

Model name: COSMOSXpress-FLATBAR
Study name: COSMOSXpressStudy
Plot type: Design Check Plot4
Criterion : Max von Mises Stress
Red < FOS = 1   < Blue

Advanced Modeling Theory and Analysis

27. Click **Next>**. The Optimize tab is displayed. The default option is Yes.

28. Select **No**.

29. Click **Next>**. The Results tab is displayed. The default option is Show me the stress distribution in the model. Accept the default option.

💡 Although COSMOSXpress calculates displacements, strains, and stresses, it only allows you to view stresses and displacement.

30. Click **Next>**. The Stress plot is displayed. You can play, stop, or save the animation of the plot in this section. View the stress plot.

31. Click **Close** from the Results tab.

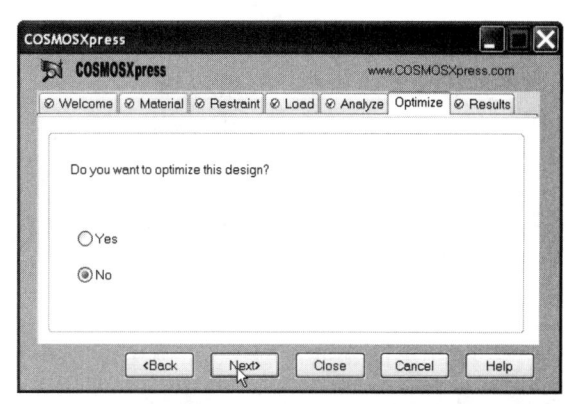

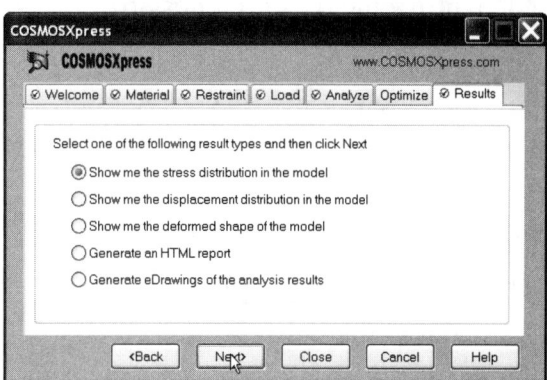

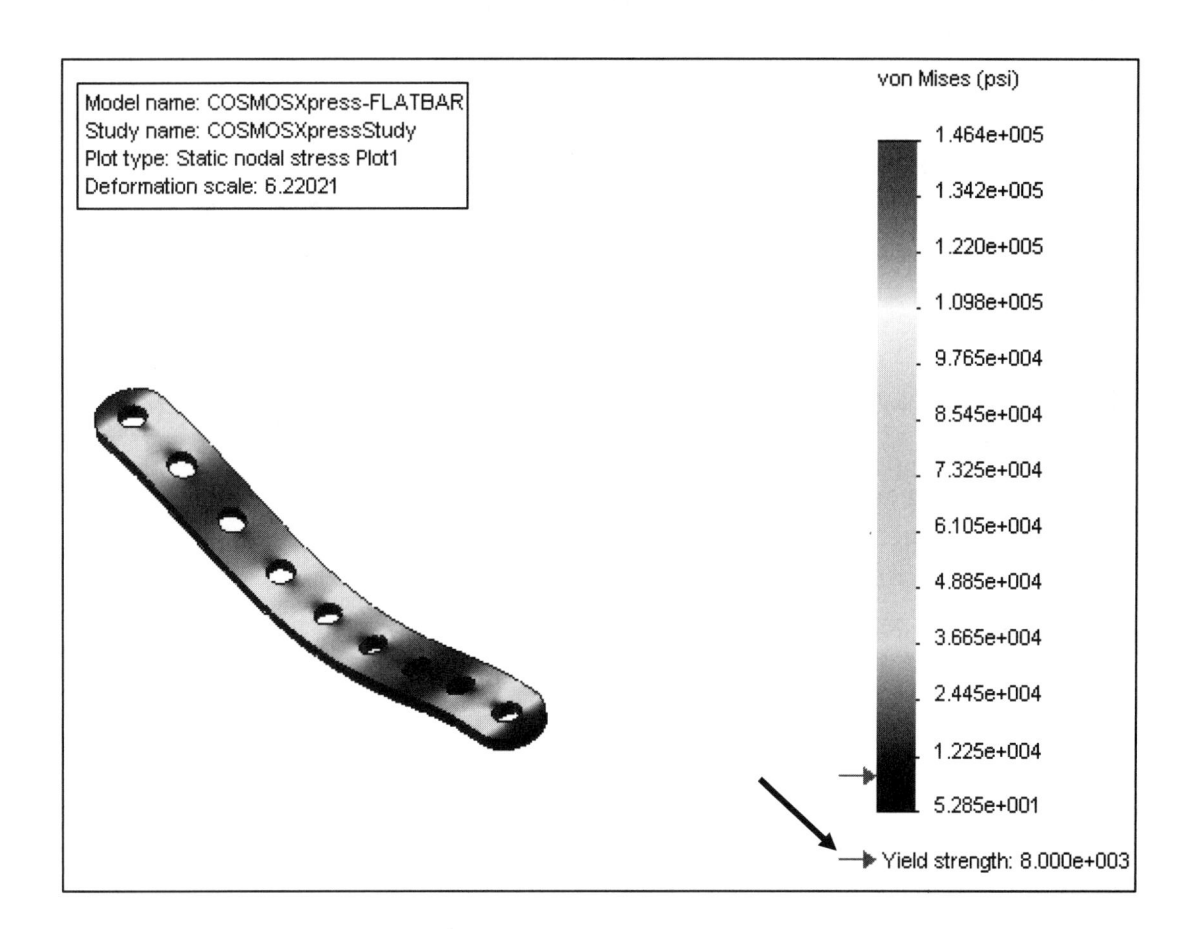

Model name: COSMOSXpress-FLATBAR
Study name: COSMOSXpressStudy
Plot type: Static nodal stress Plot1
Deformation scale: 6.22021

von Mises (psi)

1.464e+005
1.342e+005
1.220e+005
1.098e+005
9.765e+004
8.545e+004
7.325e+004
6.105e+004
4.885e+004
3.665e+004
2.445e+004
1.225e+004
5.285e+001

Yield strength: 8.000e+003

Advanced Modeling Theory and Analysis

32. Click **Yes** to save the COSMOSXpress data. The FeatureManager displays the 6061 Alloy material.

Modify the Flatbar part thickness and re-run COSMOXpress.

33. Double-click **Extrude1** from the FeatureManager.

34. **Modify** the part thickness from .090in to .250in.

35. **Activate** COSMOSXpress and re-run the analysis using the new part thickness.

36. Click the **Update** button.

37. **View** the updated FOS. The FOS is still less than one.

38. Accept the default for the plot. Click the **Show me** button.

39. **View** the Max von Mises Stress plot. The plot displays in red, where the FOS is less than 1. The plot displays in blue, where the FOS is greater than one. Red is displayed on the selected faces that are restrained.

40. Click the **Close** button.

41. Click **Yes** to save the updated information.

42. Edit the assigned material in the FeatureManager. Modify the material to **Plain Carbon Steel**.

43. **Activate** COSMOSXpress.

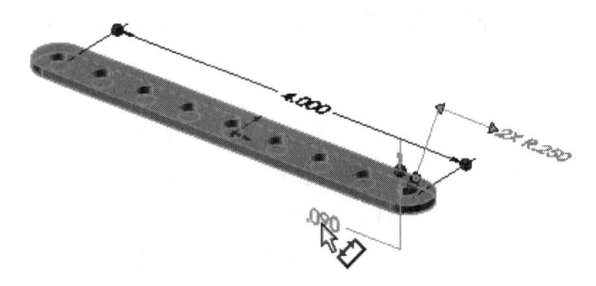

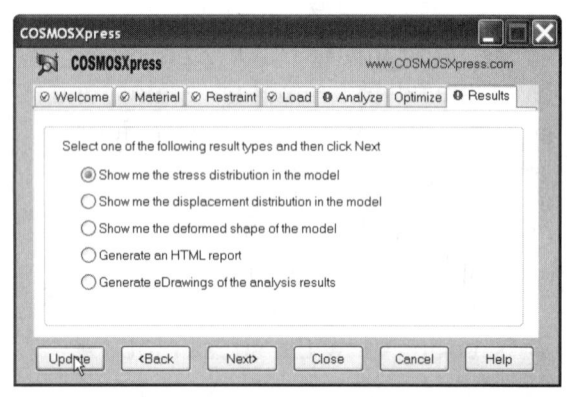

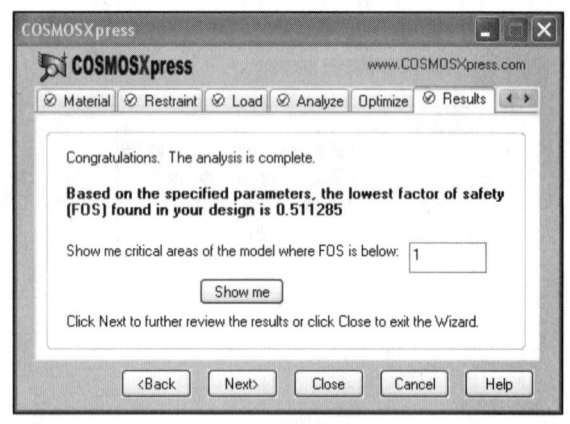

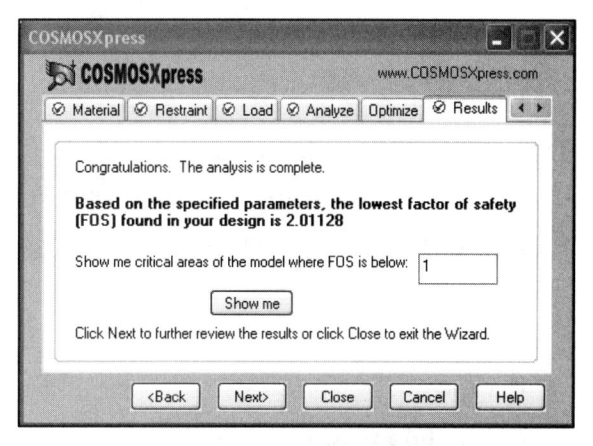

Advanced Modeling Theory and Analysis

44. Click the **Update** button.

45. **View** the new FOS. The new FOS is approximately 2.01.

46. Click **Close**.

47. Click **Yes** to save the updates. The FeatureManager is displayed.

All questions on the exam are in a multiple choice single answer or fill in the blank format. In this category, an exam question could read:

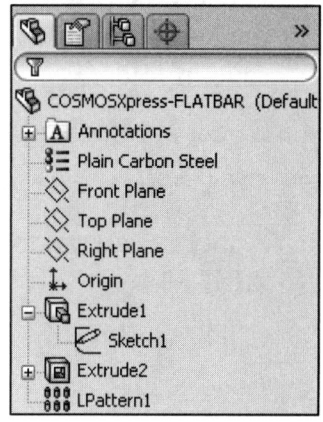

Question 1: COSMOSXpress is used to analyze?

- A = Drawings.
- B = Parts and Drawings.
- C = Assemblies and Parts.
- D = Parts.
- E = All of the above.

The correct answer is D.

Question 2: Under what COSMOSXpress menu tab do you set system units?

- A = Material tab.
- B = Restraint tab.
- C = Welcome tab.
- D = Load tab.

The correct answer is C. Either SI or the IPS system units.

Questions 3: Under what COSMOSXpress menu tab can you modify the Mesh period of the part?

- A = Restraint tab.
- B = Material tab.
- C = Analyze tab.
- D = Welcome tab.

The correct answer is C.

In the next section, perform a first-pass analysis on the COVERPLATE part. View the static displacement plot.

Tutorial: COSMOSXpress 6-2

Close all parts, assemblies, and drawings.

1. **Open** the COSMOSXpress-COVERPLATE part from the SolidWorks CSWA Folder\Chapter6 location.

2. **Activate** COSMOSXpress from the Menu bar menu. The Welcome box is displayed.

3. Click the **Options** button to select the system units and to specify a save in folder.

4. Select **IPS** for unit system.

5. Click **Next>**.

6. Apply a material to the part. Select **AISI 304**.

7. Click **Apply**.

8. Click **Next>**. Select the fixed points for the part.

9. Click **Next>**. The Restraint tab is highlighted. Restraint1 is the default restraint name for the first set of restraints.

10. Rename Restraint1 to **COVERPLATE**.

11. **Select** the two illustrated faces of the COVERPLATE to be restrained. Face1 and Face2 are displayed.

12. Click **Next>**. This box provides the ability to add, delete, or edit a restraint to the part.

13. Click **Next>**. The Load tab is highlighted.

14. Click **Next>**. Select the Load type. The default is Force. Accept the default Force Load type.

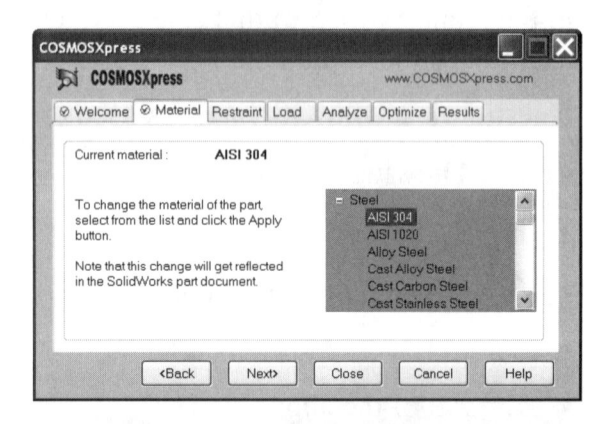

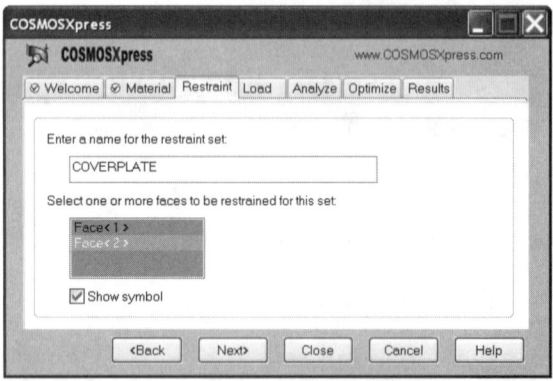

Advanced Modeling Theory and Analysis

15. Click **Next>**.

16. Enter **Force** for the load set name.

17. Select the illustrated **face** of the COVERPLATE.

18. Click **Next>**. The direction of the applied force is displayed. The direction is downward and is normal to the selected face.

19. Enter the applied **50 lb** force value.

20. Click **Next>**. This box provides the ability to add, edit, or delete a load set.

21. Click **Next>**. The Analyze tab is displayed.

22. Accept the default, Yes. Click **Next>**.

💡 As an exercise, select No and explore the settings.

23. Click the **Run** button to run the analysis.

24. **View** the results. The Factor of safety is approximately 16.53.

25. Click **Next>**. The Optimize tab is displayed.

26. Select **No**.

27. Click **Next>**.

💡 As an exercise, select the Optimize tab. The Optimize tab can perform optimization analysis after completing stress analysis on the Analyze tab.

28. Select **Show me the displacement distribution in the model**.

29. Click **Next>**.

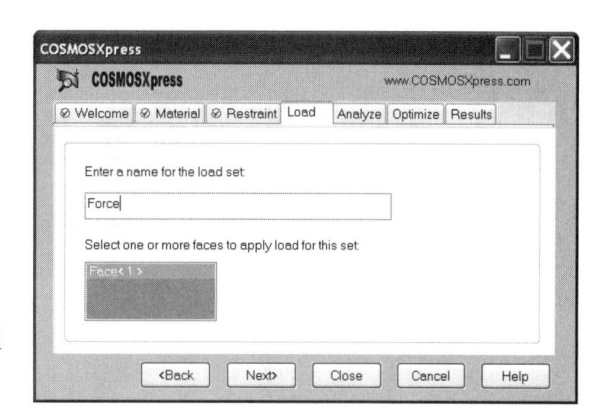

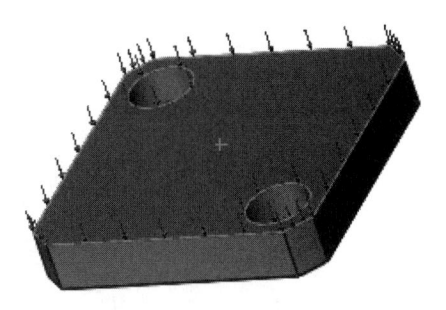

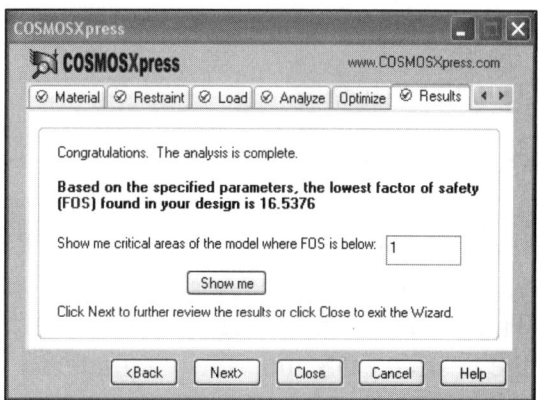

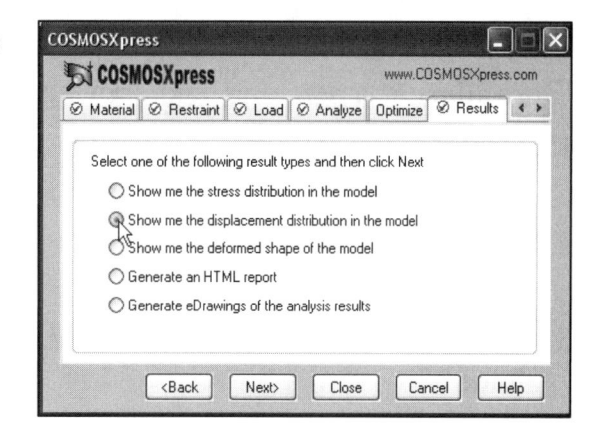

Page 6 - 18

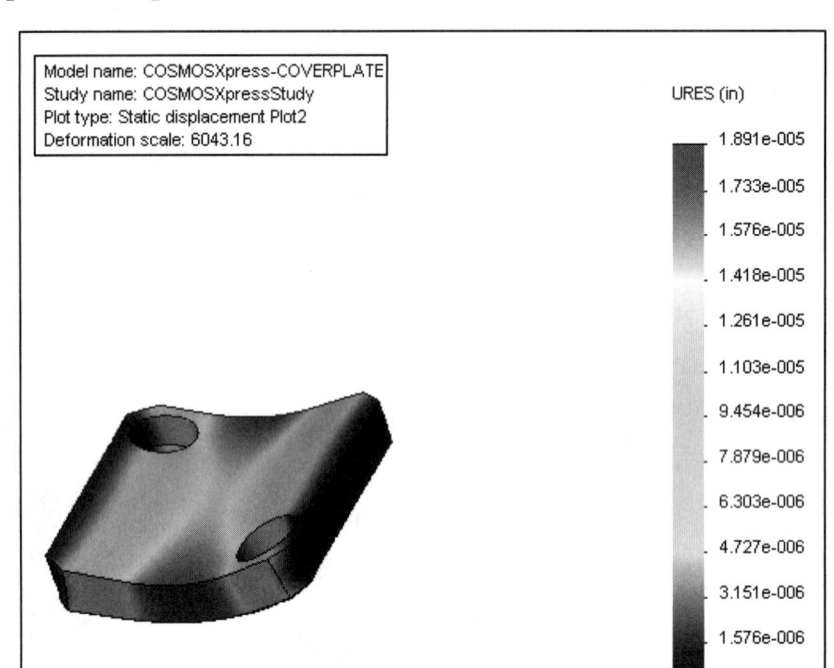

30. View the static displacement plot. Click **Next>**.

31. Click **Close**.
32. Do not save the data. Click **No**.

In the next section, perform a first-pass analysis on the LBRACKET part and assess the Factor of Safety. Then run the Optimization feature, set the desired FOS and view the material thickness change.

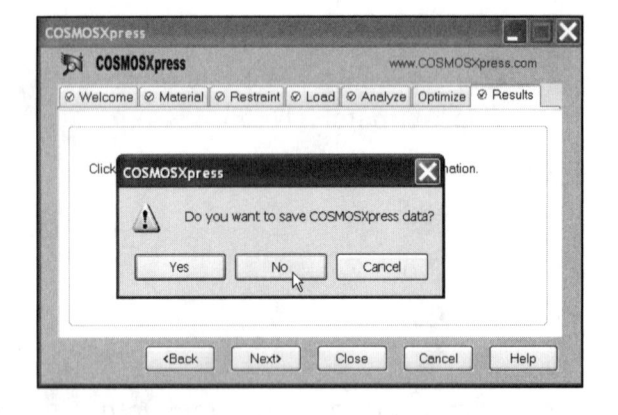

Tutorial: COSMOSXpress 6-3

Close all parts, assemblies, and drawings.

1. **Open** the COSMOSXpress-LBRACKET part from the SolidWorks CSWA Folder\Chapter6 location.

2. **Activate** COSMOSXpress from the Menu bar menu. The Welcome box is displayed.

3. Click the **Options** button to select the system units and to specify a save in folder.

4. Select **IPS** for unit system.

Advanced Modeling Theory and Analysis

5. Click **Next>**.

6. Apply a material to the part. Select **Copper**.

7. Click **Next>**. You can specify multiples sets of restraints for a part. Each set of restraints can have multiple faces.

8. Click **Next>**. The Restraint tab is highlighted. Restraint1 is the default restraint name for the first set of restraints. **Rename** Restraint1 to LBRACKET.

9. **Select** the two illustrated faces of the LBRACKET to be restrained. Face1 and Face2 are displayed.

10. Click **Next>**. This box provides the ability to add, delete, or edit a restraint to the part.

11. Click **Next>**. The Load tab is highlighted. The load section provides the ability to input information of the loads acting on the part. You can specify multiple loads either in forces or pressures.

12. Click **Next>**. Select the Load type. The default is force.

13. Select **Pressure**.

14. Click **Next>**. The Load tab is highlighted. Enter **Pressure** for the load set name.

15. Select the **face** of the LBRACKET as illustrated. Note: The edges of the LBRACKET are fixed.

16. Click **Next>**. The direction of the applied force is displayed. The direction is downward and is normal to the selected face.

17. Enter **15 psi** for value.

18. Click **Next>**. This box provides the ability to add, edit, or delete a load set.

19. Click **Next>**.

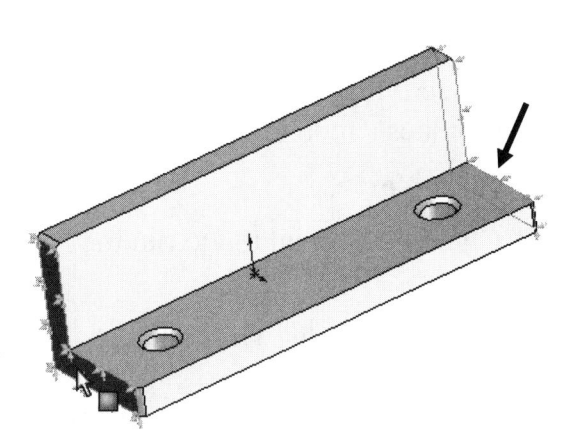

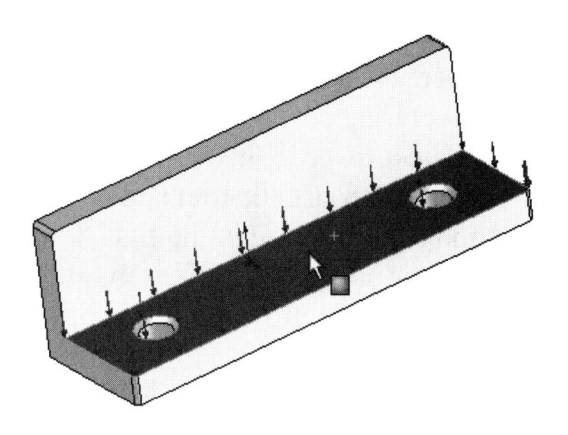

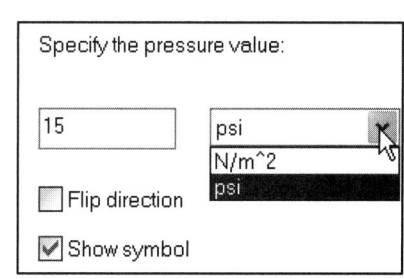

20. Select **No, I want to change the settings**.
21. Click **Next>**.
22. Modify the element size and element tolerance during the mesh period. Click and drag the **slider** to the right as illustrated.

💡 Specifying a smaller element size provides a more accurate result.

23. Click **Next>**.
24. Click the **Run** button to run the analysis.
25. **View** the results. The Factor of safety for the specified parameters is approximately 34.03. The part will not fail with the specified parameters.

Is this part over designed? Can you decrease the material and save manufacturing cost and still have a safe part? The answer to this question is yes. You can either use the trial and error method that you used in the first Tutorial or use the Optimize tab option that SolidWorks provides.

💡 The Optimize tab by default will automatically modify your part and save any changes.

26. Click **Next>**. Accept the default setting.
27. Click **Next>**. View the available criteria for the optimized part. Accept the default FOS criteria.
28. Enter **4** for FOS.

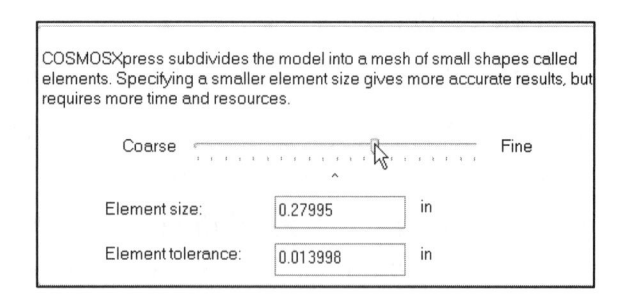

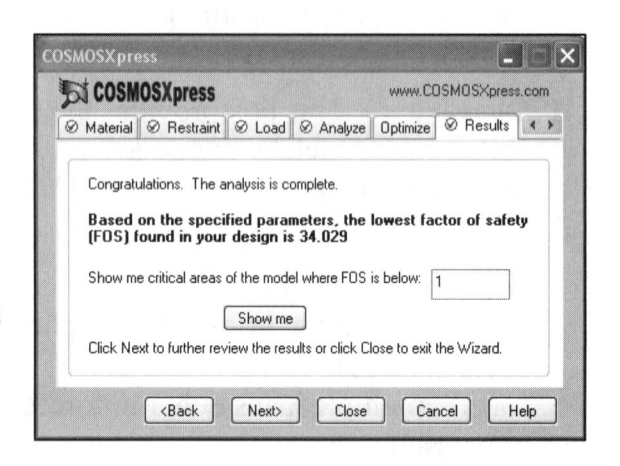

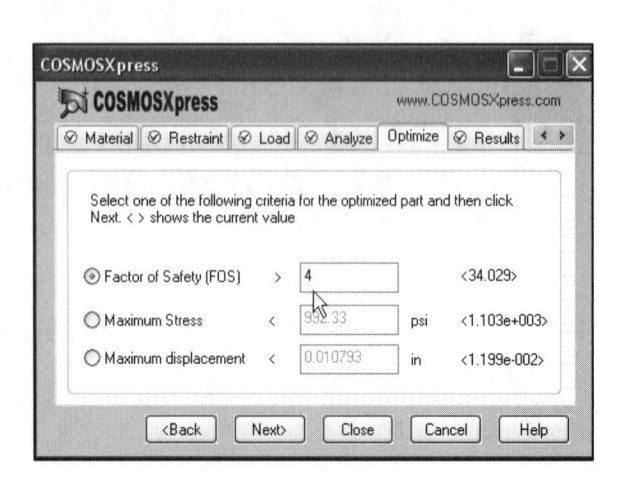

Page 6 - 21

Advanced Modeling Theory and Analysis

29. Click **Next>**.

30. Select the dimension you want to modify. Select the **.650** dimension as illustrated from the Graphics window. To obtain a FOS of 4, the lower bound is .325in, and the upper bound is .975in.

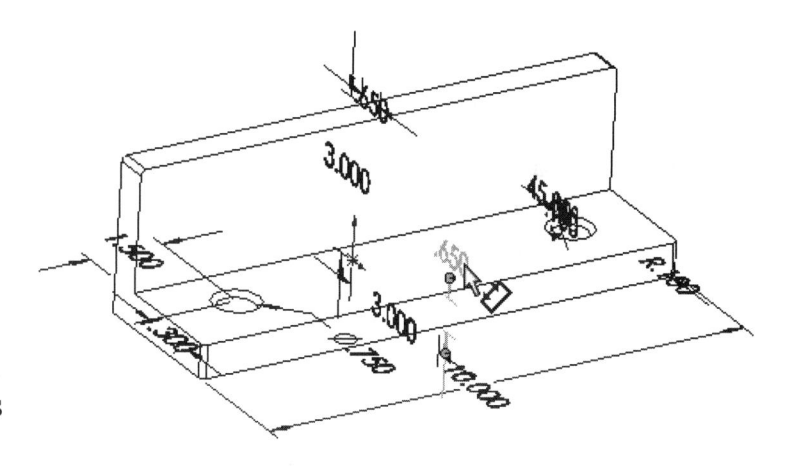

31. Click **Next>**.

32. Click the **Optimize** button. The calculation can take a few minutes.

33. **View** the results. The new design weights 21.51% less than the initial design with a FOS of 4. The thickness of the bottom face was modify from .650in to .325in.

34. Click **Next >**. Accept the default setting.

35. **Close** the model.

36. Click **No**.

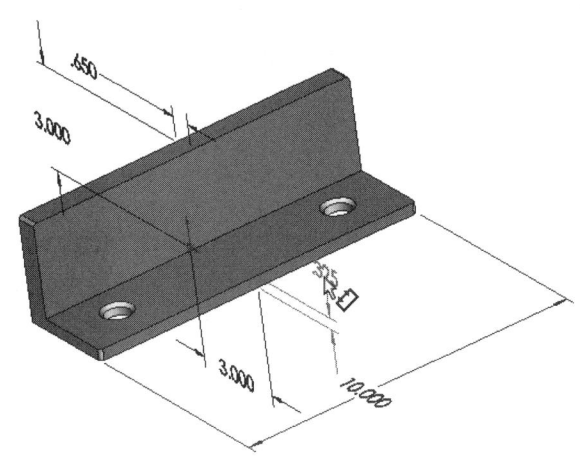

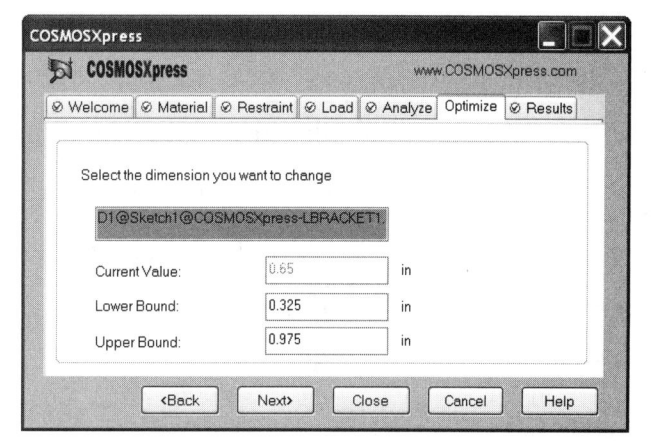

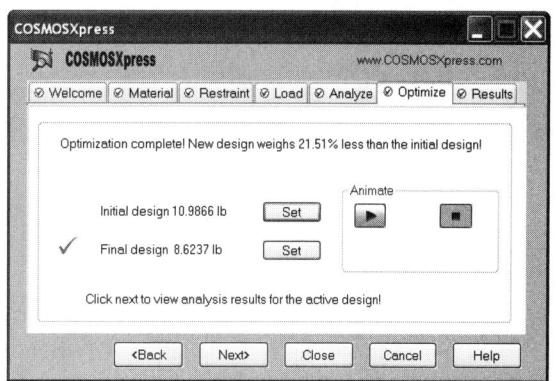

All questions on the exam are in a multiple choice single answer or fill in the blank format. In this category, Advanced Modeling Theory and Analysis an exam question could read:

Questions 1: Under what COSMOSXpress menu tab can you set the direction for a Load?

- A = Restraint tab.
- B = Load tab.
- C = Analyze tab.
- D = None of the above.

The correct answer is B.

Questions 2: An increase mesh period for a part will.

- A = Decrease calculation accuracy and time.
- B = Increase calculation accuracy and time.
- C = Have no effect on the calculation.
- D = Change the FOS below 1.

The correct answer is B.

Questions 3: COSMOSXpress uses the von Mises Yield Criterion to calculate the Factor of Safety of many ductile materials. According to the criterion:

- A = Material yields when the von Mises stress in the model equals the yield strength of the material.
- B = Material yields when the von Mises stress in the model to 5 times greater that the minimum tensile strength of the material.
- C = Material yields when the von Mises stress in the model is 3 times greater than the FOS of the material.
- D = None of the above.

The correct answer is A.

Questions 4: COSMOSXPress calculates structural failure on:

- A = Buckling,
- B = Fatigue
- C = Creep.
- D = Material yield.

The correct answer is D.

Advanced Modeling Theory and Analysis

Questions 5: Identify the loading types supported by COSMOSXPress?

- A = Force with respect to a selected reference plane .
- B = Force normal to a selected face.
- C = Pressure normal to a selected face.
- D = Imported loads on a part after running the Physical Simulation.

The correct answer is A, B, C, and D.

Questions 6: Apply a uniform total force of 200 lbs on two faces of a model. The two faces have different areas. How do you apply the load using COSMOSXPress?

- A = Select the two faces and input a normal to direction force of 200 lbs on each face.
- B = Select the two faces and a reference plane. Apply 100 lbs on each face.
- C = Apply equal pressure to the two faces. The force on each face is the total force divided by the total area of the two faces.
- D = None of the above.

The correct answer is C.

Questions 7: COSMOSXPress provides the ability to select a _____ to apply loads and restraints.

- A = Faces and Edges.
- B = Faces, Edges and Vertices.
- C = Edges only.
- D = Faces only.

The correct answer is D.

Questions 8: Can you apply a material to the part directly in COSMOSPress?

- A = Yes.
- B = No.

The correct answer is A

Questions 9: A material was applied to a part in SolidWorks. You apply a new material in COSMOSXPress. What happens to the material properties?

- A = A material can't be directly applied in COSMOSXPress.
- B = The material applied in COSMOSXPress is used only in the analysis. The material applied in SolidWorks stays the same.

- C = The part material in SolidWorks changes to the material type that you applied in COSMOSXPress.
- D = None of the above.

The correct answer is C

Questions 10: Maximum and Minimum value indicators are displayed on Stress and Displacement plots in COSMOSXPress.

- A = True.
- B = False.

The correct answer is A.

Questions 11: You can store your analysis results from COSMOSXPress.

- A = True.
- B = False.

The correct answer is A.

Question 12: Where are the analysis results stored in COSMOSXpress?

- A = They are not stored.
- B = By default, in the active SolidWorks model folder.
- C = A temporary directory folder.
- D = In a specified folder location under, the (Welcome, Options, Results Location) tools.

The correction answer is D.

Additional Information: COSMOSWorks Designer

COSMOSXpress is an introductory version of COSMOSWorks. COSMOSWorks is a design analysis application that is fully integrated with SolidWorks. SolidWorks provides two product offerings: COSMOSWorks Designer, and COSMOSWorks Professional.

COSMOSWorks Designer utilizes the SolidWorks FeatureManager and many of the same mouse and keyboard commands, so anyone who can design a part in SolidWorks can analyze it without having to learn a new interface. COSMOSWorks Designer contains the most frequently used design validation tools, offering stress, strain, and displacement analysis capabilities for both parts and assemblies.

With COSMOSWorks Designer, you can:

- Compare alternative designs so you can choose the optimal design for final production
- Study interaction between different assembly components
- Simulate real-world operating conditions to see how your model handles stress, strain, and displacement
- Interpret results with powerful and intuitive visualization tools
- Collaborate and share results with everyone involved in the product development process

COSMOSWorks reads motion loads directly from the COSMOSMotion database.

COSMOSWorks Professional

COSMOSWorks Professional offers expanded analysis capabilities over COSMOSWorks Designer including: Thermal, Frequency, Buckling, Optimization, Fatigue, and Drop Test Simulation.

With COSMOSWorks Professional, you can:

- Understand the effects of temperature changes. Temperature variations encountered by mechanical parts and structures can greatly influence the performance of a design.
- Evaluate natural frequencies or critical buckling loads and their corresponding mode shapes. Often overlooked, inherent vibration modes in structural components or mechanical support systems can shorten the life of a product and cause unexpected failures.
- Optimize design based on defined criteria. Design optimization automatically determines the most optimal design based on a specified criteria.
- Simulate virtual drop tests on a variety of surfaces. In the event that a part or assembly might be dropped, find out whether or not it can survive the fall intact.
- Study the effects of cyclic loading and fatigue operation conditions. View the effects of fatigue on the overall lifecycle of a part or assembly to find out how long it will last, and what design changes can extend its working life.

At this time, there are no questions on the CSWA exam about COSMOSWorks Professional, COSMOSMotion, or COSMOSFloWorks.

Advanced Modeling Theory and Analysis

COSMOSMotion

COSMOSMotion is design software for mechanical system simulation that enables engineers to ensure that a design works before it is built. With COSMOSMotion, you can:

- Provides confidence that an assembly performs as expected without parts colliding while the components move.
- Increases the efficiency of a mechanical design process by providing mechanical system simulation capability within the familiar SolidWorks environment.
- Uses a single model, without transferring geometry and other data from application to application.
- Eliminates the expense caused by design changes late in the manufacturing process.
- Speeds the design process by reducing costly design change iterations.

COSMOSFloWorks

COSMOSFloWorks provides the user insight into parts or assemblies related to fluid flow, heat transfer, and forces on immersed or surrounding solids.

With COSMOSFloWorks, you can:

- Analyze a wide range of real fluids such as air, water, juice, ice cream, honey, plastic melts, toothpaste, and blood, which makes it ideal for engineers in nearly every industry.
- Simulate real-world operating conditions.
- Automate fluid flow tasks.
- Interpret results with powerful and intuitive visualization tools.
- Collaborate and share analysis results.

☼ COSMOSFloWorks uses a wizard interface to setup the analysis thereby making it easy and intuitive to solve the problem. The toolbars and dialog boxes are very similar to SolidWorks interface thereby making the experience of using COSMOSFloWorks very similar to SolidWorks.

☼ In SolidWorks 2009 the following name changes will occur: COSMOSXpress to SolidWorks® SimulationXpress, COSMOSWorks to SolidWorks Simulation,

COSMOSFloWorks to SolidWorks Flow Simulation, and COSMOSMotion to SolidWorks Motion.

Summary

Advanced Modeling Theory and Analysis is one of the five categories on the CSWA exam. This chapter covered the general terminology used in Engineering analysis and provided knowledge of COSMOSXpress which is aligned to the exam.

There are two questions on the CSWA exam in this category. Each question is worth five points. The two questions are in a multiple choice single answer or fill in the blank format.

As in Chapter 2, in this category of the exam, you are not required to perform an analysis on a part or assembly, but are required to understand general Engineering analysis terminology and how to apply the COSMOSXpress tool.

Spend no more than 10 minutes on each question in this category. This is a timed exam. Manage your time. Apply the SolidWorks Help Topics tool during the exam, if needed to obtain answers to this section. Also use the Help button in the COSMOSXpress dialog box.

Key terms

- *Buckling*. Is a failure mode characterized by a sudden failure of a structural member subjected to high compressive stresses, where the actual compressive stresses at failure are smaller than the ultimate compressive stresses that the material is capable of withstanding. This mode of failure is also described as failure due to elastic instability.

- *Coefficient of Thermal Expansion*. Is defined as the change in length per unit length per one degree change in temperature (change in normal strain per unit temperature).

- *COSMOSFloWorks*. Provides the user insight into parts or assemblies related to fluid flow, heat transfer, and forces on immersed or surrounding solids.

- *COSMOSMotion*. Enables engineers to size motors/actuators, determine power consumption, layout linkages, develop cams, understand gear drives, size springs/dampers, and determine how contacting parts behave.

- *COSMOSWorks Designer*. Contains the most frequently used design validation tools, offering stress, strain, and displacement analysis capabilities for both parts and assemblies.

- *COSMOSWorks Professional.* In addition to the design validation capabilities included inside COSMOSWorks Designer, COSMOSWorks Professional offers motion simulation, drop test, design optimization, thermal heat transfer, thermal stress, vibration, buckling, and fatigue analysis.

- *COSMOSXpress.* Simulates the design cycle and provides stress results. Displays critical areas and safety levels at various regions in a part.

- *Creep.* A term used to describe the tendency of a solid material to slowly move or deform permanently under the influence of stresses. It occurs as a result of long term exposure to levels of stress that are below the yield strength or ultimate strength of the material. Creep is more severe in materials that are subjected to heat for long periods, and near the melting point.

- *Degrees of Freedom.* Are the set of independent displacements and/or rotations that specify completely the displaced or deformed position and orientation of the body or system. This is a fundamental concept relating to systems of moving bodies in mechanical engineering, aeronautical engineering, robotics, structural engineering, etc. There are six degrees of freedom: Three translations and three rotations.

- *Density.* Is mass per unit volume. Density units are lb/in3 in the English system, and kg/m3 in the SI system. Density is used in static, nonlinear, frequency, dynamic, buckling, and thermal analyses. Static and buckling analyses use this property only if you define body forces (gravity and/or centrifugal).

- *Ductile.* A mechanical property which describes how able the material lends itself to be formed into rod-like shapes before fracture occurs. Examples of highly ductile metals are silver, gold, copper, and aluminum. The ductility of steel varies depending on the alloying constituents. Increasing levels of carbon decreases ductility, i.e. the steel becomes more brittle.

- *Elastic Modulus.* For a linear elastic material, the elastic modulus is the stress required to cause a unit strain in the material. In other words stress divided by the associated strain. The modulus of elasticity was first introduced by Young and is often called the Young's Modulus.

- *Factor of Safety.* Ratio between the material strength and the calculated stress.

- *Fatigue.* Is the progressive and localized structural damage that occurs when a material is subjected to cyclic loading. The maximum stress values are less than the ultimate tensile stress limit, and may be below the yield stress limit of the material.

- *Fixed Restraint.* For solids this restraint type sets all translational degrees of freedom to zero. For shells and beams, it sets the translational and the rotational degrees of freedom to zero. For truss joints, it sets the translational degrees of freedom to zero. When using this restraint type, no reference geometry is needed.

- *Force.* Is a push or pull upon an object resulting from the object's interaction with another object. Whenever there is an interaction between two objects, there is a force upon each of the objects. When the interaction ceases, the two objects no longer experience the force. Forces only exist as a result of an interaction. For example, if you select 3 faces and specify a 50 lb force, COSMOSXpress applies a total force of 150 lbs (50 lbs on each face).

- *Linear Static Analysis.* Linear static analysis allows engineers to test different load conditions and their resulting stresses and deformation. What is stress? Stress is a measure of the average amount of force exerted per unit area. It is a measure of the intensity of the total internal forces acting within a body across imaginary internal surfaces, as a reaction to external applied forces and body forces. Deformation is a change in shape due to an applied force. This can be a result of tensile (pulling) forces, compressive (pushing) forces, shear, bending or torsion (twisting). Deformation is often described in terms of strain.

- *Material Strength.* In materials science, the strength of a material refers to the material's ability to resist an applied force.

- *Mohr-Columb Stress Criterion.* The Mohr-Coulomb stress criterion is based on the Mohr-Coulomb theory also known as the Internal Friction theory. This criterion is used for brittle materials with different tensile and compressive properties. Brittle materials do not have a specific yield point and hence it is not recommended to use the yield strength to define the limit stress for this criterion.

- *Orthotropic Material.* A material is orthotropic if its mechanical or thermal properties are unique and independent in three mutually perpendicular directions. Examples of orthotropic materials are wood, many crystals, and rolled metals. For example, the mechanical properties of wood at a point are described in the longitudinal, radial, and tangential directions. The longitudinal axis (1) is parallel to the grain (fiber) direction; the radial axis (2) is normal to the growth rings; and the tangential axis (3) is tangent to the growth rings.

- *Poisson's Ratio.* Extension of the material in the longitudinal direction is accompanied by shrinking in the lateral directions. For example, if a body is subjected to a tensile stress in the X-direction, then Poisson's Ratio NUXY is defined as the ratio of lateral strain in the Y-direction divided by the longitudinal strain in the X-direction. Poisson's ratios are dimensionless quantities. If not defined, the program assumes a default value of 0.

- *Restraints.* Restraints and loads define the environment of the model. Each restraint can contain multiple faces. The restrained faces are constrained in all directions. You must at least restrain one face of the part to avoid analysis failure due to rigid body motion. Loads and restraints are fully associative and automatically adjusted to changes in geometry.

- *Shear Modulus.* Also called modulus of rigidity, is the ratio between the shearing stress in a plane divided by the associated shearing strain. Shear Moduli are used in static, nonlinear, frequency, dynamic and buckling analyses.

- *Thermal Conductivity.* Indicates the effectiveness of a material in transferring heat energy by conduction. It is defined as the rate of heat transfer through a unit thickness of the material per unit temperature difference. The units of thermal conductivity are Btu/in sec oF in the English system and W/m oK in the SI system. Thermal conductivity is used in steady state and transient thermal analyses.

- *Tensile Strength.* Tensile strength is the maximum load sustained by the specimen in the tension test, divided by the original cross sectional area.

- *von Mises yield Criterion.* A scalar stress value that can be computed from the stress. In this case, a material is said to start yielding when its von Mises stress reaches a critical value known as the yield strength. The von Mises stress is used to predict yielding of materials under any loading condition from results of simple uniaxial tensile tests. The von Mises stress satisfies the property that two stress states with equal distortion energy have equal von Mises stress.

- *Yield Strength.* The stress at which the metal yields or becomes permanently deformed is an important design parameter. This stress is the elastic limit below which no permanent shape changes will occur. The elastic limit is approximated by the yield strength of the material, and the strain that occurs before the elastic limit is reached is called the elastic strain. The yield strength is defined in three ways, depending on the stress-strain characteristics of the steel as it begins to yield. The procedures in SAE J416, ASTM E8, and ASTM A370.

Check your understanding

1: COSMOSXpress is used to analyze?

- A = Drawings.
- B = Parts and Drawings.
- C = Assemblies and Parts.
- D = Parts.
- E = All of the above.

2: At what percent Strain is Yield Strength normally determined.

- A = 0.002%
- B: 0.20%
- C: 0.02%
- D: 0.01%

3: Under what COSMOSXpress menu tab do you set system units?

- A = Material tab.
- B = Restraint tab.
- C = Welcome tab.
- D = None of the above.
- E = All of the above.

4: Under what COSMOSXpress menu tab can you modify the Mesh period of the part?

- A = Restraint tab.
- B = Material tab.
- C = Analyze tab.
- D = None of the above.

5: An increase Mesh period for a part will?
- A = Decrease calculation accuracy.
- B = Increase calculation accuracy.
- C = Have no effect.
- D = Change the FOS below 1.

6: How many degrees of freedom does a physical structure have?
- A = Zero.
- B = Three – Rotations only.
- C = Three – Translations only.
- D = Six – Three translations and three rotations.

7: Brittle materials has little tendency to deform (or strain) before fracture and does not have a specific yield point. It is not recommended to apply the yield strength analysis as a failure criterion on brittle material. Which of the following failure theories is appropriate for brittle materials?
- A = Mohr-Columb stress criterion.
- B = Maximum shear stress criterion.
- C = Maximum von Mises stress criterion.
- D = Minimum shear stress criterion.

8: A material is orthotropic if its mechanical or thermal properties are not unique and independent in three mutually perpendicular directions.
- A = True.
- B = False.

9: You are performing an analysis on your model. You select three faces and apply a 40lb load. What is the total force applied to the model?
- A = 40lbs.
- B = 20lbs.
- C = 120lbs.
- D = Additional information is required.

The correct answer is C.

10. COSMOSXpress supports the analysis of the following?
- A = Solid, single-body part.

- B = Assemblies.
- C = Surface models.
- D = Multi-body parts.

11: What are the available system units in COSMOSXpress?

- A = SI.
- B = IPS.
- C = RPT.
- D = SI and IPS.

12: Under what COSMOSXpress menu tab can you modify the Element size and Element tolerance for the Mesh period?

- A = Restraint tab.
- B = Material tab.
- C = Analyze tab.
- D = None of the above.

13: The Maximum normal stress criterion is also known as the?

- A = Mohr-Coulomb's criterion.
- B = Maximum von Mises stress criterion.
- C = Maximum shear stress criterion.
- D = None of the above.

14: COSMOSXpress uses the vonMises Yield Criterion to calculate the Factor of Safety of many ductile materials. According to the criterion:

- A = Material yields when the vonMises stress in the model equals the yield strength of the material.
- B = Material yields when the vonMises stress in the model to 5 times greater that the minimum tensile strength of the material.
- C = Material yields when the vonMises stress in the model is 3 times greater than the FOS of the material.
- D = None of the above.

The correct answer is A.

15: COSMOSXPress calculates structural failure on:

- A = Buckling.
- B = Fatigue.
- C = Creep.
- D = Material yield.

The correct answer is D.

16: In an engineering analysis, you select a face to restrain. What is the affect?

- A = The face will not translate but can rotate.
- B = The face will rotate but can not translate.
- C = You can not apply a restraint to a face.
- D = The face will not rotate and will not translate.

17: Is it possible using COSMOSXpress to know the X,Y, Z coordinate location where the minimum and maximum stress or the displacements occur on the model?

- A = No.
- B = Yes. Export the results in HTML report format. The X,Y,Z locations are displayed in the Stress and Displacement section.

Notes:

Appendix: Check your understanding answer key

This appendix contains the answers to the Check your understanding review questions at the end of each chapter.

Chapter 2

1. Identify the following Feature icon .

    The correct answer is B: Mirror feature.

2. Identify the following Feature icon .

    The correct answer is C: Circular Pattern feature.

3. Identify the following Feature icon .

    The correct answer is A: Hole Wizard feature.

4. Identify the following Feature icon .

    The correct answer is D: Rib feature.

5. Identify the following Feature icon .

    The correct answer is D: Revolved Cut feature.

6. Question 1: Identify the number of instances in the illustrated model.

    The correct answer is C: 9 instances.

7. Identify the Sketch plane for the Extrude1 feature.

    The correct answer is B: Front Plane.

8. Identify the following Sketch Entities icon .

    The correct answer is B: Tangent Arc tool.

9. Identify the following Sketch Entities icon .

    The correct answer is C: 3 Point Arc tool.

10. Identify the following Sketch Entities icon .

    The correct answer is A: Centerpoint Arc tool.

Appendix

11. A fully defined sketch is displayed in what color?

The correct answer is B: Black.

12. What symbol does the FeatureManager display before the Sketch name in an under defined sketch?

The correct answer is A: (-).

13. Which is not a valid drawing format in SolidWorks?

The correct answer is C: * dwgg.

14. Which is a valid assembly format in SolidWorks?

The correct answer is D: * stl.

15. Identify the following Drawing View icon.

The correct answer is D: Crop View.

16. Identify the following Drawing View icon.

The correct answer is D: Aligned Section View

17. Identify the following Drawing View icon.

The correct answer is B: Standard 3 View.

18. Identify the illustrated Drawing view.

The correct answer is B: Alternative Position View

19. Identify the illustrated Drawing view.

The correct answer is B: Break View.

20. Identify the illustrated Drawing view.

The correct answer is A: Section View.

21. Identify the view procedure. To create the following view, you need to insert a:

The correct answer is B: Closed Profile: Spline.

22. Identify the view procedure. To create the following view, you need to insert a:

The correct answer is B: Closed Profile: Spline.

23. Identify the illustrated view type.

The correct answer is A: Crop View.

Chapter 3

1. Calculate the overall mass of the part, volume, and locate the Center of mass with the provided information using the provided Option 1 FeatureManager.
   - Overall mass of the part = 1105.00 grams
   - Volume of the part = 130000.00 cubic millimeters
   - Center of Mass Location: X = 43.46 millimeters, Y = 15.00 millimeters, Z = -37.69 millimeters

2. Calculate the overall mass of the part, volume, and locate the Center of mass with the provided information using the provided Option 3 FeatureManager.
   - Overall mass of the part = 269.50 grams
   - Volume of the part = 192500.00 cubic millimeters
   - Center of Mass Location: X = 35.70 millimeters, Y = 27.91 millimeters, Z = -146 millimeters

3. Calculate the overall mass of the part, volume, and locate the Center of mass with the provided information.
   - Overall mass of the part = 1.76 pounds
   - Volume of the part = 17.99 cubic inches
   - Center of Mass Location: X = 0.04 inches, Y = 0.72 inches, Z = 0.00 inches

4. Calculate the overall mass of the part, volume, and locate the Center of mass with the provided illustrated information.
   - Overall mass of the part = 1280.91 grams
   - Volume of the part = 474411.54 cubic millimeters
   - Center of Mass Location: X = 0.00 millimeters, Y = -29.17 millimeters, Z = 3.18 millimeters

5. Calculate the overall mass of the part, volume, and locate the Center of mass with the provided information.
   - Overall mass of the part = 248.04 grams
   - Volume of the part = 91868.29 cubic millimeters
   - Center of Mass Location: X = -51.88 millimeters, Y = 24.70 millimeters, Z = 29.47 millimeters

Appendix

Chapter 4

1. Calculate the overall mass of the part, volume, and locate the Center of mass with the provided information.
   - Overall mass of the part = 1.99 pounds
   - Volume of the part = 6.47 cubic inches
   - Center of Mass Location: X = 0.00 inches, Y = 0.00 inches, Z = 1.49 inches

2. Calculate the overall mass of the part, volume, and locate the Center of mass with the provided information.
   - Overall mass of the part = 279.00 grams
   - Volume of the part = 103333.73 cubic millimeters
   - Center of Mass Location: X = 0.00 millimeters, Y = 0.00 millimeters, Z = 21.75 millimeters

3. Calculate the overall mass of the part, volume, and locate the Center of mass with the provided information.
   - Overall mass of the part = 1087.56 grams
   - Volume of the part = 122198.22 cubic millimeters
   - Center of Mass Location: X = 44.81 millimeters, Y = 21.02 millimeters, Z = -41.04 millimeters

4. Calculate the overall mass of the part, volume and locate the Center of mass with the provided information.
   - Overall mass of the part = 2040.57 grams
   - Volume of the part = 755765.04 cubic millimeters
   - Center of Mass Location: X = -0.71 millimeters, Y = 16.66 millimeters, Z = -9.31 millimeters

5. Calculate the overall mass of the part, volume and locate the Center of mass with the provided information. Create Coordinate System1 to locate the Center of mass for the model.
   - Overall mass of the part = 2040.57 grams
   - Volume of the part = 755765.04 cubic millimeters
   - Center of Mass Location: X = 49.29 millimeters, Y = 16.66 millimeters, Z = -109.31 millimeters

6. Calculate the overall mass of the part, volume and locate the Center of mass with the provided information.
    - Overall mass of the part = 37021.48 grams
    - Volume of the part = 13711657.53 cubic millimeters
    - Center of Mass Location: X = 0.00 millimeters, Y = 0.11 millimeters, Z = 0.00 millimeters
7. Calculate the overall mass of the part, volume and locate the Center of mass with the provided information.
    - Overall mass of the part = 37021.48 grams
    - Volume of the part = 13711657.53 cubic millimeters
    - Center of Mass Location: X = 225.00 millimeters, Y = 70.11 millimeters, Z = -150.00 millimeters

Chapter 5

1. Calculate the overall mass and volume of the assembly. Locate the Center of mass using the illustrated coordinate system.
    - Overall mass of the assembly = 843.22 grams
    - Volume of the assembly = 312304.62 cubic millimeters
    - Center of Mass Location: X = 30.00 millimeters, Y = 40.16 millimeters, Z = -53.82 millimeters
2. Calculate the overall mass and volume of the assembly. Locate the Center of mass using the illustrated coordinate system.
    - Overall mass of the assembly = 19.24 grams
    - Volume of the assembly = 6574.76 cubic millimeters
    - Center of Mass Location: X = 40.24, Y = 24.33, Z = 20.75
3. Calculate the overall mass and volume of the assembly. Locate the Center of mass using the illustrated coordinate system.
    - Overall mass of the assembly = 19.24 grams
    - Volume of the assembly = 6574.76 cubic millimeters
    - Center of Mass Location: X = 40.24, Y = -20.75, Z = 24.33

Appendix

Chapter 6

1. COSMOSXpress is used to analyze?

The correct answer is D: Parts. COSMOSXpress supports the analysis of a single solid body. For multibody parts, you can analyze one body at a time. For assemblies, you can analyze the effect of Physical Simulation one component at a time. Surface bodies are not supported.

2. At what percent Strain is Yield Strength normally determined?

The correct answer is B: 0.20%. View the graph on page 6-4.

3. Under what COSMOSXpress menu tab can you set system units?

The correct answer is C. Welcome tab. Allows you to set default analysis units; SI or IPS (English) and to specify a folder for saving analysis results.

4. Under what COSMOSXpress menu tab can you modify the Mesh period of the part?

The correct answer is C: Analyze tab. Select to analyze with the system default settings or change the settings.

5. An increase Mesh period for a part will?

The correct answer is B: Increase calculation accuracy. COSMOSXpress subdivides the model into a mesh of small shapes called elements. Specifying a smaller element size provides a more accurate result, but requires additional time and resources.

6. How many degrees of freedom does a physical structure have?

The correct answer is D: Six – Three translations and three rotations.

7. Brittle materials has little tendency to deform (or strain) before fracture and does not have a specific yield point. It is not recommended to apply the yield strength analysis as a failure criterion on brittle material. Which of the following failure theories is appropriate for brittle materials?

The correct answer is A: Mohr-Columb stress criterion.

8. A material is Orthotropic if its mechanical or thermal properties are not unique and independent in three mutually perpendicular directions.

The correct answer is B: False.

9. You are performing an analysis on your model. You select three faces and apply a 40lb load. What is the total force applied to the model?

The correct answer is C: 120lbs.

Appendix

10. COSMOSXpress supports the analysis of the following.

The correct answer is A: Solid, single-body parts.

11. What are the available system units in COSMOSXpress?

The correct answer is D: Select SI (International System of Units), or IPS (English). Setting the preferred system of units does not restrict you from entering data in other units, the dialog boxes of loads and boundary conditions let you override the default preferred units. The preferred units will be used by default to display the results, but you can choose to display the results in other units. For example you can choose the SI system as your default unit system, but still can apply pressure in psi and displacements in millimeters.

12. Under what COSMOSXpress menu tab can you modify the Element size and Element tolerance for the Mesh period?

The correct answer is C: Analyze tab. Select to analyze with the system default settings or change the settings.

13. The Maximum normal stress criterion is also known as?

- A = Mohr-Coulomb's criterion
- B = Maximum von Mises stress criterion
- C = Maximum shear stress criterion
- D = None of the above

The correct answer is A: Mohr-Coulomb's criterion. The Mohr-Coulomb stress criterion is based on the Mohr-Coulomb theory also known as the Internal Friction theory. This criterion is used for brittle materials with different tensile and compressive properties. Brittle materials do not have a specific yield point and it is not recommended to use the yield strength to define the limit stress for this criterion.

14. COSMOSXpress uses the vonMises Yield Criterion to calculate the Factor of Safety of many ductile materials. According to the criterion:

The correct answer is A: Material yields when the vonMises stress in the model equals the yield strength of the material.

15. COSMOSXpress calculates structural failure on:

The correct answer is D: Material yield

16. In an engineering analysis, you select a face to restrain. What is the affect?

The correct answer is D: The face will not rotate and will not translate.

17: Is it possible using COSMOSXpress to know the X,Y, Z coordinate location where the minimum and maximum stress or the displacements occur on the model?

The correct answer is A: Yes. Export the results in HTML report format. The X,Y,Z locations are displayed in the Stress and Displacement section.

Notes:

Index

2D Sketch	2-14	Angle Distance Chamfer feature	4-8
2D Sketching	3-24		
3 Hole Link part	5-10, 5-11	Angle distance chamfer feature	4-23
3 Point Arc Sketch tool	2-36		
3 Point Arc Sketch tool	4-4	Angle Mate	5-3, 5-15, 5-25
3 Standard views	2-42		
3D Drawing View	1-8	Angular units	2-53
3D Modeling techniques	2-3	Annotation tool	2-44
3D Sketch	2-13	ANSI	2-15
3D Sketch tool	2-35, 3-25, 3-26, 3-35	Ansi Metric	5-29
		Anti-Aligned option	5-8
3D Sketching	3-24, 3-26, 3-35	Arc Properties First arc condition	4-13
3DSketch 3-1 part	3-25	Arc PropertyManager	4-4
5 Hole Link part	5-10, 5-12	ASME Y14.3M Standard	2-15
		Assem1	5-13
A		Assembly modeling	5-1
Add Configurations	2-34	Assembly Modeling 5-1	5-18
Add Material	2-22	Assembly Modeling 5-2	5-25
Add Relations dialog box	2-18	Assembly Modeling 5-3	5-35
Add Relations Sketch tool	2-37	Assembly modeling techniques	5-2
Adding features	1-3		
Add-Ins	1-13	Assign material	3-5
AdjustingPin part	5-38	Associativity	1-4, 2-3
Advanced Part 4-1	4-5	Auxiliary view	2-46, 2-49
Advanced Part 4-2	4-8	Axis of revolution	3-42, 5-30
Advanced Part 4-3	4-11	Axle part	5-10, 5-11
Advanced Part 4-3 MMGS System	4-11	Axle40 part	5-27
Advanced Part 4-4	4-16	**B**	
Advanced Part 4-4 Modify	4-16	Base feature	2-21
		Base sketch	1-2, 3-7
Advanced Part 4-5	4-23	Base100 part	5-38
Advanced Part 4-5A	4-25	Basic concepts in SolidWorks	1-2
Advanced Part 4-5B	4-26		
Advanced Part 4-6	4-32	Basic Tolerance type	3-33
Advanced Part 4-6A	4-33	Begin Assembly PropertyManager	5-13, 5-22, 5-32
Advanced Part 4-7	4-38		
A-Landscape	2-43	Bilateral Tolerance type	3-31
Aligned option	5-8	Bottom-up assembly modeling	5-2
Aligned Section view	2-46, 2-50		
Alternate Position view	2-46, 2-52	Bracket100 part	5-27

Index

Brittle material	6-7	Rebuild	2-31
Broken reference	2-32	Resolved assembly	2-31
Broken view	2-46	Resolved part	2-30
Broken-out Section view	2-51	Smart Component	2-31
Buckling	6-1	Suppressed	2-30
Build a simple part	3-4	Concentric mate	5-3, 5-5, 5-14

C

		Confirmation Corner	1-7
Callout value	3-31	Consolidated flyout tool	1-6
Center of gravity	2-25	Constant radius fillet	3-38
Center of mass	2-25, 3-12, 3-14	Constraints	1-4
		Construction geometry	3-13, 3-42, 3-45, 3-47
Centerline Sketch tool	2-36		
Centerpoint Arc Sketch tool	2-36, 4-4	Contents tab	2-41
		Control area	1-12
Centroid	2-25	Convert Entities Sketch tool	2-37, 3-42, 4-13, 4-14, 4-18
Chamfer feature	4-5, 4-8, 4-23, 4-38, 5-30		
		Coordinate System	5-18
Circle Sketch tool	2-35	Coordinate system location	4-17, 4.23, 4-24
Circular Pattern feature	3-45, 3-46, 4-7, 4-22		
		Coordinate System PropertyManager	4-18, 4-19, 4-24, 5-18, 5-25
Circular Pattern PropertyManager	4-7, 4-22		
Clevis part	5-10	Coordinate System tool	4-18, 4-19
Coefficient of Thermal Expansion	6-1	Coordinate System1	4-18, 4-19
		Coordinates	4-17
Coincident mate	5-3, 5-5, 5-6	Coradial Relation	4-7
		COSMOSFloWorks	6-27
Collapse items	1-17	COSMOSMotion	6-27
Collar	5-10, 5-13	COSMOSWorks	6-25
Color – Sketch status	2-28	COSMOSWorks - Designer	6-25
CommandManager	1-9		
Assembly document	1-11	COSMOSWorks Professional	6-26
Drawing document	1-10		
Part document	1-9	COSMOSXpress	6-8
Component	5-2	COSMOSXpress – COVERPLATE part	6-17
Component State	2-30		
Hidden	2-30	COSMOSXpress – Flatbar part	6-10
Hidden Lightweight	2-30		
Hidden Smart Component	2-31	COSMOSXpress – LBRACKET part	6-19
Hidden, Out-of-Date	2-30	COSMOSXpress – User Interface	6-9
Lightweight part	2-30		
Out-of-Date	2-30	Analyze tab	6-10
Lightweight		Load tab	6-10

Index

Material tab	6-9	Copy Scheme	1-18
Optimize tab	6-10	Show Tolerance Status	1-18
Restraint tab	6-9	TolAnalyst Study	1-18
Results tab	6-10	Displacement	6-3
Start Over button	6-10	Display grid	1-8
Update button	6-10	Display Pane status	2-33
Welcome tab	6-9	Color	2-33
COSMOSXpress Wizard	6-8	Display Mode	2-33
Counterbore	3-33	Hide/Show	2-33
Countersink	3-33	RealView	2-33
Create a new coordinate system location	4-18	Texture	2-33
		Transparency	2-33
Create a new drawing	2-43	Display Style	1-8
Create an assembly	5-13, 5-22, 5-32	Hidden Lines	1-8
		Hidden Lines Removed	1-8
Creating mates	5-4	Shaded	1-8
Creep	6-2	Shaded With Edges	1-8
Crop view	2-46, 2-47, 2-51	Visible	1-8
		Wireframe	1-8
Customize CommandManager	1-12	Display/Delete Relations Sketch tool	2-37
Customize FeatureManager	1-17	Distance Chamfer feature	4-5
		Distance mate	5-3, 5-8
		Document Precision	3-30
D		Document Properties	2-52
Deactivate planes	1-8	Document Recovery	1-15
Default Sketch planes	2-14	Drawing Annotation	2-42
Degrees of Freedom	6-2	Drawing name views	2-46
Density	6-2	Drawing scale	2-43
Depth/Deep symbol	3-33, 3-35	Drawing sheet	2-42
Design Intent	2-18, 5-2	Drawing toolbar	2-46
Assembly	2-20	Drawing view mode option	2-43
Drawing	2-20		
Feature	2-19	Drawing views	2-42
Part	2-20	Drawings	1-4
Sketch	2-18	Drop-down menu	1-5
Design Library	1-13	Ductile	6-2
Detail view	2-46, 2-47, 2-50		
		E	
Dimension PropertyManager	3-31	Edit a feature	2-10, 3-3
		Edit a sketch	2-12, 3-3
Dimension text symbols	3-33	Edit material	2-22
Dimensioning standard	2-53	Edit Sheet Format mode	2-45
DimXpertManager tab	1-18	Edit Sheet mode	2-45
Auto Dimension Scheme	1-18	Edit Sketch Plane	2-16, 3-2
		Editing features	1-3

Index

Elastic Modulus	6-2	Previous View	1-7
Engineering document	3-2	Rotate view	1-8
External reference	2-32	Section View	1-7
Broken	2-32	View Orientation	1-8
Locked	2-32	Zoom to Area	1-7
Out of Context	2-32	Zoom to Fit	1-7
Extruded Base feature	3-5, 3-9	Help Topics	2-41
Extruded Boss feature	3-11, 3-51	Hidden - Component State	2-31
Extruded Cut feature	3-9, 3-48, 4-22	Hidden Lightweight - Component State	2-31
Extrude-Thin feature	2-16, 3-51, 4-15, 4-21	Hidden Smart Component - Component State	2-31
		Hole Wizard feature	5-29
F, G		Hole Wizard PropertyManager	5-29, 5-30
Factor of safety	6-6, 6-14		
Fatigue	6-2		
FeatureManager Design Tree	1-3, 1-16	**I, J, K**	
ConfigurationManager	1-16	Identify dimensions	2-10
DimXpertManager	1-16	Identify parameters	2-10
PropertyManager	1-16	In-context assembly modeling	5-2
Features	1-2		
File Explorer	1-13	Index tab	2-41
File formats	2-38	Inertia	6-4
Fillet feature	3-38, 3-48, 3-52, 4-5, 4-8	Insert a sheet	2-44
		Insert component	5-14, 5-15, 5-16, 5-34
Fillet Sketch tool	4-3	Insert Components PropertyManager	5-14, 5-15, 5-16, 5-34
FilletXpert	4-32		
First Angle projection	2-15	Instances	2-11, 2-12
Fix component	5-14	Intersection relation	3-53
Fixed Component	2-31	Invalid Solution Found - Sketch	2-28
Fixed Restraint	6-2		
Float	2-31	IPS Unit system	3-8
Float component	5-7, 5-14	Isometric drawing view	2-45
Force	6-3		
Front Plane	2-13	**L**	
Full round fillet feature	4-21	Length units	2-53
Fully defined - Sketch	2-28	Lightweight part - Component State	2-30
H		Limit Tolerance type	3-31
Heads-up View toolbar	1-7	Line Sketch tool	2-35
3D Drawing View	1-8	Line1 part	3-2
Apply Scene	1-8	Line2 part	3-6
Display Style	1-8	Linear Pattern feature	5-28
Hide/Show Items	1-8	Linear Static Analysis	6-3, 6-6

Index

Locked reference	2-32

M

Machined Brackets	5-39
Mass Properties dialog box	2-22, 2-26, 3-5, 4-18
Mass Properties tool	3-5, 4-18
Mass/Section Property Options dialog box	3-5
Mass-Volume 3-10 part	3-36
Mass-Volume 3-10A part	3-36
Mass-Volume 3-11 part	3-38
Mass-Volume 3-11-MMGS part	3-38
Mass-Volume 3-12 part	3-40
Mass-Volume 3-13 part	3-41
Mass-Volume 3-13A part	3-41
Mass-Volume 3-14 part	3-43
Mass-Volume 3-15 part	3-44
Mass-Volume 3-16 part	3-46
Mass-Volume 3-5 part	3-20
Mass-Volume 3-6 part	3-21
Mass-Volume 3-6-MMGS part	3-21
Mass-Volume 3-7 part	3-23
Mass-Volume 3-7-IPS part	3-23
Mass-Volume 3-8 part	3-27
Mass-Volume 3-8-MMGS part	3-27
Mass-Volume 3-9 part	3-30
Mate 5-1 assembly	5-5
Mate 5-2 assembly	5-6
Mate 5-3 assembly	5-7
Mate PropertyManager	5-3
Material	2-22
Material icon	3-5
Material Strength	6-4
Mates	5-3
Mates folder	5-5, 5-8
MateXpert	5-6
Max von Mises Stress plot	6-13
Maximum variation	3-32
Measure dialog box	2-23
Measure tool	2-23

Menu bar menu	1-5
Edit	1-5
File	1-5
Help	1-5
Insert	1-5
tool	1-5
Window	1-5
Menu bar toolbar	1-4
New	1-4
Open	1-4
Options	1-4
Print	1-4
Rebuild	1-4
Save	1-4
Undo	1-4
Mid Plane End Condition	2-21
Midpoint Relation	2-19
Minimum Variation	3-32
Mirror Component PropertyManager	5-17, 5-35
Mirror Components	5-17, 5-35
Mirror Entities Sketch tool	2-37
Mirror feature	3-30, 3-48, 3-57, 4-23, 5-22
Mirror PropertyManager	3-30, 3-48, 5-22
Mirror Sketch PropertyManager	3-7
Mirror Sketch tool	3-7, 3-13, 4-3, 4-7
MMGS Unit system	3-8
Model	1-2
Model View PropertyManager	2-43
Mohr-Columb Stress Criterion	6-4
Mohr-Columb theory	6-4
Motion Study	1-15
Multiple views	2-43

N

New assembly	5-13, 5-22, 5-32
New Motion Study	1-16

Index

New tool	1-4	Poisson's Ratio	6-5
No Solution Found - Sketch	2-28	Primary datum plane	2-14
		Principal moments of inertia	2-25
Number of Instances	4-7	Print tool	1-4
		Projected view	2-46

O

Offset Entities Sketch tool	2-38		
Offset Start Condition	4-35	**R**	
Options tool	1-4	Reaction force	6-3
Origin	2-17, 2-21, 3-3	RealView	1-14
		Rebuild - Component State	2-31
Orthographic projection	2-14		
Orthotropic Material	6-5	Rectangle Sketch tool	2-35
Out of Context	2-32	Reference configuration	2-34
Out-of-Date Lightweight part - Component State	2-30	Reference Plane	4-10, 4-13
		Reference Planes	2-13, 3-53, 3-54
Output coordinate system	4-18		
Over defined - Sketch	2-28	Refining the design	1-3
		Reordering features	1-3
P,Q,U		Resolved assembly - Component State	2-31
Parallel mate	5-3, 5-7		
Parallel Plane at Point	4-30	Resolved part - Component State	2-30
Parallel relation	4-10		
Part-Modeling 3-1 part	3-49	Resolved state	2-32
Part-Modeling 3-1-Modify part	3-49	Revolve feature	5-30
		Revolved Base feature	3-42, 5-30, 5-31
Part-Modeling 3-2 part	3-52		
Part-Modeling 3-2-Modify part	3-52	Revolved Boss feature	3-44
		Revolved Cut feature	5-32
Part-Modeling 3-3 part	3-55	Revolved feature	3-42
Part-Modeling 3-3-Modify part	3-55	Rib feature	3-55
		Rib PropertyManager	3-55
Part-Modeling 3-4 part	3-58	Ribs	3-55
Perpendicular mate	5-3	Right Plane	2-13
Perpendicular relation	4-10	Right-click Pop-up menus	1-6
Pin part	5-20	Rollback bar	1-3
Pin-4 part	5-27	Rotate view	1-8
Pin-5 part	5-39		
Plane feature	4-30	**S**	
Plane PropertyManager	3-51, 3-54, 4-10, 4-13	Save as Copy	5-12
		Save As tool	1-5
Plane tool	3-51, 3-54, 4-10, 4-13	Save As type	2-38
		Save tool	1-4
Plate-3 part	4-17	Scale	2-43
Plate-X-Y-Z part	4-19	Search	1-13
Point Sketch tool	2-36	Search tab	2-41

Secondary datum plane	2-14	Mate alignment	5-3
Section view	2-46, 2-49	Parallel	5-3
Seed feature	2-11, 3-45, 4-6, 4-22	Perpendicular	5-3
		Tangent	5-3
Seed sketch	2-11	Standard view	2-46
Shear Modulus	6-5	Standard Views toolbar	1-7
Short cut keys	1-18	Start Condition: Extrude feature	4-35
Simple-Cut 3-1 part	3-17		
Single view	2-43	strains	6-3
Sketch	2-13	stress	6-3
Sketch Entities	2-35	Stress-strain curve	6-6
Sketch Fillet Sketch tool	2-36, 4-21	Suppressed - Component State	2-30
Sketch plane	2-13, 2-15, 2-17		
		Symbol library dialog box	3-34
Sketch Plane – Front Plane	2-17	Symmetric relation	4-3
		Symmetry	2-19, 3-28
Sketch Plane – Top Plane	2-17, 2-21	System feedback	1-6
Sketch Plane PropertyManager	2-16	Dimension	1-6
		Edge	1-6
Sketch Relations	2-21	Face	1-6
Sketch states	2-28	Vertex	1-6
Fully defined	2-28	System feedback symbols	5-8
Invalid Solution Found	2-28		
No Solution Found	2-28	**T**	
Over defined	2-28	Tangent	4-10
Under defined	2-28	Tangent Arc Sketch tool	2-36, 4-22
Sketch toolbar	2-35	Tangent mate	5-3, 5-6
Sketch tools	2-35	Task Pane	1-12
SketchXpert	2-29	Design Library	1-13
Smart Component - Component State	2-31	Document Recovery	1-15
		File Explorer	1-13
Smart Dimension tool	2-35	RealView	1-14
SolidWorks	1-2	Search	1-13
SolidWorks Help	2-41	SolidWorks Resources	1-12
SolidWorks model	1-2	View Palette	1-14
SolidWorks Resources	1-12	Temporary Axes	3-55, 3-57
SolidWorks Toolbox	1-13	Tensile Strength	6-5
Spline Sketch tool	2-36	Tertiary datum plane	2-14
Split FeatureManager	1-17	Thermal Conductivity	6-5
Square block part	5-19	Third Angle projection	2-15
Standard mates	5-3	Tip of the Day	1-12
Angle	5-3	Tolerance Precision	3-32
Coincident	5-3	Tolerance type	3-31
Concentric	5-3	Top Plane	2-13
Distance	5-3	Top-down assembly modeling	5-2
Lock	5-3		

Index

Trim Entities Sketch tool	2-38, 4-37
Trim Sketch tool	3-42, 4-14

U

U-Bracket part	5-19
Under defined - Sketch	2-28
Undo tool	1-4
Unit system	2-53
Units	2-15, 2-16, 2-53, 3-8
Up To Next End Condition	4-21, 4-22
Up-To Vertex End Condition	3-54
User Interface, (UI)	1-4

V

Vertex	4-19, 4-25
View Layout toolbar	2-46
View Mate Errors	5-6
View Mates tool	5-4
View Orientation	1-8
Bottom	1-8
Dimetric	1-8
Four view	1-8
Front	1-8
Isometric	1-8
Left	1-8
Right	1-8
Single view	1-8
Top	1-8
Trimetric	1-8
Two view	1-8
View Palette	1-14
View toolbar	1-7
Volume-Center of mass 3-2 part	3-12
Volume-Center of mass 3-2-IPS partC	3-12
Von Mises plot	6-14
Von Mises yield Criterion	6-5

W, X, Y, Z

What is SolidWorks	1-2
What's Wrong dialog box	2-28
Wheel1 part	5-27
WheelPlate part	5-27
Yield Strength	6-6
Yoke part	5-38

Cengage Learning has provided you with this product for your review and, to the extent that you adopt the associated textbook for use in connection with your course, you and your students who purchase the textbook may use the Materials as described below.

IMPORTANT! READ CAREFULLY: This End User License Agreement ("Agreement") sets forth the conditions by which Cengage Learning will make electronic access to the Cengage Learning-owned licensed content and associated media, software, documentation, printed materials, and electronic documentation contained in this package or made available to you via this product (the "Licensed Content"), available to you (the "End User"). BY CLICKING THE "I ACCEPT" BUTTON AND/OR OPENING THIS PACKAGE, YOU ACKNOWLEDGE THAT YOU HAVE READ ALL OF THE TERMS AND CONDITIONS, AND THAT YOU AGREE TO BE BOUND BY ITS TERMS, CONDITIONS, AND ALL APPLICABLE LAWS AND REGULATIONS GOVERNING THE USE OF THE LICENSED CONTENT.

1.0 SCOPE OF LICENSE

1.1 <u>Licensed Content.</u> The Licensed Content may contain portions of modifiable content ("Modifiable Content") and content which may not be modified or otherwise altered by the End User ("Non-Modifiable Content"). For purposes of this Agreement, Modifiable Content and Non-Modifiable Content may be collectively referred to herein as the "Licensed Content." All Licensed Content shall be considered Non-Modifiable Content, unless such Licensed Content is presented to the End User in a modifiable format and it is clearly indicated that modification of the Licensed Content is permitted.

1.2 Subject to the End User's compliance with the terms and conditions of this Agreement, Cengage Learning hereby grants the End User, a nontransferable, nonexclusive, limited right to access and view a single copy of the Licensed Content on a single personal computer system for noncommercial, internal, personal use only, and, to the extent that End User adopts the associated textbook for use in connection with a course, the limited right to provide, distribute, and display the Modifiable Content to course students who purchase the textbook, for use in connection with the course only. The End User shall not (i) reproduce, copy, modify (except in the case of Modifiable Content), distribute, display, transfer, sublicense, prepare derivative work(s) based on, sell, exchange, barter or transfer, rent, lease, loan, resell, or in any other manner exploit the Licensed Content; (ii) remove, obscure, or alter any notice of Cengage Learning's intellectual property rights present on or in the Licensed Content, including, but not limited to, copyright, trademark, and/or patent notices; or (iii) disassemble, decompile, translate, reverse engineer, or otherwise reduce the Licensed Content. Cengage reserves the right to use a hardware lock device, license administration software, and/or a license authorization key to control access or password protection technology to the Licensed Content. The End User may not take any steps to avoid or defeat the purpose of such measures. Use of the Licensed Content without the relevant required lock device or authorization key is prohibited. UNDER NO CIRCUMSTANCES MAY NON-SALEABLE ITEMS PROVIDED TO YOU BY CENGAGE (INCLUDING, WITHOUT LIMITATION, ANNOTATED INSTRUCTOR'S EDITIONS, SOLUTIONS MANUALS, INSTRUCTOR'S RESOURCE MATERIALS AND/OR TEST MATERIALS) BE SOLD, AUCTIONED, LICENSED OR OTHERWISE REDISTRIBUTED BY THE END USER.

2.0 TERMINATION

2.1 Cengage Learning may at any time (without prejudice to its other rights or remedies) immediately terminate this Agreement and/or suspend access to some or all of the Licensed Content, in the event that the End User does not comply with any of the terms and conditions of this Agreement. In the event of such termination by Cengage Learning, the End User shall immediately return any and all copies of the Licensed Content to Cengage Learning.

3.0 PROPRIETARY RIGHTS

3.1 The End User acknowledges that Cengage Learning owns all rights, title and interest, including, but not limited to all copyright rights therein, in and to the Licensed Content, and that the End User shall not take any action inconsistent with such ownership. The Licensed Content is protected by U.S., Canadian and other applicable copyright laws and by international treaties, including the Berne Convention and the Universal Copyright Convention. Nothing contained in this Agreement shall be construed as granting the End User any ownership rights in or to the Licensed Content.

3.2 Cengage Learning reserves the right at any time to withdraw from the Licensed Content any item or part of an item for which it no longer retains the right to publish, or which it has reasonable grounds to believe infringes copyright or is defamatory, unlawful, or otherwise objectionable.

4.0 PROTECTION AND SECURITY

4.1 The End User shall use its best efforts and take all reasonable steps to safeguard its copy of the Licensed Content to ensure that no unauthorized reproduction, publication, disclosure, modification, or distribution of the Licensed Content, in whole or in part, is made. To the extent that the End User becomes aware of any such unauthorized use of the Licensed Content, the End User shall immediately notify Cengage Learning. Notification of such violations may be made by sending an e-mail to infringement@cengage.com.

5.0 MISUSE OF THE LICENSED PRODUCT

5.1 In the event that the End User uses the Licensed Content in violation of this Agreement, Cengage Learning shall have the option of electing liquidated damages, which shall include all profits generated by the End User's use of the Licensed Content plus interest computed at the maximum rate permitted by law and all legal fees and other expenses incurred by Cengage Learning in enforcing its rights, plus penalties.

6.0 FEDERAL GOVERNMENT CLIENTS

6.1 Except as expressly authorized by Cengage Learning, Federal Government clients obtain only the rights specified in this Agreement and no other rights. The Government acknowledges that (i) all software and related documentation incorporated in the Licensed Content is existing commercial computer software within the meaning of FAR 27.405(b)(2); and (2) all other data delivered in whatever form, is limited rights data within the meaning of FAR 27.401. The restrictions in this section are acceptable as consistent with the Government's need for software and other data under this Agreement.

7.0 DISCLAIMER OF WARRANTIES AND LIABILITIES

7.1 Although Cengage Learning believes the Licensed Content to be reliable, Cengage Learning does not guarantee or warrant (i) any information or materials contained in or produced by the Licensed Content, (ii) the accuracy, completeness or reliability of the Licensed Content, or (iii) that the Licensed Content is free from errors or other material defects. THE LICENSED PRODUCT IS PROVIDED "AS IS," WITHOUT ANY WARRANTY OF ANY KIND AND CENGAGE LEARNING DISCLAIMS ANY AND ALL WARRANTIES, EXPRESSED OR IMPLIED, INCLUDING, WITHOUT LIMITATION, WARRANTIES OF MERCHANTABILITY OR FITNESS FOR A PARTICULAR PURPOSE. IN NO EVENT SHALL CENGAGE LEARNING BE LIABLE FOR: INDIRECT, SPECIAL, PUNITIVE OR CONSEQUENTIAL DAMAGES INCLUDING FOR LOST PROFITS, LOST DATA, OR OTHERWISE. IN NO EVENT SHALL CENGAGE LEARNING'S AGGREGATE LIABILITY HEREUNDER, WHETHER ARISING IN CONTRACT, TORT, STRICT LIABILITY OR OTHERWISE, EXCEED THE AMOUNT OF FEES PAID BY THE END USER HEREUNDER FOR THE LICENSE OF THE LICENSED CONTENT.

8.0 GENERAL

8.1 <u>Entire Agreement.</u> This Agreement shall constitute the entire Agreement between the Parties and supercedes all prior Agreements and understandings oral or written relating to the subject matter hereof.

8.2 <u>Enhancements/Modifications of Licensed Content.</u> From time to time, and in Cengage Learning's sole discretion, Cengage Learning may advise the End User of updates, upgrades, enhancements and/or improvements to the Licensed Content, and may permit the End User to access and use, subject to the terms and conditions of this Agreement, such modifications, upon payment of prices as may be established by Cengage Learning.

8.3 <u>No Export.</u> The End User shall use the Licensed Content solely in the United States and shall not transfer or export, directly or indirectly, the Licensed Content outside the United States.

8.4 <u>Severability.</u> If any provision of this Agreement is invalid, illegal, or unenforceable under any applicable statute or rule of law, the provision shall be deemed omitted to the extent that it is invalid, illegal, or unenforceable. In such a case, the remainder of the Agreement shall be construed in a manner as to give greatest effect to the original intention of the parties hereto.

8.5 <u>Waiver.</u> The waiver of any right or failure of either party to exercise in any respect any right provided in this Agreement in any instance shall not be deemed to be a waiver of such right in the future or a waiver of any other right under this Agreement.

8.6 <u>Choice of Law/Venue.</u> This Agreement shall be interpreted, construed, and governed by and in accordance with the laws of the State of New York, applicable to contracts executed and to be wholly preformed therein, without regard to its principles governing conflicts of law. Each party agrees that any proceeding arising out of or relating to this Agreement or the breach or threatened breach of this Agreement may be commenced and prosecuted in a court in the State and County of New York. Each party consents and submits to the nonexclusive personal jurisdiction of any court in the State and County of New York in respect of any such proceeding.

8.7 <u>Acknowledgment.</u> By opening this package and/or by accessing the Licensed Content on this Web site, THE END USER ACKNOWLEDGES THAT IT HAS READ THIS AGREEMENT, UNDERSTANDS IT, AND AGREES TO BE BOUND BY ITS TERMS AND CONDITIONS. IF YOU DO NOT ACCEPT THESE TERMS AND CONDITIONS, YOU MUST NOT ACCESS THE LICENSED CONTENT AND RETURN THE LICENSED PRODUCT TO CENGAGE LEARNING (WITHIN 30 CALENDAR DAYS OF THE END USER'S PURCHASE) WITH PROOF OF PAYMENT ACCEPTABLE TO CENGAGE LEARNING, FOR A CREDIT OR A REFUND. Should the End User have any questions/comments regarding this Agreement, please contact Cengage Learning at Delmar.help@cengage.com.